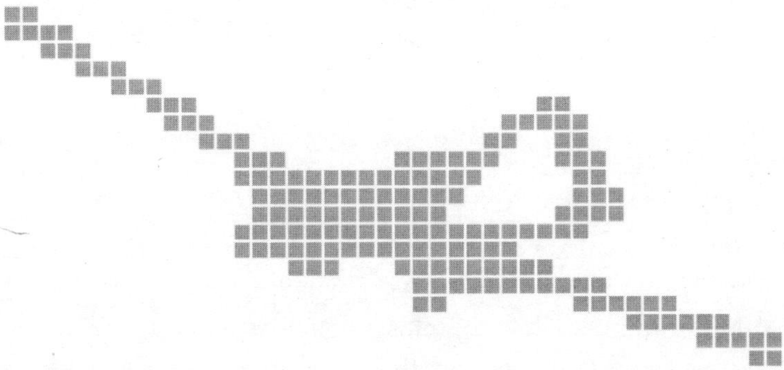

BYTES AND BULLETS

Also by Steve Feldstein

The Rise of Digital Repression

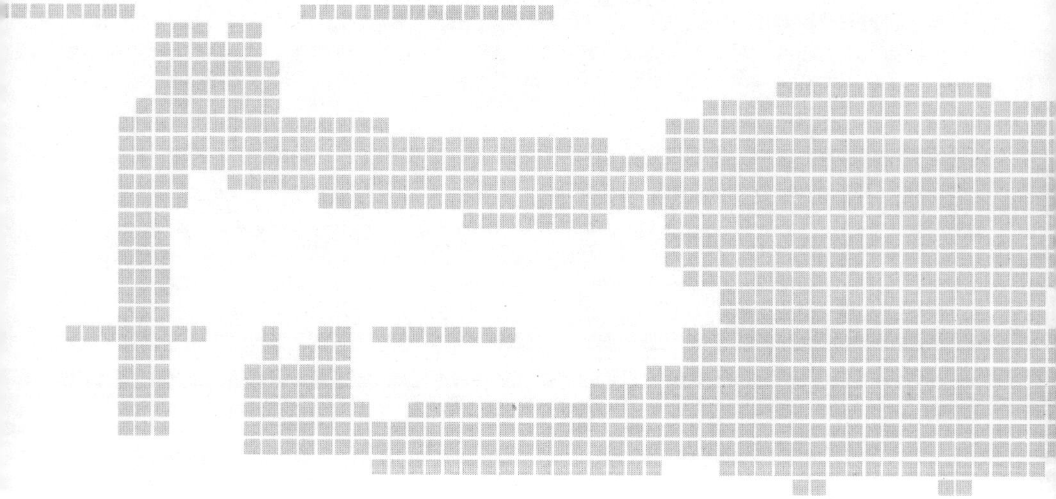

BYTES AND BULLETS

Global Rivalries, Private Tech, and the Future of Modern Warfare

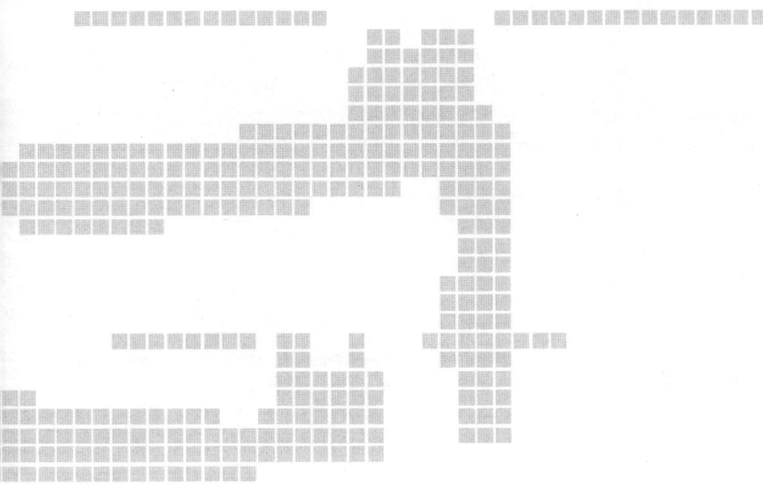

Steve Feldstein

monoray

First published in Great Britain in 2026 by Monoray, an imprint of
Octopus Publishing Group Ltd
Carmelite House
50 Victoria Embankment
London EC4Y 0DZ
www.octopusbooks.co.uk

An Hachette UK Company
www.hachette.co.uk

The authorized representative in the EEA is Hachette Ireland,
8 Castlecourt Centre, Dublin 15, D15 XTP3, Ireland (email: info@hbgi.ie)

Published in the United States by St. Martin's Press,
an imprint of St. Martin's Publishing Group.

Hardback ISBN: 978-1-80096-286-6
Trade Paperback ISBN: 978-1-80096-287-3
eISBN: 978-1-80096-289-7

A CIP catalogue record for this book is available from the British Library.

Printed and bound in Great Britain.

10 9 8 7 6 5 4 3 2 1

Designed by Jonathan Bennett

This FSC® label means that materials used for the product have been responsibly sourced.

FSC
www.fsc.org
MIX
Paper | Supporting responsible forestry
FSC® C104740

To Elizabeth

CONTENTS

Introduction 1

Part I: Global Rivalry in the Digital Age

1. The End of American Unipolarity 15
2. Rival Powers Compete for Technology 27
3. Rising Tech Powers 36

Part II: The Private Tech Revolution

4. Snowball Units and Starlink Terminals 49
5. From the Last Supper to the "Pro-Military Arm of Silicon Valley" 65
6. "A Feeling of a Thousand Snipers in the Sky" 71
7. Defense Tech Gold Rush 78

Part III: America's Tech War

8. Rise of Economic Warfare 95
9. What Is a Strategic Technology? 102
10. "We Cannot Let China Get These Chips. Period." 109
11. Supercharging American Innovation 119
12. The Pitfalls of "Damming Half the River" 127
13. Punishing Russia's War Machine 138

Part IV: China Seizes the Future, Russia Adapts

14. China's Quest to Become a Tech Superpower 151
15. Seizing the "Crown Jewels of U.S. Innovation" 161
16. "We Must Be Prepared for Worst-Case and Extreme Scenarios" 171
17. Will China's Strategy Work? 188
18. An Axis of Disruptors? 201
19. "We Are Strong People and Can Cope with Any Challenge" 206

Part V: Innovators and Disruptors: Rising Powers

20. Military Innovations Spread Across the Globe 221
21. Israel's Culture of Innovation 229
22. Drones and Geopolitics 238
23. The Emiratis' Bid to Become an AI Power 251

Part VI: Consequences

24. Power Struggle 265
25. Tech War in the Age of Trump 278
26. Managing the Future 293

Appendix I: Rising Tech Powers Analysis 305
Appendix II: China's Progress Overcoming Tech Choke Points 309
Appendix III: Data Tables for Drone Strikes and Fatalities 311
Acknowledgments 313
Notes 315
Index 375

BYTES AND BULLETS

Introduction

Just after 10:00 a.m. on January 3, 2019, Beijing's Chang'e-4 spacecraft accomplished a major breakthrough in space exploration. The vessel, named after a Chinese moon goddess, successfully landed in the Von Kármán crater on the far side of the moon. None of the previous twenty-seven lunar missions in human history had even attempted this; the challenge of landing on the side of the moon that is perpetually hidden from the Earth's view was immense. The moon itself is a massive obstacle, blocking direct communications between Earth and its far side. To facilitate signal exchanges, six years earlier, Chinese scientists had positioned a communications relay satellite into orbit. While they were confident that their efforts would pay off, Beijing's fear of losing face if the mission failed kept it shrouded in secrecy. Chinese researchers were exultant after the landing. "The Chinese people have done something that the Americans have not dared try," said Zhu Menghua, a scholar with the Macau University of Science and Technology, who has worked closely with the space agency.[1]

In 2024, China pulled off the feat again, sending the Chang'e-6 spacecraft to collect and deliver far-side rock samples back to Earth—another first. These unprecedented accomplishments have fueled a growing sentiment that China's space program has closed the gap with NASA. Beijing's next goals are ambitious: to put astronauts on the moon by 2030 and build a lunar base in partnership with Russia by 2035.[2]

Needless to say, American lawmakers have become alarmed by China's rapid progress. Beijing has built a vibrant commercial space industry that threatens to overtake the United States with over 150 companies developing satellites, pursuing research, and launching rockets. NASA has struggled to match these results. Early in 2024, America's first attempted moon landing in fifty years ended with the spacecraft toppling sideways moments after coming into contact with the lunar surface, compromising critical functions. Another planned mission never even made it to the moon due to glitches with its propulsion system shortly after reaching orbit.[3]

The pressure is on for NASA to show that it can keep pace with China's achievements. Through its Artemis program, the United States intends to launch multiple moon landings within the decade and to eventually build a permanent lunar base on the moon's surface. But Artemis has not gone smoothly; it has faced delays and cost overruns while relying on a complicated mix of government workers and private personnel.

Beijing's space effort is highly symbolic. "It's the equivalent of building cathedrals in the twenty-first century," says Alanna Krolikowski, a professor at the Missouri University of Science and Technology. But unlike cathedrals, China's space program comes with an obvious military dimension. According to US intelligence, China already fields counterspace capabilities, such as directed-energy weapons, electronic warfare systems, and anti-satellite missiles that are capable of destroying US equipment. Experts fear that China's growing capacity is creating ripe conditions for a "new space arms race" with the United States.[4]

Nearly four thousand miles away from Beijing, Ukraine offered a very different lesson in geopolitical competition. In the fall of 2023, Russian missiles were pummeling Ukraine's cities and destroying its power stations. Ukraine's defenders felt powerless to stop the onslaught. They lacked aircraft that could defend against Russia's bombers. Their air defenses had been bolstered by American Patriot missiles, but these were expensive to use and in short supply. For every Russian missile that was intercepted, many others penetrated Ukraine's defenses. President Volodymyr Zelensky summoned his spy chief, Lt. Gen. Vasyl Maliuk, and asked him: "How can we fight back?"[5]

Maliuk, a career officer in the Security Service of Ukraine (SBU), was known for running daring operations on a shoestring budget. He had already directed a punishing drone offensive using aerial and seaborne systems that forced Russia's navy to withdraw most of its Black Sea fleet. He also spearheaded a creative bombing campaign against Russia's interior, targeting gas plants and defense factories. Now, Zelensky asked Maliuk to accomplish something much more difficult: neutralize Russia's long-range bombers, which were launching powerful missiles far beyond the range of Ukraine's defenses. Most of these planes were stationed at air bases deep inside Russia, some as far away as Siberia—thousands of miles from the Ukrainian border.

The Ukrainians met in secret in an operation directly supervised by Zelensky. Their mission was so closely held that not even the Americans were aware of what the SBU was planning. Over many months, they developed an audacious plan called "Spider's Web," resembling "a modern version of the Trojan horse." Their gambit was risky: find a way to bring dozens of

Ukrainian drones close to the airfields housing Russia's strategic bombers in the hopes of catching the Kremlin by surprise. Ukrainian operatives assembled purpose-built drones equipped with AI machine learning capabilities that could take out Russia's most valuable aircraft. Their engineers used mothballed Soviet bombers displayed in a museum in central Ukraine to teach their systems how to pinpoint structural plane weaknesses, such as missile-mounting points beneath the wings and fuel tank seams.[6]

The operation then moved to the infiltration phase. Maliuk's team broke down the drones into their component parts for later assembly. They also smuggled in custom prefabricated cabins equipped with solar panels and EcoFlow batteries. Later, the cabins would serve as launch pads for the drones; the solar panels and batteries would keep the drones charged and allow Ukraine to maintain communication with the devices throughout the operation. Getting the cabins into Russia wasn't easy; Moscow had banned items such as solar panels from entering the country, forcing the Ukrainians to "bribe Russian customs agents," Maliuk said. Once the equipment was safely inside Russia, the Ukrainians then needed someone who could painstakingly reassemble the unmanned aerial vehicles (UAVs for their final dispatch to Russia's far-flung airfields.[7]

For this stage of the operation, the SBU tapped Artem Tymofeyev, a former DJ, and his wife, Kateryna Tyofeyeva, a tattoo artist, based in the industrial center of Chelyabinsk, about 900 miles east of Moscow. They had both moved to the city in 2018 so that Artem could go into business with his father, who operated a flour mill there. The Tymofeyevs had a history of activism and had participated in 2014 street protests in Kyiv that ousted the pro-Russian president, Viktor Yanukovych. But as Ukraine's economy worsened, they decided to move to Russia to seek better opportunities. Artem and Kateryna "ticked all the boxes" for the SBU, which approached them about taking part in the operation. After a rigorous vetting process, the SBU felt comfortable proceeding and the couple threw themselves into the mission. Artem set up a logistics company, renting a warehouse close to a local FSB field office. He started acquiring trucks and recruiting drivers—placing an ad on a Russian website and interviewing more than twenty applicants. They spent hours assembling 150 drones and eight cabins while receiving remote instructions from the SBU via phone and video calls. By April 2025, their preparations were complete, although the operation wouldn't launch until the following month.[8]

Between May 23 to 26, five different trucks departed from Chelyabinsk with the houses and drones aboard, their drivers unaware of what they were transporting. Several times, the operation nearly unraveled. Once, the roof of a prefabricated cabin slid off, revealing the illicit drone cargo. The driver of the truck called Artem, telling him, "This is some kind of bullshit, there are drones under the roof." Thinking quickly, Artem played dumb—"What the fuck?" he said. SBU operatives became nervous, reported *The Wall Street Journal*. What if the trucker couldn't reinstall the roof? What if the FSB had the truck under surveillance? What if the trucker's wife found out, and then told her friends? They concocted an excuse for Artem to relay back to the driver: the drones were for hunting—to track animals across a wide area. After a short delay, the driver texted Artem back with a photo of the roof correctly placed on the cabin with a single world: "closed."[9]

Each driver was instructed to arrive at specific times and to park their trucks at a designated location close to one of five strategic Russian air bases. Every participant in the operation only had access to essential information. "The people who made the drones did not know that they were for Spiderweb. The people who made the houses did not know that these would be for aerial vehicles. . . . Each was responsible for his own extremely important, separate direction of work," said Maliuk.[10]

At 1:06 p.m. on June 1, 2025, Ukrainian commanders issued the strike order. Remote operators simultaneously slid open the container roofs and released scores of UAVs. One hundred seventeen first-person view (FPV) drones streamed toward their targets. They were equipped with a "unique double warhead," allowing the first 1.5-pound charge to penetrate the skin of the aircraft and the second explosive to detonate inside. Explosions began lighting up Russia's airfields, sending towering columns of smoke into the sky. Intelligence sources began sharing direct footage of the attacks. One compelling video taken from a drone camera panned across the airfield, revealing multiple Russian bombers engulfed in flames. The drone then plunged toward the ground, slamming into yet another sitting aircraft.[11]

The operation dealt a crushing blow to the Russians. While their exact losses remained uncertain, forty-one Russian planes were struck, and at least twelve aircraft were irreparably damaged, including seven TU-95 heavy strategic bombers, which are capable of carrying nuclear weapons. Several TU-22 long-range bombers were also destroyed, along with at least one Beriev A-50 early-warning and control aircraft, a rare plane worth an estimated $330 million. All told, using domestically produced Osa quadcopters—just thirteen to fifteen inches long and costing six hundred to one thousand dollars each—Ukraine wiped out about 20 percent of Russia's operational long-range bomber fleet, worth billions of dollars. Moscow's equipment was irreplaceable; Russia no longer makes the 1950s-era Tupolev planes, compromising the Kremlin's ability to project power beyond its borders.*[12]

Analysts were stunned by the daring operation. Some experts compared Kyiv's strikes to the Imperial Japanese Navy's attack against Pearl Harbor more than eighty years prior, an operation that transformed combat and marked the rise of aircraft carriers as naval warfare's dominant weapon. "The Ukrainians rewrote the rules of warfare again," wrote military commentator Max Boot, signaling "the obsolescence of once-dominant weapons systems: battleships in 1941, manned aircraft today."[13]

Ukraine's capacity to strike at the heart of Russian power illustrated a profound shift in geopolitics: big states no longer have a monopoly on the tools of war. Smaller states—with the right engineers and programmers—can now develop sophisticated and often devastating tools to challenge much larger rivals.†[14]

■ ■ ■

The world is experiencing a significant amount of turbulence. Since the end of the Cold War, the United States stood unchallenged as the "uncontested Rome" of the modern world. But conditions have changed. The rise of China and the aggressive revanchism of Russia are reconfiguring the global order, and emerging states are demanding a greater role in the international system.[15] Meanwhile, President Donald Trump's second term has brought radical disruption to the United States. At home, his agenda has gutted government institutions, put scientific research funding at risk, and deeply altered public life. Overseas, America is abandoning decades-long alliances, shirking international law, initiating new conflicts, and rejecting the post–World War II international order it designed and built. US global leadership, once taken for granted, now hangs by a thread.

---

* While Russia makes newer planes that are updated and more agile, they lack essential qualities of the destroyed bombers, notably range and payload capacity.

† Ukraine's ability to achieve its goals was also heavily dependent on support from the United States. The Pentagon stood up a mission command center in Wiesbaden, Germany, that served as "the backbone of Ukrainian military operations that, by U.S. counts, have killed or wounded more than 700,000 Russian soldiers." This partnership aided Kyiv in planning its counteroffensives, offered intelligence to guide overall battle strategy, and provided precise targeting information for Ukrainian drones in the sky and soldiers in the field. Without the United States' provision of over $66 billion worth of weaponry—"more than a half-billion rounds of small-arms ammunition and grenades, 10,000 Javelin antiarmor weapons, 3,000 Stinger antiaircraft systems, 272 howitzers, 76 tanks, 40 High Mobility Artillery Rocket Systems, 20 Mi-17 helicopters and three Patriot air defense batteries"—Ukraine would have struggled to mount long-term resistance against Moscow's forces.

As US power ebbs, the world is experiencing rising insecurity, growing fragmentation, and political uncertainty. War has returned to the global stage—the number, intensity, and length of conflicts stand at their highest levels since the Cold War. Adversaries are seeking new technologies to gain an advantage on the battlefield, leading to dangerous and unpredictable outcomes. In 2026, the *Bulletin of the Atomic Scientists* moved the doomsday clock to 85 seconds to midnight—"closer to catastrophe than ever before"—citing declining international cooperation, leadership failure, and escalating risks from nuclear war, climate change, and disruptive technologies. European leaders have urged their citizens to stockpile food and other essential items to sustain them for at least seventy-two hours in case of crisis. They warn that the bloc faces a "new reality marred with risk and uncertainty"—including Russia's war in Ukraine, rising geopolitical tensions, sabotage of critical infrastructure, and threats from electronic warfare.[16]

Conflict is taking on an economic dimension as well. China and the United States are pursuing stringent measures meant to deny their rivals access to their markets and to keep critical technologies under their exclusive control. They have erected trade barriers and restricted the flow of services and goods. Beijing and Washington have slapped on tariffs, penalized rival companies, and authorized billions of subsidies to protect their domestic industries. In contrast, global financial institutions first set up to promote stability, security, open trade, and prosperity—from the World Bank and International Monetary Fund to the World Trade Organization—have been marginalized as nation-states downgrade collaboration and treat the world as a zero-sum zone of competition.

In short, geopolitical competition is intensifying in all domains. I have spent many years studying how technology impacts geopolitics, democracy, and national security. For more than two decades, my foreign policy career spanned the federal government, nongovernmental organizations, think tanks, and universities. That experience showed me just how seriously nations around the world take the competition for technology, and the critical role they expect innovation to play in international affairs. In 2021, I wrote my first book—*The Rise of Digital Repression: How Technology Is Reshaping Power, Politics, and Resistance*—which examined how a suite of technologies was helping dictators maintain power in their countries. I found that in countries like China and Russia, intrusive tools were giving autocrats like Xi and Putin a powerful means to crush dissent and remove all obstacles to their rule. More recently, I've turned my attention to the impact of technology on geopolitics and global affairs.

In the heady period following the fall of the Iron Curtain, American power reigned supreme. Throughout the 1990s and 2000s, the United States set the pace on scientific research, innovation, and development—thereby acquiring crucial geopolitical benefits. Back then, it faced no serious rivals. "The defining and unique feature of world politics after the Cold War was the absence of balancing against the United States," writes Thomas Wright in *All Measures Short of War*. "The absence of balancing did not mean that other countries always agreed with the United States. They did not. Rather, it meant there were limits to how they would express their opposition."[17] But America's unipolar moment would not last. Over time, both China and Russia chafed at American power and pursued other options.

▪ ▪ ▪

This book tells the story of the rebirth of geopolitical competition in the twenty-first century—a contest centered around technology. We face a constant barrage of dramatic headlines: little-known Chinese startup DeepSeek releases an AI model that shocks the world and causes tech stocks to lose billions in valuation. Beijing debuts a mysterious sixth-generation stealth fighter with potential breakthrough capabilities that challenge US air superiority. Israel deploys a machine-driven AI targeting system, seemingly out of science fiction, to hunt for Hamas militants in the ruins of Gaza. Ukrainian soldiers fighting on the eastern front lines convert basic quadcopter drones into devastating weapons that pummel Russian tanks and equipment worth millions of dollars. How can we gauge the significance of these innovations? What is their larger impact on global affairs and the waging of war? What strategies is the United States pursuing to protect its interests and maintain its tech leadership? And what steps are America's rivals taking to secure their own objectives?

To answer these questions and others, I have organized the book into six sections. Part I explains how nations compete for power in the digital age. Throughout history, opposing countries have vied to control technology, and America's rivalry with China is a continuation of that dynamic. As American unipolarity has waned, Beijing has invested extraordinary effort into narrowing its technological gap with the United States, and to a large degree it has succeeded. Now, Beijing is accelerating its modernization plans—remaking its military and building cutting-edge AI tools. US policymakers are suitably alarmed and pursuing strategies to slow China's rise. But America's concerns extend beyond Beijing. From Russia's ongoing aggression to threats stemming from an emerging group of middle-power states, the United States faces mounting challenges to its technological leadership.

Part II describes how America's transition to an open innovation system at the end of the Cold War set the stage for a dramatic expansion of private tech power. Today, private actors control most cutting-edge research and development. Commercial labs, backed by private capital, are driving innovation forward and weakening government influence as a result. These forces are changing the game and shifting the balance of power between private companies and governments.

Indeed, it is now impossible to segregate civilian technologies from military ones—the line between the two has already collapsed. Partly due to the open tech revolution, capabilities developed for consumers are swiftly being repurposed for military use and vice-versa. We need look no further than the Ukraine war to see how the boundaries between civilian and military tech have blurred—geotagged photos from smartphones are used to call in strikes against Russian tanks; Telegram chat groups coordinate cyberattacks against enemy infrastructure. To paraphrase author Rosa Brooks, everything has become war, and the military has become everything.[18] Accordingly, it makes little sense to maintain a clear division between military and civilian tech innovations, and I have intentionally included examples of both types throughout the book.

Part III steps back to examine the larger tech war between the United States and China. Both countries are squaring up for a protracted fight—restricting key technologies, investing heavily in the innovations of the future, and recruiting allies to support their goals. I describe how successive US presidents have unleashed economic warfare against their rivals, and dig into the details behind America's unremitting export-controls push. Part IV then analyzes China's and Russia's responses to these efforts. What strategies is Xi Jinping using to offset America's tech restraints and win the future? What countermeasures has Vladimir Putin implemented to keep his war machine humming? In Part V, I shift the aperture to middle-power states and examine how they are using technology to gain political influence and stature, discussing how new centers of innovation—from Singapore and Tel Aviv to Abu Dhabi—are increasingly contesting US tech leadership.

Finally, in Part VI, I probe the consequences of the new age of geopolitics. In the coming years, major powers will intensify their efforts to stop advanced technologies from proliferating to their opponents; if history is a guide, these strategies will expand tensions and hasten confrontation. We are entering a risky period. The combination of expanding US tech export controls, volatility from Donald Trump's White House, and retaliatory countermeasures from China could create an escalatory spiral. Of course, many countries caught in the middle are hesitant to align with either bloc. Countries like Brazil, India, and Saudi Arabia will hedge—developing independent relationships with different great powers so they can preserve their options. As tech becomes a crucial point of contestation and particularly as AI becomes more central to militaries, there is a growing risk of a race to the bottom when it comes to developing ever more powerful tools. In the face of these challenges, is confrontation inevitable, or can leaders find a way to manage competition and uphold stability?

■ ■ ■

My motivation for writing *Bytes and Bullets* is simple: I believe that advanced technology like artificial intelligence, robotics, spacecraft and satellites, leading-edge semiconductors, and quantum computing has become a powerful tool of competition and a primary focal point of geopolitical rivalry. Other authors have provided deep accounts of how these technologies work in isolation, but most have missed the bigger picture. This book shows how the latest tech innovations are now becoming crucial pieces in a geopolitical chess match spanning regions and continents. They have the potential to change how wars are fought, to transform economies, and to upend diplomatic alliances.

Consequently, nations are in a race to subsidize new innovations while denying access to their rivals. The book's perspective is rooted first and foremost in US interests, reflecting my expertise. But since all sovereign nations are making similar calculations and following comparable logic, these insights apply broadly. Moreover, given America's outsized global footprint, I would argue that Washington's policy shifts deserve particular attention. In response to this burgeoning competition, the United States has deployed tools of economic war to curb China's tech capacity—and following Russia's attack on Ukraine in 2022, to check Moscow's access to advanced technologies as well. These measures reflect an important reality: tech has become an essential component of power, and US leaders believe it is vital to keep leading innovations out of the hands of their rivals. But this tech-control strategy faces inherent limitations; for a variety of reasons, Washington's controls will yield diminishing returns over time.

America's allies are increasingly reluctant to back its tech war. European states like France and Germany, as well as East Asian countries such as Japan, South Korea, and Taiwan, remain economically interdependent with China. They are loath to cut ties and damage their economies. During Joe Biden's presidency, they started to band together to deter China and Russia; many agreed upon mutual tech restrictions to check Russian aggression, imposed limited semiconductor controls on China, and called out both countries for violations of international law. But those efforts weren't uniform. Some nations, like Germany, maintained distance from America's approach. Even more forward-leaning countries like the Netherlands and Japan had to be cajoled to curb their exports of high-end chip equipment. With Trump back in the Oval Office, this approach has been effectively shelved. He has leveled harsh words against America's allies, imposed tariffs against their products, and left their leaders fuming. Their doubts have grown. Turbulence in the White House doesn't mean that countries will exit treaty alliances like NATO (unless, of course, Trump decides to unilaterally withdraw from them); these structures are deeply entrenched and offer numerous advantages. But it could cause other instruments of alliance-building, such as the Quadrilateral Security Dialogue (an arrangement made up of Australia, India, Japan, and the United States that cooperates on security, economic, and global health matters), to fall by the wayside.

History tells us that it is exceedingly difficult to curb technological diffusion—the outward flow of expertise, innovation, research, and ideas. It is already too late to curtail China's expanding tech capabilities; it is the world's leading manufacturing power, has invested billions in catching up to the United States, and is now a global leader in many technology fields. Inevitably, Chinese researchers and scientists will come up with creative ways to bypass semiconductor controls or AI model restrictions—it is just a matter of time.

Even if the United States can successfully curb the spread of a small group of cutting-edge technologies for a limited period, many other tools—from autonomous drones and satellite networks to generative AI applications—are already proliferating widely. An array of rising powers is pursuing independent foreign policies alongside powerful tech capabilities. Countries like Iran, Israel, Saudi Arabia, Türkiye, and the United Arab Emirates seek greater influence in a competitive global environment and are using technology as a means to get there.

Finally, Washington's ability to control the development of cutting-edge technologies is increasingly supplanted by America's own private companies.

Large tech corporations, from SpaceX to Microsoft, have gained newfound sway historically reserved to nation-states, allowing them to shape diplomatic and military outcomes. They are spearheading the race to achieve artificial general intelligence (AGI). They control critical infrastructure—from data centers and underwater sea cables to satellite internet networks—and they are much more focused on short-term profits than diplomatic consequences or international stability. Simply put, American tech firms now wield significant power, further complicating Washington's strategic options.

The consequence of these trends is that the United States will find itself engaged in an intensifying contest as rival countries implement hard-nosed policies to preserve their advantages, isolate their opponents, and develop ever more powerful systems. The outcome of this competition will redefine the balance of global power and international relations for generations to come.

PART I

Global Rivalry in the Digital Age

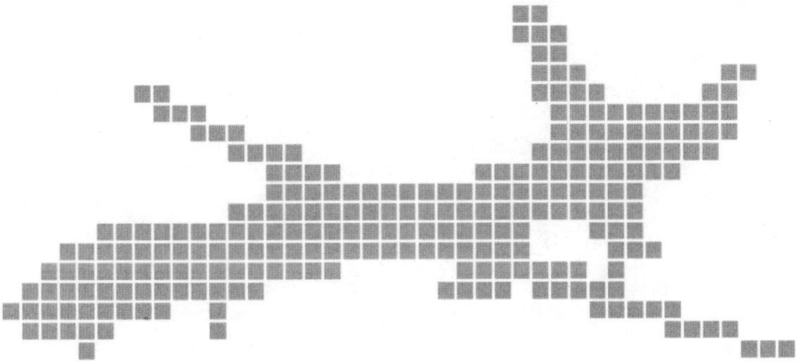

1

The End of American Unipolarity

The upheaval racking the world today seemed far-fetched until recently. In the 1990s, after decades of Cold War tensions, the fall of the Soviet Union gave policymakers renewed hope that global peace was not only possible, but had already been achieved. Scholars like Francis Fukuyama proposed that we were at the "end of history" and were witnessing an "unabashed victory of economic and political liberalism." Experts broadly agreed on the idea of "convergence," which offered an appealing argument: now that America had firmly entrenched itself as the world's dominant power, major countries had more to gain from competing over business than from fighting wars. Governments would cease treating each other as adversaries and instead work cooperatively and "converge" to tackle common challenges. It was in everyone's best interests to play by the rules of the international system—created and overseen by the United States—rather than cause problems or create alternative structures.[1]

For nearly a quarter century, no major powers sought to balance against American unipolarity. Neither China nor Russia was capable or interested in pushing back against US dominance. When they had disagreements with America, they expressed their opposition in UN Security Council votes or in occasional "spasms of anger," such as China's fury at NATO's 1999 bombing of the Chinese embassy in Belgrade.[2] But this extraordinary period of convergence began to unravel in the late 2000s.

A significant milestone was President George W. Bush's response to the September 11, 2001, terrorist attacks. His administration articulated a grand strategy positioning America as the world's central provider of order and security. "The United States would be the global Leviathan," writes Princeton academic G. John Ikenberry. It would stand above other states, unconstrained by multilateral rules and institutions. "The United States would have an open-ended license to deploy power and intervene around the world—and do so with fewer checks and balances." The Iraq War marked the debut of Washington's new global security doctrine—with all of its consequences. Bush's strategy ultimately failed on two fronts: it was not operationally sustainable, and it triggered countries to reassess their participation in American hegemony.[3]

Russia's public break with the United States came in 2007, at the annual Munich Security Conference. In the Hotel Bayerischer Hof, Russian President Vladimir Putin stood in front of an international assembly of leaders and diplomats and for thirty minutes upbraided America for trying to establish a "unipolar world," with "one center of authority, one center of force, one center of decision-making." He warned that the country's unconstrained use of force would plunge the world into an "abyss of permanent conflicts." This critique was not particularly original, but Putin's signaling that the US liberal order was no longer of "interest or value" to Russia was a serious charge.[4]

American policymakers had spent substantial time working with Russia over the previous decade, in the belief that Moscow would eventually integrate into the post–Cold War system of international stability. At first, it seemed likely that the two countries would foster a close relationship—perhaps even an alliance. In 1992, President Boris Yeltsin paid his first visit to the United States as Russia's head of state. President George H. W. Bush hosted him at Camp David. The primary focus of the meeting was nuclear reduction, but toward the end of the conversation Yeltsin asked a revealing question:

> YELTSIN: Are we still adversaries or not?
>
> BUSH: No, we are not.
>
> YELTSIN: You have nothing here [in the joint statement] which says that we are no longer adversaries and are moving to be allies.
>
> SECRETARY OF STATE JAMES BAKER: It says friendship.
>
> YELTSIN: No, no. We should say that we are moving from a stage of adversaries to allies. This gives a new quality.

Bush held firm, and the joint statement did not include the term "allies" because Bush didn't "want to act like all our problems are solved."[5] Still, such an outcome didn't appear far off.

In the ensuing years, however, Russia shifted its stance from cooperation to competition. Its break with the West didn't happen all at once; rather, it occurred in fits and starts. After succeeding Yeltsin, Putin initially believed that a mutually beneficial partnership was possible, too. He envisioned forming a "common front" in the post-9/11 War on Terror, in exchange for American acceptance of Russia's special influence in its backyard and an agreement not to interfere with its domestic politics. But such a deal was "never in the cards." America had no interest in trading anything for a partnership with Russia, a country it viewed in decline.[6]

William Burns, the former US ambassador to Russia, recounted a meeting he held with Putin and Secretary of State Condoleezza Rice in 2006. Putin invited the Americans to his presidential compound outside of Moscow. After keeping them waiting for some three hours—a common ploy he used to unnerve visitors—he ushered them into his quarters. In front of a "roaring fire," talk turned to Georgia and its pro-Western, liberal president, Mikheil Saakashvili. Putin expected Saakashvili to show deference to Moscow—which Saakashvili was loath to do. Putin began to lose his cool. As Burns recollects, "Putin's intimidating aura is often reinforced by his controlled mannerisms, modulated tone, and steady gaze. But he can get quite animated if he wants to drive home a point, his eyes flashing and his voice rising in pitch." Standing in front of the fireplace, Putin "wagged his index finger" and warned the Americans that "if Saakashvili starts something, we will finish it." At this point, Rice also got up, "looming several inches taller than Putin in her heels." Putin said sharply, "Saakashvili is nothing more than a puppet of the United States. You need to pull back on the strings before there's trouble." While tensions in the room eventually dissipated, disagreements about how to deal with Georgia and larger differences about Russia's place in the world were never resolved.[7]

American policymakers wanted to keep Russia involved in international-security initiatives, such as nuclear arms control, but they were unwilling to cede any influence to Moscow. To that end, America continued supporting pro-democracy groups operating in Russia, as well as dangling the prospect of NATO membership to its neighbors (and former USSR satellites), Georgia and Ukraine. Putin in turn became embittered against the United States—blaming America for fomenting "color revolutions" in Georgia, Ukraine, and Kyrgyzstan—and reacted with fury to any signals about NATO enlargement. In 2008, for the first time since the breakup of the Soviet Union, Russian troops marched into Georgia. In a five-day war, Russian forces came within striking distance of the capital of Tbilisi, signaling the reemergence of Russian military power. This operation paved the way for Russia's annexation of Crimea from Ukraine in 2014, which effectively "symbolized the end of the post–Cold War era and the reemergence of a new East-West divide in Europe."[8]

China underwent a similar shift in the 2000s. Until 2008, China and the United States forged a remarkably stable partnership despite the differences in their political systems. Both countries benefited from good trade relations. China's economy soared as foreign investment flowed into the country, and it became the manufacturing hub of the world. American corporations enthusiastically opened up shop and pushed Washington to maintain friendly relations. The mantra from Washington was for Beijing to be a "responsible stakeholder" in the international system—as Deputy Secretary of State Robert Zoellick called for in an oft-cited 2005 speech.[9]

But the 2008 financial crisis changed everything. Seemingly overnight, Wall Street banks and insurance companies collapsed, the stock market plunged, and housing prices plummeted. It represented the biggest economic downturn in the United States since the Great Depression. The crisis caused Chinese leaders to reevaluate the strength of the West. They concluded that America was far weaker than they had previously thought. This stimulated demand among elites and the public for a more "ambitious" foreign policy, driven by the belief that China's time had come.[10] Beijing began staking assertive positions, from pursuing territorial claims in the South China Sea to increasing its investments in military technologies. In diplomatic meetings, China went from accommodating to bullying. In a 2010 meeting of Southeast Asian countries, for example, Chinese Foreign Minister Yang Jiechi flew off the handle. He lambasted delegates who had spoken in favor of collective arrangements to manage maritime claims—something that had become contentious due to China's newfound aggression. Yang said, "China is a big country and other countries are small countries, and that's just a fact." He then scolded Secretary of State Hillary Clinton for "conniving" with the other leaders against China.[11]

Domestically, Beijing went from decades of restraint to a new era of overreach. The Chinese Communist Party (CCP) instituted policies at home that matched its bellicose rhetoric abroad. Under President Xi Jinping, authorities began implementing technology-fueled repression. They enacted mass surveillance in "smart cities" and built Orwellian digital gulags in outlying provinces like Xinjiang. It became impossible for American policymakers to ignore that China's grand strategy had shifted. Rather than continue following Chinese leader Deng Xiaoping's maxim of "hide your strength, bide your time"—a principle that had guided Chinese policy for decades—Xi moved the country in a very different direction. Under his leadership, China emerged as a principal rival to Western democracies, with the aim of "undermining the Western liberal order while reaching for PRC [People's Republic of China] hegemony in Asia and the expansion of Chinese influence worldwide." By 2010, the United States began to respond in kind. Obama administration officials organized Southeast Asian countries to push back against China's maritime claims, they offered more weaponry to Taiwan, and they announced a strategic pivot to Asia in reflection of these tensions.[12]

Today, the post–Cold War convergence toward the Western liberal order has collapsed, seemingly bringing an end to what Charles Krauthammer famously termed America's "unipolar moment." Even Fukuyama himself acknowledges that "we've seen frightening reversals to the progress of liberal democracy over the past 15 years" (although he maintains that the liberal democratic order can endure if "people are willing to struggle on its behalf").[13] The world appears headed toward greater fragmentation and a wider dispersion of power.

One consequence of this tumult is a startling increase in violence and war. Research shows that in periods where leaders have uncertainty about their present or future security, the risk of conflict grows.[14] Not since the Cold War has the world faced such volatility.

In 2022, for example, nearly 277,000 battle-related deaths were recorded, a number unmatched since 1972. Casualties largely stemmed from two conflicts—the Russia-Ukraine war and armed conflict in Ethiopia between Tigrayan insurgents and government forces. Yet battlefield deaths tell only part of the story. Data collected by Oslo's Peace Research Institute shows that the number of wars or "conflict events" has also risen dramatically; 2024 registered the highest number of conflict events since 1946. The only other period since World War II with so many state-based conflicts occurred in the early 1990s—an era that included the first Gulf War and civil war in Rwanda. Notably, 2024 saw an increase in interstate war—conflict between sovereign states—something that had become a rarity after World War II. Analysts recorded four interstate conflicts between Russia and Ukraine, Iran and Israel, Afghanistan and Pakistan, and the United States/United Kingdom and Yemen.[15] Figure 1 illustrates these grim trends—the line graph shows the number of conflict events per year; the vertical bar graph depicts annual battle-related deaths.

Figure 1. Trends in Conflict, 1946–2024[16]

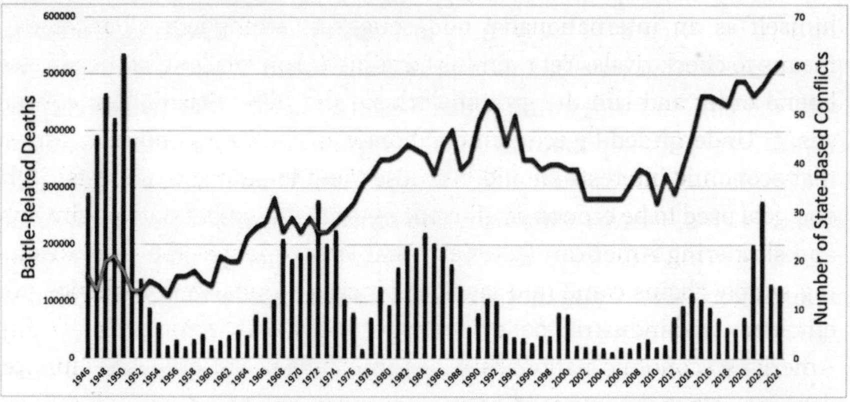

The present state of turmoil has caused some analysts to argue that history is reversing course, "pushing the world back in the direction of geopolitical and ideological groupings."[17] Historian Hal Brands, for instance, analogizes current conditions to the interwar period of the 1930s. He writes that between China's massing of military power in the Pacific, Russia's assault in Ukraine, and Iran's interventions through its proxies in the Middle East, many Americans "don't realize how close the world is to being ravaged by fierce, interlocking conflicts."[18]

But it is not just armed conflict that has risen to the fore. Economic warfare has returned as well, displacing the previous era of global collaboration and integration. The United States has undergone one of the biggest reversals on this front. America once stood as the chief proponent of open markets, free trade, and globalized manufacturing. The Treasury Department sent officials around the world to spread the gospel of privatization, lower taxes, and the benevolent power of markets. The benefits of doing so were many. Foreign exports enriched American companies; consumers profited from accessing cheap goods produced in China's Pearl River Delta (Walmart's business model would be completely upended without inexpensive products from China). The world accrued immense gains from economic integration. In a twenty-year span, beginning in 1990, 1.3 billion people were lifted out of extreme poverty. A child today is twice as likely to survive into adolescence, while malnutrition and illiteracy rates have plummeted.[19] Walling off trade and innovation behind national borders harkens back to the protectionism of the interwar period, to the 1920s and '30s, and rejects years of accumulated wisdom about protecting the free flow of commerce and information.

But the landscape has changed. By all accounts, President Biden saw himself as an internationalist and sought to strengthen alliances as a means to check rivals. Yet even his administration challenged the old neoliberal order and laid the groundwork for the "new Washington consensus."[20] Undergirded by economic nationalism, the new consensus argued that economic interests should be subservient to strategic interests. If the end goal used to be economic efficiency—even if it meant outsourcing jobs and shuttering American factories—that was no longer the case. Reshoring supply chains came into fashion, officials emphasized resilience over efficiency, and industrial policy no longer was a dirty word. Under Trump, America's economic nationalism has intensified. Not even one hundred days into office, he imposed 10 percent across-the-board tariffs for all imports to the United States and supplemented them with additional taxes tailored to individual countries.[21]

There is some wisdom and a lot of folly to America's course correction. Based on the data, there's little doubt that China has become the world's sole manufacturing superpower. In 2008, China supplanted America as the globe's largest producer—a feat it accomplished in about fifteen to twenty years (far quicker than it took the United States to replace the UK as the world's leading manufacturer prior to World War I). China hasn't looked back. In the intervening years, Beijing has "more than doubled its world share" while America's portion has fallen by three percentage points. At this stage, China has established an unassailable industrial advantage over all of its competitors; its production "now exceeds that of the next [seven] largest manufacturers combined."[22]

While the United States cannot realistically overcome China's manufacturing advantage, that doesn't mean it is out of options. For one, not all manufacturing is worth the same. A substantial portion of China's economy remains centered in low-end manufacturing, such as textiles, ceramics, and steel. These are labor-intensive industries that account for a diminishing percentage of future economic output. A smarter strategy would be to subsidize investments in critical sectors linked to high-tech industries (and to constrain competitors from doing the same). These sectors matter quite a bit to countries' security; the race among great powers to develop ever more powerful microchips, hypersonic missiles, and AI systems is fierce. Therefore, imposing protectionist measures in these areas—even if the result is lower productivity and higher prices—could be a worthwhile trade-off. Such moves, of course, come with risks. A big danger is that political leaders will continue broadening protectionist measures to cover an expanding range of high-tech products. With Trump as president, there is no telling how far these trade restrictions will extend.

What America's changing economic preferences come down to is whether its leaders believe the country will obtain greater benefits from keeping the door open and maintaining technological interdependence, or whether they are convinced that closing the door to imports and focusing on building up domestic industries is a better course. Their outlook will be heavily influenced by their expectations about the changing balance of power. To get more clarity about where trends are headed, let's take a look at global forecasting projections.

Changing Balance of Power

The balance of power is shifting in the world, but not always in the ways people think. The University of Denver's Pardee Center created a "global power index" to measure countries' relative share of power worldwide. Its model uses a weighted combination of military, demographic, economic, technological, and diplomatic factors. From the end of the Cold War through 2022, figure 2 shows a gradual erosion of American power—in large part due to China gaining ground.

US influence remains high relative to other states. As of 2022, America still controlled nearly a quarter of the global share of power despite having under 5 percent of the world's population. When the United States' portion is combined with the European Union's, their global share of power exceeds 40 percent. In contrast, China continues to lag behind. Other would-be challengers, such as India, Japan, or Russia, barely exceed the 3 percent threshold. But longer trends paint a less optimistic picture; America's global power has clearly slipped over the decades.

Looking to the future, the signs are muddled. Most experts agree that the foremost geopolitical challenge of the twenty-first century involves managing a rising China seeking to become a dominant power and a declining United States trying to preserve its influence. But consensus breaks down from there. One school of thought holds that China will overtake the United States and claim "power predominance" in as little as a decade. Experts argue that only a dramatic shift by Washington can stem China's rise. Other analysts fall into an "optimist" camp, unconvinced that China will overtake the United States (and even if this transpires, they argue, China will be incentivized to pursue relatively moderate policies). Their biggest worry is that the two countries will enter a destabilizing "spiral of hostility" fueled by high-tech export restrictions and military buildups in East Asia.[23]

Figure 2. Global Power Index, Historical Trends (1990–2022)[24]

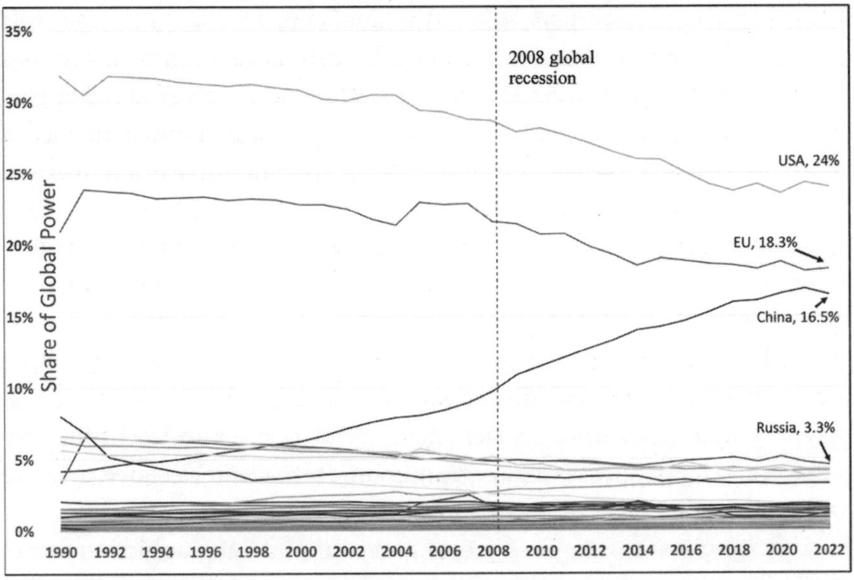

Despite forecasters' best efforts, it is impossible to anticipate how these trends will play out. Even deciphering the present data is difficult. China's statistics are deliberately obscure and its officials often release misleading data or "present only facets of the reality they want the world to see," making it hard to know China's actual power and status, let alone what they will look like in ten or fifteen years' time.[25] The best we can hope for, then, is to obtain a deeper understanding of underlying trends that will shape future policies in the United States and China. On this basis, three insights are relevant about America's role in the world.

First, while China's rise will undoubtedly continue and its material share of global power could surpass the United States in the next decade, this doesn't mean that China will supplant the United States as a superpower. China will encounter bumps along the road. Hal Brands, for example, argues that the drivers of China's growth are already stagnating or in decline. From the 1970s to the 2000s, China possessed abundant food, water, and energy resources. It enjoyed "the greatest demographic dividend in history," with ten working-age adults for every senior citizen. Those factors led China to become the global center of manufacturing and allowed the nation to cut poverty at a stunning rate. Now, the situation has reversed. China is running out of water and is a net importer of energy and food, making economic growth more costly and fostering an unhealthy reliance on external trade. Its declining demographic picture also raises alarm bells—China is losing population at a rapid clip. Thanks to its one-child policy, which was implemented in 1980 to curb overpopulation and ended in 2016, China's population fell by 2.08 million, with new births dropping by 6 percent in 2023 alone. A new trend is taking hold: outward migration. More than three hundred thousand Chinese annually are leaving for other countries, twice as many as a decade ago.[26]

Looking at China's economic productivity also offers insights. Historically, two factors are crucial to determining whether a rising state will overtake an existing great power: the size of the economy and its productivity (roughly calculated as the ratio of a country's GDP to total hours worked). If a country is economically efficient but has a small economy—such as Norway or Switzerland—then its lack of size renders it noncompetitive. Conversely, if a country possesses a large economy but runs it inefficiently—like India—then this also nullifies its advantages.

China's economy is large. By some measures, its GDP is equal to that of the United States. In 2022, its GDP stood at $17.96 trillion, close behind America's GDP of $25.44 trillion; when purchasing-power parity is factored in, China's economy slightly exceeds the United States'. But Beijing faces a major obstacle: lagging productivity. From 2000 to 2017, China's total productivity "never surpassed 43 percent of the US," writes Jeffrey Ding, a scholar at George Washington University. Moreover, China's future productivity growth is uncertain. Due to inefficient infrastructure outlays, China's productivity growth declined from 2.8 percent in the early 2000s to 0.7 percent in the decade after (2009–2018).[27] As a result, Beijing faces a major productivity challenge in the years ahead. Unless China can dramatically improve its efficiency from new technologies and manufacturing products, it will struggle to make up ground with the United States.

For the sake of argument, suppose that China somehow found a winning formula that allowed it to dramatically increase its productivity and surpass America's economic output as well as develop new military technologies. That still wouldn't make China a military peer of the United States. Developing new weapons and tactics "depends on decades of expensively accumulated technological and tactical experience," writes MIT scholar Barry Posen.[28] Successful war fighting requires extensive training, experience in combat, and effective force deployment. Military dominance is not something that magically happens once a country's economy reaches a certain threshold of size or a country's army attains a specific number of planes or tanks.

Second, it is unlikely that any middle-power state—or group of middle powers—will challenge the United States or China for dominance in the near future. Aside from the EU (which, as a regional grouping of countries, belongs in a slightly different category), the closest state competitor is India. But even by 2050, its expected share of global power will remain low. While India will accrue greater influence due to its large population and emerging industrial base, on its own it will be unable to tilt the global balance of power. However, it would be unwise to underestimate the growing heft of the "hedging middle." Countries in the Global South will be intent on diversifying their relationships and optimizing their choices. As William Burns warns, they will see little benefit to "sticking to monogamous geopolitical relationships with either the United States or China." Instead, more countries will seek open-ended relationships with multiple larger powers—perhaps following America's lead on certain issues while fostering ties with China on other matters.[29]

The growing relevance of the BRICS grouping, of which China is a major anchor, is instructive. The acronym stands for the group's founding five members: Brazil, Russia, India, China, and South Africa. It has since expanded to include Egypt, Ethiopia, Iran, and the UAE (Saudi Arabia is also a participant, but it is not a formal member). Many in the West predicted the group's rapid demise when it launched in 2009 but its trajectory has gone in the opposite direction. Its ten current members represent over 35 percent of global GDP in purchasing power parity—the exchange rate at which money buys the same amount of goods and services in different countries—exceeding the G7's global GDP share of 30 percent. In the coming years, the BRICS will likely expand further, "with more than 40 countries expressing interest in joining." Its October 2024 summit in Kazan, Russia, was telling. The Kremlin crowed that the meeting was the biggest foreign policy event ever hosted by Russia and represented the "literal embodiment of a revisionist approach to the transformation of the international order." Attendees lavished praise on one other. Prime Minister Abiy Ahmed of Ethiopia, once the toast of the West (and the ill-advised winner of the Nobel Peace Prize) gushed to Putin, "This period was not easy for Russia, but under your leadership you have succeeded to maintain the economic resilience which might be exemplary for most of us."[30]

Third, the influence of the West (anchored by the United States and Europe) will continue to wane. America's diminished ability to "command the commons"—to exert ongoing and continuous superiority over air, sea, and space—will open up space for rival actors.[31] Arguably, Trump's anti-globalization instincts are speeding up this process. As China catches up to the United States, this will inject more uncertainty into the international system. Already, the long post–Cold War peace guaranteed by American power has ended. If policymakers once expected that major powers would stop treating each other as rivals and would instead compete over economic investment and trade, events in the last decade have refuted these assumptions. Russia is engaged in a protracted hot war against Ukraine. China is arming itself with alacrity and pursing menacing actions in the South China Sea and toward Taiwan. Iran, Israel, and Türkiye have developed sophisticated weapons industries whose proliferation is triggering more instability. Meanwhile, Trump is steering the United States in a more confrontational, unilateral, and illiberal direction.

What does all this add up to? More than anything, these trends mean there will be greater unpredictability in the international system. Within a decade, no country will be a fully dominant power. As leaders contemplate these projections, they must confront an unsettling reality: they are steering their nations into security waters more treacherous than any navigated in recent decades.

2

Rival Powers Compete for Technology

Rival powers have competed to control technology throughout history. As far back as the sixteenth century BCE, adversaries raced to develop the chariot as a weapon, which altered the balance of power in favor of Egypt and ultimately sparked "a centuries-long arms race from Anatolia to Mesopotamia." Superior technology has often helped a weaker army attain victory even when faced with long odds, like King Henry V's deployment of the longbow to inflict a devastating defeat upon the numerically superior French cavalry at the Battle of Agincourt in 1415, or Britain's use of the Maxim gun (the world's first fully automatic machine gun) to wipe out thousands of attacking warriors during the First Matabele War in South Africa in 1893. Even today, the popular imagination maintains a belief that a decisive technology can change the course of war and remake geopolitics.

America's rivalry with China for technological dominance reflects this dynamic. Over the last twenty years, Beijing has sought to narrow the gap with the United States. Its authorities have channeled immense resources into research and development (R&D), invested in high-tech manufacturing, enticed labs and companies to set up shop in-country, and acquired cutting-edge technologies (sometimes legitimately but often illicitly, through IP theft and forced IP transfers). In 2019 alone, China reportedly spent close to $250 billion to support its industrial policies. These efforts have led to impressive gains: today, China's R&D spending as a percentage of GDP largely matches that of the United States, China's output of patent applications has grown exponentially, and its volume of high-technology exports significantly exceeds that of the United States.[1]

Some experts argue that China not only has built the foundations to be a science and technology superpower but is challenging the United States as the world's leading innovator. The Australian Strategic Policy Institute, for instance, concluded that in sixty-six out of seventy-four critical technologies—spanning defense, space, robotics, energy, biotechnology, and AI—China had taken a significant lead in research generation and development. The National Security Commission on Artificial Intelligence reached a similar finding, noting that for the first time since World War II, "America's technological predominance—the backbone of its economic and military power—is under threat." They assessed that if trends persist, China could surpass America as the global leader in AI within a decade.[2]

Beijing is not limiting its innovation push to civilian technologies; China is also undertaking a major military modernization. It has built up the world's largest navy in terms of sheer numbers and just completed its third aircraft carrier, the *Fujian*. Beijing is investing heavily in boosting its nuclear stockpile; the Pentagon estimates that China is set to "quadruple" its nuclear supply by the end of the decade to at least one thousand warheads. China has also committed significant resources to developing "intelligentized" warfare—which relies on AI and other disruptive technologies to wage an integrated campaign in the land, sea, air, space, cyber, and cognitive arenas against its adversaries. And Xi has explicitly called for integrating China's civilian and defense economies through the concept of "military-civil fusion." Fundamentally, this initiative involves sharing research, talent, resources, and technology between the two ecosystems (allowing the military to capitalize on China's world-class innovation economy).[3] These combined efforts have not gone unnoticed; rather, they have spurred alarm in US policy circles.

It has become an article of faith among US leaders that technological innovation grants states a decisive edge in strategy and war. In major speeches, policymakers warn that advances in science and technology "have the potential to reshape the global balance of power, spark entirely new industries, and revolutionize the way we live and work." These ideas provide the intellectual scaffolding for America's tech war against China. Simply put, if technology is the key to power in the twenty-first century, then it is a "national security imperative for the United States to achieve and maintain unquestioned and unchallenged global technological dominance."[4] In practice, this entails imposing onerous tech export controls against America's rivals, limiting the proliferation of military technologies, and prioritizing security concerns over economic interests.

But is this view correct? How decisive is technology in matters of geopolitics and war?

I would argue that historically, the connection between technological innovation and achieving of geopolitical interests is not quite so linear as proponents suggest. The spread of firearms in Europe in the sixteenth century is a good illustration. Gunpowder and firearms were not exclusive to Europe in this era—many empires from Mughal India to Ottoman Turkey had access to these weapons. But Europe became the global leader in producing and developing them. Once firearms technology became reliable, Europeans quickly ended the dominance of Central Asian horse archers, who were little match against infantrymen equipped with guns. European armies then turned their firearms against each other, changing the offense-defense balance on the battlefield. Formerly impregnable stone fortresses were now vulnerable to artillery assaults; the new weapons made short work of many "old stone castles scattered around Europe." As political theorist Niccolò Machiavelli himself wrote, "No walls exist, however thick, that artillery cannot destroy in a few days."[5]

But the apparent triumph of firearms proved illusory. Soon, the best minds of the day were devising adaptations meant to blunt the offensive advantages of firepower. Leading lights in Italy, such as Michelangelo and Leonardo da Vinci, came up with an answer. Max Boot describes how they designed "thick, low-slung earthwork fortresses bristling with artillery ingeniously positioned in angled bastions to command interlocking fields of fire." This style of fortress became known as *trace italienne,* and it proved pivotal. These forts were not only impenetrable to most contemporary armies, but their proliferation turned the tides of conquest itself: no invader "could hope to unify all of Europe" as a result.[6] Within a decade, the initial advantages conferred by gunpowder were supplanted by new adaptations.

Countries that dominate leading industries—such as Germany and Japan's mastery of the steam engine in the early twentieth century, or Russia and the United States leaping into the atomic age in the aftermath of World War II—can gain crucial military and geopolitical benefits for a period of time. But there are limits to how much technological innovation can propel military victory or foster long-term geopolitical dominance, for several reasons.

First, rivals do not stand still. As historian Paul Kennedy writes, the race for new technologies "never remains constant."[7] Countries that are first movers—those that initially figure out how to use a new military innovation—soon face emulation by their adversaries. One week into its 2022 invasion of Ukraine, Russia had massed a forty-mile mechanized column to attack Kyiv from the north. Due to a combination of miscalculation and ineptitude, the column had grown to include one thousand tanks, twenty-four hundred mechanized vehicles, and fifteen thousand troops—all largely organized in a single-file line. Within days, Ukrainian troops blew out key bridges and destroyed vehicles sitting at the head of the convoy, forcing it to halt. Then, a mix of special forces and volunteer operators equipped with night-vision goggles and thermal-imaging drones wreaked havoc against Moscow's troops. Said one soldier, "We are like a hive of bees. One bee is nothing, but if you are faced with a thousand, it can defeat a big force. We are like bees, but we work at night."[8] Heavy pressure from Ukrainian drone strikes and artillery brought the convoy to a standstill. After four weeks, the Russians were forced to withdraw.

For months afterward, Ukraine deployed drones to great effect, recapturing lost territory and delivering immense losses to Russia's troops. But Moscow began to adapt. It implemented electronic countermeasures that caused Ukrainian drones to crash harmlessly. Its military acquired its own UAVs, thousands of Iranian Shahed one-way attack drones that terrorized Ukrainian cities. Later, Russian factories started mass-producing "first-person view" drones to seize the advantage in the trenches. As Luke Mogelson wrote in *The New Yorker*, this innovation-adaptation pattern was "emblematic of a vicious cycle in which Russia absorbs Ukrainian ingenuity and turns it back against Ukraine, spurring further lethal ingenuity."[9] As a result, Russia reduced Ukraine's qualitative edge on drone warfare to a draw.

This leads to a second point: unlike in prior eras, technological development today is characterized by open innovation. In the Cold War, states hoarded their capabilities and limited the availability of their technologies. "In a period of closed military innovation, states can largely control access to major technological developments such as nuclear weapons, jet fighters, or precision-guided munitions," writes Carnegie Mellon scholar Audrey Kurth Cronin. "These technologies are hard to develop, require a high level of expertise, and are carefully protected." The closed-innovation model dominated for most of the twentieth century through the end of the Cold War. As changes swept through the world and US leaders saw the Soviet Union crumble before their eyes, a post–Cold War "euphoria" about the possibilities of a new world order took hold. The government opened up its innovations and made them publicly accessible. ARPANET became the internet; GPS and smartphone technology were commercialized. Emerging technologies not only became publicly available for use—giving individuals and companies free rein to "buy, use, and distribute them"—but the changed mindset also allowed for experimentation and for people to come up with new ideas, forms, and purposes for them.[10]

Yet the open innovation model also had its shortcomings. It meant that the bulk of research and new innovations would no longer come from government labs or from public funding. Instead, the private sector became the primary source of technological innovation. Now, new technologies are far more likely to be invented by private companies, financed through private sources of capital, and marketed to clients around the world with fewer regulations. Technologies are often dual use, harboring both military and civilian applications. While private companies have arguably given rise to a new age of innovation, they have also led to painful trade-offs: governments are compromising their ability to control the spread of advanced products that strengthen their geopolitical advantages and militaries.

Tech giants like Apple, Google, and Microsoft wield influence once reserved to nation-states, giving them an outsized ability to impact diplomatic strategies and military decisions. Certain industry leaders, like Elon Musk, have secured extraordinary sway. Tech companies are at the forefront of the contest toward attaining artificial general intelligence (AGI). They control vital infrastructure—cloud computing platforms, telecommunications networks, large data centers, and high-speed satellite networks. Companies are leading the development of more powerful and innovative weapons. Private tech platforms control the information space, setting the rules for how individuals and even governments can communicate with one another.

Third, technological innovation is a necessary but not sufficient condition for military or geopolitical success. Scholarship tells us that how a country uses its innovations—the manner in which it incorporates new technologies into its military structures and institutions or the way these innovations diffuse—makes all the difference. In his book *The Diffusion of Military Power*, Michael Horowitz, a political scientist at the University of Pennsylvania, argues that whether a country can successfully adopt and use an innovation is dependent on resources and organizational capacity. No matter how "intrinsically compelling a new innovation may seem, it will not diffuse throughout the system" if the costs are too high and the capacity requirements are too onerous."[11]

Stephen Biddle, a prominent military historian, analyzed sixteen interstate wars fought between 1956 and 1992. He found that technological superiority was "no better than a coin flip for predicting victory and defeat." Instead, the real differentiator between victory and defeat was force employment. Militaries that mastered the "modern system" of warfare—which in the offense involved "cover, concealment, dispersion, small-unit independent maneuver, suppression, and combined arms integration" and for defense entailed "depth, reserves, and counterattack"—reduced their vulnerability to twenty-first century weapons and sensors and allowed them to overcome the impact of technological change.[12]

Political scientist Jeffrey Ding makes a similar point. He notes that influential theorists have long argued that "dominating cycles of innovation in key technologies propels a rising power to global leadership." But Ding contends that being the first to introduce a novel technology offers limited benefits. First mover advantages "can be offset by other factors that affect the rate at which a country adopts new innovations at scale," from the strength of communication channels to the density of relationships that connect research institutions and companies. In other words, without requisite organizational capacity, resources, or diffusion, the innovation itself only carries so much weight.[13]

As we approach a potential moment of technological transformation, it is useful to keep in mind a simple idea: geopolitical and military outcomes are influenced by a complex array of factors; technological innovation is just one piece of a larger puzzle.

An AI Arms Race?

Today, the globe faces an intensifying technological competition led by China and the United States. A key question is whether the world has entered a new age of technology more akin to the first industrial revolution (which pioneered steam engines, machine tools, and the use of basic materials like iron and steel), the second industrial revolution (which saw the advent of the internal combustion engine, electrification, railways, the telegraph, and more efficient manufacturing), or perhaps the information revolution, which ushered in the age of computers and the internet. What makes this age of technological innovation distinctive and how does it compare to prior eras? Answering this question is critical—for it will dictate how governments and companies allocate resources and investments.

The current landscape is dominated by a fascination with artificial intelligence technology. The momentum behind AI is rooted in three developments: increased availability of data from public and private sources, enhanced machine learning and algorithmic approaches, and advanced computer processing. Breakthroughs in AI are occurring with regularity but are dependent on exponentially greater demands for computing power (or "compute" for short). One study suggests that in the past thirteen years, "the amount of compute used to train leading AI systems has increased by a factor of 350 million."[14] Advanced computing is tied to a range of objectives: supporting scientific breakthroughs, enhancing intelligence collection, and protecting critical infrastructure from cyberattacks, just to name a few. It is also foundational to military applications of AI, an issue that has seized policymakers' attention.

How might AI affect military outcomes? Henry Kissinger, Eric Schmidt (the former CEO of Google), and Daniel Huttenlocher (a dean at MIT) outlined a thought-provoking line of reasoning. In the past, strategic technologies possessed one or two of the following characteristics: (1) carried both civilian and military purposes, (2) spread easily and widely, (3) exhibited massive destructive potential. For example, railways possessed both civilian and military attributes; they could carry troops to the front lines of battle or bring goods to market. Railways were also easily adopted by other states (at least those with the requisite resources and organizational capacity). But railroads offered no destructive potential. In a similar vein, nuclear technologies are also dual use and can generate immense destructive power, but they do not proliferate easily (at last count, only nine states officially possessed nuclear weapons).

AI, on the other hand, "breaks this paradigm." It is clearly a dual-use technology, it proliferates easily and widely, and its applications have tremendous destructive potential. Not only can AI-enabled weapons help adversaries launch digital assaults with exceptional speed, but "a state may effectively have no time to evaluate the signs of an incoming attack. Instead, it may need to respond immediately or risk disablement." Imagining a world where AI technology is paired with, say, nuclear weapons underscores the destructive implications of this technology.[15]

However, we should be careful not to simply assume that AI will bring about such revolutionary transformations. Much of the concern involves AGI systems, which analysts suggest would have the potential to equal or surpass humans "across a wide range of cognitive tasks—the equivalent of millions of brilliant minds working tirelessly at the top of their fields at machine speed." Some experts contend that not only will AGI unlock scientific discoveries and enhance economic productivity, but it will confer decisive national security benefits, such as inventing a "wonder weapon" that would completely disable an opponent through a "splendid cyber strike," or developing an advanced autonomous weapon that would lead to military dominance.[16]

But such outcomes are atypical. "History suggests that technological breakthroughs rarely yield wonder weapons that provide an immediate, decisive impact on military balances or national security," write Jim Mitre and Joel B. Predd for RAND. "Except for rare examples, such as nuclear weapons, cultural and procedural factors drive an institution's technological adoption capacity and are more consequential than being the first to achieve a scientific or technological breakthrough."[17] Given the low odds of the wonder weapon scenario materializing, it makes little sense to spend excessive time or resources preparing for that eventuality.

Aside from wonder weapons, other disruptive possibilities could emerge from AGI. For example, it could recalibrate the instruments of national power, destabilizing the current global order. As AGI became rooted in societies, it could uplift certain military capabilities, such as autonomous warfare, and provide big advantages to the side best able to master it. Alternatively, it could reshape national economies through a massive boost in productivity or a major disruption of the labor market.[18] But history indicates such scenarios take time to emerge and are contingent on a host of economic, political, and organizational factors, rather than guaranteed by the invention of a unique, breakthrough technology.

In my own research, I have explored the impact of technology on political repression. Dictators have indeed acquired sophisticated digital tools; they can use facial recognition to surveil their populations, deploy spyware to hack into opponents' phones and peer into personal photos and messages, and install filtering software to censor what citizens can read and access online. Given the plethora of new techniques available to autocratic leaders, the "rise of digital repression" would appear to be a game changer for authoritarians. But advanced technology has not yet enabled autocrats to extinguish democracy globally. Around the world, even in semi-closed political environments, democratic opposition remains robust; there are copious examples of protest movements mobilizing citizens onto the street to overturn despotic regimes.

When people ask whether digital technologies—including AI systems used for repression—are fundamentally tilting the playing field in favor of authoritarian stakeholders, my general answer is no. In high-capacity repressive states, like China or Russia, digital tools can have a transformative effect. But those states are the exception, not the rule. In countries that have lower institutional capacity and diverse political incentives, or where democratic safeguards still exist, there are limits to how much digital technologies can support government coercion.[19]

It is difficult to say whether AI marks a revolutionary new era of technology; at this early stage, we simply cannot know whether it represents a paradigm shift or instead will bring more incremental progress. There are strong indications that AI technologies—and a complement of other emerging technologies, such as robotics, quantum, biotechnologies, and clean energy—have the *potential* to transform certain aspects of global affairs. But their impact remains speculative.[20] What we can acknowledge is that significant technological change is occurring at a fraught geopolitical moment. Competition between the United States and China (as well as Russia) has intensified; by necessity, this makes the current innovation race important even if we don't yet know if AI will be transformational.

The challenges to America's technological dominance originate from sources beyond Beijing and Moscow alone. Smaller nations and even nonstate actors are quietly changing the global order as well. While these actors rarely take on great powers directly, their maneuvers are eroding larger states' influence.

3

Rising Tech Powers

In the fall of 2020, in a largely overlooked war, Azerbaijan achieved a decisive victory against Armenian-backed forces in the disputed enclave of Nagorno-Karabakh. Dating back to the waning days of the Soviet Union, the territory had witnessed periodic heavy fighting and major displacement. Nagorno-Karabakh is a landlocked mountainous region located between Armenia and Azerbaijan in the southern Caucasus. Its geographic area is small, roughly the size of Delaware, with a population numbering under 150,000. Like many disputes linked to the former Soviet Union, the original point of tension lay far in the past—in this case, Joseph Stalin's decision in 1921 to designate the region as part of the Soviet Social Republic of Azerbaijan rather than Armenia. At the time, the bulk of Nagorno-Karabakh's population was ethnically Armenian; his decision set the stage for decades of clashes. Already, the two sides had fought a brutal war between 1988 and 1994. The dueling armies pummeled each other and took out their fury on the civilian population. Eventually, Armenia emerged as the victor but at a heavy cost, sustaining thirty thousand civilian and military casualties. Azerbaijan faced even worse consequences, suffering eighty thousand civilian and military casualties as well as the loss of Nagorno-Karabakh and adjacent territories. The warring parties never signed a peace treaty after the fighting stopped. Rather, Nagorno-Karabakh became a "frozen conflict" where the two armies would remain deadlocked for the foreseeable future.[1]

The 2020 war broke this pattern. In the short fighting that ensued, Azerbaijani forces decimated Armenia's military thanks to high-tech drones and missiles supplied by Israel and Türkiye. While Türkiye was a longtime backer of Azerbaijan, Israel's alliance was more recent. Starting around 2016, Israel reportedly supplied Baku with nearly 70 percent of its arsenal, giving Azerbaijan a major edge while boosting Israel's defense industry. Israeli Harop loitering munitions—small triangular suicide drones that issue a "shrill scream" when diving toward a target—caused panic among

Armenia's troops. When soldiers heard the Harop streaming toward them, they knew they had "seven seconds" to take cover before it struck."[2]

Türkiye's Bayraktar TB2 drones played an equally destructive role in the war. A couple of months before hostilities started, Azerbaijan purchased about two dozen TB2s. Reportedly, they were deployed so quickly that there wasn't enough time to train Azeri pilots to operate them—Turkish pilots had to step in instead. As the battle raged, footage was broadcast showing TB2s systematically picking off Armenian radar systems, missile batteries, and tanks. On the first day of the war alone, the Armenians lost half their antiaircraft forces and 40 percent of their artillery in the span of fifteen minutes. Their elaborate trench lines and fortified bunkers, better suited for World War I, were no match for the modern-age systems Azerbaijan unleashed. Within days, Armenia's forces were in ruins. Azerbaijan's military destroyed or captured over 250 tanks, 70 armored fighting vehicles, 5 Russian S-300 missile defense systems, and 29 self-propelled artillery systems from their Armenian rivals. All told, Azerbaijan claimed to have wiped out over $1 billion worth of military equipment. Still, it stopped short of taking over the whole territory due to pressure from Moscow—leaving a rump state governed by the Karabakh Armenian population.[3]

After hostilities ceased, Moscow deployed two thousand troops to keep peace in the region. One of their primary responsibilities was to protect a single road linking Nagorno-Karabakh to Armenia. This vital lifeline allowed trucks to bring in food, medicine, and other critical goods to the capital city of Artsakh. Despite the Russian presence, over the next three years tensions simmered.

Then, in 2023, Azerbaijan spotted a new opportunity. Its autocratic president, Ilham Aliyev, saw that Moscow was distracted by its war in Ukraine and was loath to jeopardize ties with Azerbaijan and its backer Türkiye. Early in the year, Azerbaijani troops closed the road and imposed a blockade on Artsakh. Said a former resident, life became unbearable: "There was hunger. No cigarettes, no bread, nothing."[4] As Aliyev slowly strangled the enclave, Russian peacekeepers stood by and did nothing. By the fall, Aliyev was ready to make his final move. On September 19, he ordered a lightning strike against Artsakh and this time Azerbaijan finished the job. Once more relying on Israeli Harop drones and LORA quasi-ballistic missiles as well as Turkish Bayraktar TB2 drones, Azerbaijan forces overran the enclave in a single day. Armenian leaders had little choice but to relinquish control of the territory. "We are all in shock. Everyone understands that this is the end—the complete destruction of Artsakh," said Benyamin Poghosyan, formerly the head of the Armenian defense ministry's research unit.[5]

Azerbaijan's victory came despite the presence of Russian peacekeepers (several of whom were killed during the skirmish), as well as recent moves by Armenia to align more closely with the United States. Kremlin spokesperson Dmitry Peskov could only offer empty words about the sudden realignment. "It is hardly possible to talk about who is to blame," he insisted amid questions about Russia's failure to intervene. He described Baku's victory as "a new system of coordinates" and advised residents to "get to know the agreements on living under Azerbaijani rule."[6]

What lessons could be drawn from Azerbaijan's victory? For one, advanced weaponry from Israel and Türkiye made a critical difference on the battlefield. Careful preparation by Baku in close coordination with its allies showed the devastating effect of new precision systems. Azerbaijan's plan in both wars was to use massive precision fire to overwhelm Armenia's key defense networks—its air defense radars, electronic warfare systems, and command and control hubs—thereby blinding Armenia's forces. While their opponent staggered to reorganize, Azeri forces would go for the jugular, targeting tanks, armored vehicles, logistical sites, and troops themselves. By the time Armenia's military knew what had hit them, the war would be effectively over.

The conflict also showed that a smaller state like Azerbaijan needn't turn to a large power such as the United States, Russia, or China to achieve victory against its opponent. Weaponry from Israel and Türkiye worked just fine—and likely came with a far lower price tag. Said Ben Wallace, the UK's former defense minister, Türkiye's TB2s illustrated how other countries were now "leading the way," noting that its drones were "responsible for the destruction of hundreds of armored vehicles and even air defense systems." Middle powers were prepared to commit enormous sums to "invent, adopt, and adapt" new technologies.[7] They recognized that high-tech innovations would help them increase their military might, enhance their economic leverage, and boost their global appeal. Traditionally, experts assumed that technological innovations followed a hierarchical path: innovations first appeared in advanced countries and subsequently spread outward to less sophisticated or smaller countries. Yet accumulating evidence casts doubt on this assertion, indicating that these countries present real challenges to America's strategic interests. Countries like Azerbaijan can turn to other middle powers to achieve their security objectives, which directly undercuts Washington's influence. This leads to the question: Which specific countries are leading the charge and deserve the attention of larger states?

"Rising Tech Powers" Challenge Large States

Determining a country's standing in the global pecking order is harder than it seems. One common metric for assessing a country's relative power is to look at population size and geographic territory (along with GDP and military size). Typically, countries with less than 10 or 15 million inhabitants are considered small states—possessing commensurately low levels of influence. But this is arguably an outdated measure of power. It fails to take into consideration other factors, such as technological capacity, military and administrative capabilities, and ideational influence (the degree to which a country's ideas influence citizens in other parts of the world).[8]

Take a country like Israel. It sits on a tiny sliver of land, barely 8,630 square miles, and its population is just under 10 million. On these factors alone, Israel should classify as a small state. Yet a deeper look proves otherwise. Its military expenditures are huge relative to its size. Israel ranked fifteenth globally for its level of defense spending, allocating $27.5 billion to its military in 2023—ahead of much larger countries such as Canada and Brazil (as well as exceeding the defense spending of regional rivals like Iran and Türkiye). Israel's distinctive position extends beyond its military spending. Its economy is quite significant, possessing a GDP of $550 billion, which ranks thirtieth in the world. Its innovation capacity is also notable—the World Intellectual Property Organization (WIPO) places Israel's innovation ecosystem as fourteenth best in the world.[9] Clearly, Israel wields military and economic influence equivalent to that of a much larger state despite its small population and geographic territory. So what factors should be relevant?

For our purposes, I have identified a slate of countries that I term "rising tech powers"—states that wield an independent foreign policy, and have a high degree of innovation capacity and military expenditures alongside a robust economic base and considerable regional influence.[*10] I used the following criteria to come up with this list: My initial set of countries included

* The question about how to properly categorize a cohort of regional powers is a major topic of discussion for the US government. The Office of the Director of National Intelligence, for example, identified a set of "regional powers" in its "Global Trends 2040" report, noting that states such as "Australia, Brazil, Indonesia, Iran, Nigeria, Saudi Arabia, Turkey, and the United Arab Emirates" will seek to take on roles "previously filled by a major power" to maintain regional stability or enhance their influence. Regional powers "may seek to build their own coalitions or strengthen regional blocs to project influence and in some cases, collaborate on global challenges, but in other cases they may act more aggressively in conflicts in their region." However, the report failed to provide criteria for determining which countries would qualify for this grouping and the logic behind it.

the world's fifty largest economies by GDP. I then excluded the three large powers (China, Russia, United States), EU member states, Group of Seven (G7) members, countries in the Five Eyes intelligence sharing alliance (Australia, Canada, New Zealand, the United Kingdom, and the United States), and South Korea, as these countries are either part of an advanced economic or political grouping or are intimately involved in a common military alliance with the United States. That left twenty-four countries in the pool.

Next, I incorporated other measures associated with a country's innovation and military capacity: global AI and innovation rankings, annual military expenditures, and exports of military arms. I filtered out countries that didn't possess requisite technological capacity—those states placing outside the top seventy-five for the AI innovation rankings or for WIPO's innovation index were excluded. Then, I scrutinized military expenditures and military exports for the most recent year of data (2024)—countries that didn't rank in the top sixty for either category were likewise eliminated. This produced a set of ten countries, including two special cases: India and South Africa.

India easily met all the criteria. But as the world's most populous country and a state with growing heft on the international stage, India's capacity was disproportionately larger than the other group members', so I decided to exclude it from the cohort. Conversely, South Africa exhibited borderline numbers. It ranked low on the innovation index and for its AI capabilities. However, South Africa is also a founding member of the BRICS grouping and wields regional influence, and therefore warranted inclusion.

Country	Population (millions)	Geographic Size (sq mi)	GDP 2024 (billions)[11]	Military expenditures 2024 (billions)[12]	WIPO Innovation Ranking 2025[13]	Tortoise Global AI Index 2024[14]
Brazil	212	3,287,357	2,310	24.4	52	30
Iran	91.6	636,400	464	8.04	70	60
Israel	10	8,522	551	33.75	14	9
Saudi Arabia	35.3	830,000	1,140	71.73	46	14
Singapore	6	286	562	15.17	5	3
South Africa	64	470,900	418	2.9	61	69
Taiwan	23	13,976	814	18.86	NA	21
Türkiye	85.5	302,535	1,460	14.27	43	34
UAE	11	32,278	569	22.27	30	20

Table 1. Nine Countries Making Up the Rising Tech Powers†

† For further detail about the ranking methodology, see Appendix I.

These countries are technology makers, not just technology takers. They are invested in building advanced military capabilities but are also focused on developing strong innovation ecosystems. While rising tech powers are distributed globally, the two biggest clusters are located in the Middle East and East Asia.

Members of this group are heterogeneous. One category of states—Israel, Singapore, Taiwan—are recognized innovators that manufacture vital components for cutting-edge civilian and military systems. Some of their companies—Taiwan's TSMC, Israel's Elbit Systems, and Singapore's Sea Limited—are leading global brands. A second group of states views emerging technologies like AI as the key to their future success. Countries like Saudi Arabia and the UAE are investing billions to develop their industries and form partnerships to build data centers and network hubs around the world. A third category of countries, such as Iran and Türkiye, have established a strong niche developing "fast follower" military innovations for items like drones and precision missiles. (Iran has even taken steps to develop an AI arms industry, despite crippling sanctions, including making AI enhancements to its Shahed drone.[15]) As global instability and conflict reemerge, their products are in high demand. In short, these states have differing resources, interests, and incentives, causing them to respond uniquely to new innovations.

How Military Actors Respond to Innovations

As military and technological innovations emerge—such as autonomous drones or satellite-based weapons—countries face difficult choices about how to respond. Some countries will pursue adoption. They may attempt to replicate the innovation—either selectively or in full. Or they may adapt the innovation to suit their needs. A second possibility is non-adoption. Rather than copying advancements, countries or insurgent organizations could pursue offsetting strategies, either undercutting the original innovation or developing alternative technologies engineered to defeat it.

Türkiye's development of military drones is a useful illustration. Ankara came to drone manufacturing late; while the United States had close to eleven thousand UAVs in its arsenal by 2011, it was not until 2015 that the Turkish company Baykar debuted its TB2 drones. Yet, by 2022, Baykar had amassed seventy-eight global orders for its drone systems, each consisting of multiple UAV units, making Türkiye the most prolific global exporter of military-grade drones after the United States.[16] How did it make such dramatic inroads in such a short period of time? The answer: creative adaptation.

Selçuk Bayraktar, the inventor of the TB2 drone and Baykar's chief technology officer, realized several things about the UAV market. First, American drones were priced way too high. General Atomics' MQ-9 Reaper drone, a popular export model, costs around $16 million per unit. For many countries, this cost was simply out of range. Bayraktar priced TB2s much more cost-effectively at $1 million to $2 million per unit. Second, Bayraktar understood that countries would be responsive to that price point only if they were confident that cheaper Turkish units would be effective. After all, why would they bother purchasing lower-priced drones that were unable to precisely strike targets, or were sitting ducks for air defenses? To that end, Bayraktar outfitted TB2s with a host of capabilities—laser-guided munitions, high-altitude flight, and extensive operational range. While they were not quite as good as state-of-the-art offerings from US manufacturers, they weren't far off. Third, buyers of Türkiye's products faced far lower export barriers. While military exports from the United States and Europe must go through extensive review processes to ensure human rights and foreign policy compliance, Türkiye's government didn't impose such conditions. Bayraktar could tell prospective clients from nondemocratic states like Azerbaijan or Libya that a handshake that day would lead to a delivery within a year.[17]

Few actors have been able to adapt military innovations as successfully as Türkiye. Instead, many countries and militant groups adopt offsetting strategies instead.

On October 7, 2023, Hamas initiated a surprise attack against a variety of targets in Israel, assaulting towns, army bases, even a music festival. One of the biggest obstacles faced by Hamas fighters was the presence of the "Iron Wall," a set of technologically enhanced physical barricades running the entire forty-mile length of the Gaza border, bristling with surveillance cameras, radar systems, and underground sensors. Hamas fighters didn't have the means to directly attack the wall. But that didn't stop them. Instead, they used a combination of low-tech maneuvers and simple machinery to neutralize Israel's advantage. First, they acquired off-the-shelf commercial drones (models that any user could purchase from Amazon for about three hundred dollars), retrofitted the drones with grenades, and used them to "bomb" Israeli observation towers and blind the towers' sensors. Next, they fired more than three thousand rockets into Israel, targeting both the Iron Wall and surrounding communities. While the commotion distracted Israeli defenders, a first wave of militants flew into the territory on fan-powered hang gliders and paragliders. Hamas then tapped explosives to blow gaping holes into different sections of the wall. Finally, they used bulldozers to widen the holes so that larger vehicles could go through the breaches.[18] None of these tactics required advanced technologies or expensive military armaments. Instead, Hamas's creative use of simple tools was enough to offset Israel's technological superiority and temporarily put the Israel Defense Forces (IDF) on its back foot.

Offsetting strategies aren't confined to using cheap tools against more advanced opponents. They can also involve policy offsetting—using "export controls, arms control treaties, and other multilateral regimes" to limit the reach of new innovations. For example, in response to Iraq's continual use of chemical weapons in the 1980s against Kurdish and Iranian forces, countries such as Australia pushed for more stringent enforcement of the Chemical Weapons Convention. They wanted to avoid a chemical arms race, so they turned to treaties to constrain the spread of these weapons.[19]

States also form alliances in response to technological innovations—allowing them to access powerful technologies to defend against their opponents. In some cases, countries will align with a leading state generating the innovation (known as "bandwagoning"). In others, states will create or join alliances against a more powerful country (referred to as "balancing"). Among the rising tech powers, Iran relies heavily on bandwagoning with both China and Russia against the United States. Tehran has received sophisticated Russian military technologies, including Su-35 fighter jets, Mi-28 attack helicopters, and possibly the S-400 missile defense system. Russia's space program has even sent Iranian satellites into orbit, signifying growing collaboration in the space industry. In return, Iran has filled in needed gaps for Russia during its war in Ukraine, supplying Moscow with munitions, artillery shells, and attack drones (including building a drone facility in Russia's Tatarstan region). Iran has also established a close tech relationship with Beijing, contracting with Chinese companies such as Huawei and ZTE to build out its internet network infrastructure. In 2024, reports surfaced that Iran was chasing deals with two Chinese satellite companies that would help it carry out remote surveillance and gather intelligence for military targets.[20]

Rather than lock into a "rigid alignment" with a major power, another option is for countries to hedge. This is an insurance-seeking strategy where states cultivate independent relationships with multiple large countries in order to preserve their options. Countries such as Brazil, India, Saudi Arabia, Singapore, and the UAE demonstrate textbook hedging behavior.[21]

Brazil is a good case in point. It remains a close technology partner of China's despite American efforts to reduce their cooperation. Brasília continues to use Chinese company Huawei to administer its 5G network; President Luiz Inácio Lula da Silva personally toured a Huawei research facility during a 2023 visit to China (during the same trip he also showed an interest in building a Chinese semiconductor factory). Lula has insisted that "nobody can stop Brazil from continuing to develop its relationship with China." When reporters asked foreign policy adviser Celso Amorim about the United States' perceived displeasure with the country's relationship with China, he responded, "We are not afraid of the big bad wolf." Under Trump, relations between the two countries have taken a turn for the worse. In July 2025, the White House announced a 50 percent tariff hike for all Brazilian goods, accusing President Luiz Inácio Lula da Silva of undertaking a "politically motivated persecution" of former president Jair Bolsonaro (a rightwing populist who shares common ideological ground with the MAGA movement). In response, Lula declared that Brazil "will not be anyone's colony."[22]

Yet these developments do not mean Brasília has decisively shifted toward Beijing. Many Brazilians are concerned about China's assertive foreign policy; a plurality has negative views about Beijing and worries that the country will fall too deeply under China's sway. Brazil has formed crucial partnerships with the United States, including signing a defense cooperation agreement with America and becoming a major non-NATO ally in July 2019. Given Brazil's long history of hedging, it will continue threading a middle ground. "For Brasília, Beijing constitutes an indispensable strategic partner, but Washington, even after its recent aggressions, remains an irreplaceable global power," writes Hussein Kalout, Brazil's former special secretary for strategic affairs.[23]

Table 2 breaks down strategies adopted by middle powers and smaller countries in response to technology and military innovations.

Adoption	• Imitation (replicating an innovation in its entirety) • Adaption (modifying an innovation) • Selective emulation (replicating part of an innovation)
Non-adoption	• No response • Offsetting (methods to prevent, counter, or limit an innovation) • Developing a new innovation (in lieu of imitation or adaptation)
Policy offsetting	• Leveraging treaties or other agreements to limit innovations • Developing or joining alliances (balancing, bandwagoning) • Hedging (maximizing options without choosing sides) • Neutrality (referred to as "hiding")

Table 2: How States and Militant Groups Respond to Major Technological or Military Innovations[24]

The strategic calculations of middle-tier and smaller states—whether they pursue adaptation, offsetting, or something else—bring complications. Great powers need to guard against challenges from smaller states and ensure that new innovations won't undermine their security (Türkiye's export of drones to Ukraine and their subsequent deployment against Russia's military is a classic example of what's at risk). They must also compete with their rivals to influence smaller states. American officials recognize this dynamic and are pushing back against China's sway across the world by dangling the possibility of investments or wielding sanctions threats. Since returning to the White House, Trump has used coercion and bullying to get countries to do America's bidding. The results have been lackluster. Said one African diplomat, "Threats may get results for a short time, but over a long time, people are taking notes: We're not going to put all our eggs into America."[25]

The end of the Cold War left the United States sitting alone atop the international order. The world enjoyed relative peace as free trade and globalization took center stage. But these conditions wouldn't last. Rivals, such as China, were biding their time to challenge America's predominance. Meanwhile, regional powers were gaining influence and using asymmetrical means to compete in war and technology. This story of geopolitical competition is not unique; history is full of similar rivalries.

But something else emerged out of the era of free trade and globalization. Private tech companies became increasingly vital. As US officials pivoted to an open innovation approach, allowing the public to buy, use, adapt, and distribute emerging technologies developed in government labs, commercial firms emerged as power brokers on the world stage. "ARPANET became the internet. Tax dollars developed the Global Positioning System," writes scholar Audrey Kurth Cronin. "All of the major components of smartphones derive from US government programs."[26] Subsequently, research and innovation shifted to private hands. Today, private actors oversee the lion's share of cutting-edge R&D. New innovations largely come from private labs, are backed by private capital, and are sold to clients worldwide. These companies are beholden to shareholders and profits, and they are changing the rules of the game.

PART II

The Private Tech Revolution

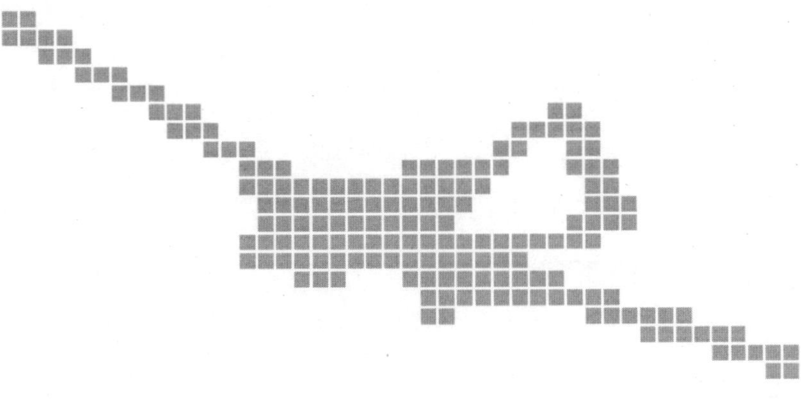

4

Snowball Units and Starlink Terminals

Mykhailo Fedorov, the youngest member of Volodymyr Zelensky's cabinet, was worried. In 2022, Vladimir Putin had ratcheted up his rhetoric against Ukraine, denouncing the country's growing relationship with the United States and its allies. In an end-of-the-year news conference, Putin warned that he needed guarantees "right away, right now," that Ukraine would never join NATO. He also demanded the withdrawal of all US nuclear weapons from Europe and insisted that troop rotations through former Warsaw Pact states (such as Poland, the Czech Republic, and Bulgaria) cease at once. Following these remarks, Putin went conspicuously silent, but Russia's actions spoke volumes. Moscow amassed thousands of troops along Ukraine's border, as well as tanks, artillery, advanced missile systems, helicopters, fighter jets, and antiaircraft batteries. Russian soldiers surrounded Ukraine on three sides, from Crimea in the south and the Donbas in the east to Belarus in the north. As January turned into February, the atmosphere was electric with tension.[1]

Fedorov, who, at the time, served as Ukraine's minister of digital transformation, dressed casually, often in a hoodie and jeans. He got just three or four hours of sleep a night, interrupted every thirty minutes or so by alerts from his iPhone. He wasn't responsible for military strategy, nor was he in charge of Ukrainian diplomacy. But he was entrusted with something just as important—safeguarding all of the nation's data and its ability to communicate internally and with the outside world. For years, Ukraine had physically stored all government records and citizen information on local servers within its borders, due to a data protection law that prohibited the government from storing data in the cloud. Fedorov knew these facilities were vulnerable to missile attacks and that their destruction could paralyze the country.[2]

As the drumbeats of war grew ever louder, Fedorov moved into action. On February 17, 2022, just days before Russian troops flooded into Ukraine, he worked with the parliament to amend the data law and allow private cloud providers to host government data in Europe. Right away, Fedorov struck deals with Amazon, Microsoft, Oracle, and Google to transfer critical information. This wasn't an easy process. As Liam Maxwell, a senior official at Amazon Web Services, recounted, they used "pen and paper" to outline a list of the most important items to move: the population registry, land and property records, tax payment receipts, bank records, education lists, and anti-corruption databases. The project involved twenty-seven Ukrainian ministries, eighteen universities, Ukraine's largest K–12 remote-learning school (which served "hundreds and thousands of displaced children"), along with dozens of businesses, such as PrivateBank, Ukraine's biggest financial institution.[3]

To get these records safely out of the country, Amazon flew in "Snowball units"—large "shock-proof gray containers"—from Dublin to Kraków, Poland. Company officials then placed the Snowballs onto trains and transported them into Ukraine. In Ukraine, they downloaded data onto secure devices and put them back into the Snowballs.

Loaded with up to eighty terabytes of encrypted data each, the Snowballs were moved back to Amazon facilities in Europe. Stress levels were high. Said Maxwell, "It's a tense moment around the baggage carousel. Here's government in a box, literally." Once the data was distributed across different cloud facilities, everyone breathed easier. Said Maxwell, "You can't take out the cloud with a cruise missile." Fedorov's concerns proved well founded. Days after the invasion, a Russian missile knocked out a facility in Kyiv housing critical information. Fortunately, said Fedorov, "all of the backups were already transported to other European countries and no damage was done."[4]

Helping Ukraine secure its data was just the beginning. As the conflict deepened, Western tech companies became more and more active supporting Ukraine's government. *The New York Times* compared Microsoft's participation to the role of Ford in World War II, "when the company converted automobile production lines to make Sherman tanks."[5] In the coming months, Fedorov worked the phones (and social media) nonstop, badgering companies to make products available for the government's use and to cease providing services to Russia. One of his most notable exchanges took place over social media with Elon Musk.

In the first week of the war, Russia launched a devastating cybersecurity attack targeting Ukraine's communications routers. The digital strike crippled networks across the country and disabled the military's command system. Without this network, Ukraine couldn't mount a coordinated defense against the Russian attack; a substitute was desperately needed. One of the few options available for the Ukrainians was to use satellite internet services—specifically Elon Musk's Starlink network. Starlink works by constantly pinging internet signals back and forth between its terminals on the ground and low-Earth-orbit (LEO) satellites in space. In order to stay afloat, LEO satellites must travel at high velocities—nearly seventeen thousand miles per hour, which means that a typical LEO satellite will circle the Earth in about ninety minutes. As a result, orbiting LEOs are only briefly in contact with the ground transmitter. To maintain an uninterrupted signal, a second satellite needs to take the place of the first before it disappears from view, so a successful system requires "thousands of satellites operating in synchronicity." Starting in 2019, Musk sent SpaceX rockets with Starlink payloads into space nearly every week. His efforts paid off; as of March 2026, he had nearly ten thousand satellites in orbit.[6]

Fedorov took to X to call for Musk's help: "While you try to colonize Mars—Russia try to occupy Ukraine! While your rockets successfully land from space—Russian rockets attack Ukrainian civil people! We ask you to provide Ukraine with Starlink stations and to address sane Russians to stand." Musk replied, "Starlink service is now active in Ukraine. More terminals en route." Two days later, five hundred Starlink terminals arrived in Kyiv.[7] The coming months would bring thousands more terminals to Ukraine, providing vital communications for its armed forces.

Fedorov's pleading wasn't limited to Starlink. Ukrainian officials were concerned that Russian soldiers were using Google Maps to identify crowds of people for missile strikes or places where Ukrainian soldiers were assembling. (The day before the invasion, a Middlebury professor posted on social media that he had located Russian troops massing near the Ukrainian border using live traffic data from Google Maps.)[8] Fedorov called Karan Bhatia, a senior Google official, and asked him to disable Russian access to live traffic data in Ukraine. Google agreed and also suspended the sales of other products and services, eventually blocking Russian state media's access to YouTube.

He even reached out to Chinese drone maker DJI, alleging that Russia's forces were using its technology to guide its missile attacks. He posted: "@ DJIGlobal are you sure you want to be a partner in these murders? Block your products that are helping Russia to kill the Ukrainians!" Within hours, DJI responded and offered to try to block drone flights by putting geofencing into place throughout the country. Said DJI's spokesperson, the company "does not support any use of our products that harms people's lives." All told, Fedorov's ministry sent more than four thousand requests to various tech companies and organizations requesting them to curb transactions with Russia and provide vital services for Ukraine.[9] And tech companies responded. Hundreds of firms agreed to stop their sales to Russia. Many offered material or financial support to Ukraine.

In fairness, the Ukraine war was hardly the first time large private companies found themselves in a major geopolitical role. The Dutch East India Company, which operated from 1602 to 1799, was valued at $7.4 trillion at its height (adjusted for inflation), worth more than Google, Apple, Meta, Amazon, Nvidia, and Microsoft combined. It possessed government-like powers, such as the right to make treaties with local sovereigns, to build forts and keep military forces, as well as to oversee administrative functions in its colonies. Its counterpart, the British East India Company, also maintained its own army—by 1800 it controlled some two hundred thousand soldiers (surpassing the British Army in manpower). Its soldiers subdued restive states in India, enforced taxation, carried out looting, and fortified Britain's empire. While both entities were chartered by their home governments, thus ensuring a nominal amount of oversight, in practice they operated with a high degree of independence.[10]

American businesses also had a long history of cooperating with the US government in times of war. During World War I, wealthy industrialists such as Henry Ford and J. P. Morgan poured substantial resources into supporting the war effort. Ford's factories manufactured war boats, military vehicles, artillery, and many other products. Morgan personally extended enormous loans to support the Allies and organized a consortium of banks to provide hundreds of millions more to France, Russia, and the UK. In World War II, Ford once again retooled its factories to build equipment for the US military, such as tanks, jet engines, and even the B-24 bomber.[11]

But Big Tech's active support of Ukraine represented a sea change for the industry. For the most part, Silicon Valley's tech giants had deliberately avoided getting involved in geopolitical matters. Their job wasn't to take sides in a conflict or get entangled in diplomatic maneuvering. They instead followed the "Californian Ideology," a worldview "that is simultaneously countercultural in lifestyle, laissez-faire in economics, and libertarian in politics." This "suit-hoodie divide" between technology companies and the government manifested itself in slogans such as Google's "Don't be evil" (a mantra it quietly dropped in 2018). The idea was simple: networks and software could empower individuals to change the world. Google should confine itself to designing cutting-edge products and not get tangled up in foreign policy debates. Governments were seen as obstacles; they'd either suffocate companies with burdensome regulations or exploit their products in harmful ways. "'Don't be evil' became part of Google's identity—internally and to the outside world. It represented a new kind of ethos for a future corporate powerhouse," wrote Shirin Ghaffary and Alex Kantrowitz in *Vox*, "one that would help shape the culture of Silicon Valley and the many tech companies that formed in the past 20 years. Google was meant to be innovating technology to make the world a better place."[12]

Companies such as Microsoft tried to formalize the idea of worldwide tech neutrality. In 2017, the firm's president, Brad Smith, proposed enacting a "digital Geneva convention." Microsoft increasingly found itself pulled in multiple directions by different governments. The United States wanted Microsoft to help it spy on its adversaries. China demanded that Microsoft make a special version of Windows that would be impenetrable to US intelligence. Russia wanted Microsoft to turn a blind eye toward its cyber-hacking operations, which were collecting valuable information on the firm's customers. Smith felt the only way out of this situation was to declare that the company would become a "trusted and neutral Digital Switzerland." In exchange for immunity from attacks from nation-states, Microsoft would agree not to help any governments in their own attacks. As the scholars Henry Farrell and Abraham Newman wrote in their book *Underground Empire*, "The term 'digital Switzerland' implied that tech companies were themselves a new kind of world-spanning country, even if they lacked a territory of their own." Smith's hope was that "geopolitics would stop devouring technology, if only the technology sector was allowed to withdraw from geopolitics."[13]

And yet Smith's push for tech neutrality would barely last five years. When Russia invaded Ukraine, almost immediately Microsoft found itself publicly on the front lines supporting Kyiv. Smith declared that Microsoft was "horrified, angered and saddened" by the war and that the company condemned Russia's unlawful invasion. This wasn't empty rhetoric, either; Smith announced a range of actions the company would immediately take, such as coordinating Ukraine's cybersecurity, pledging over $107 million worth of services to the country, and "working in lockstep" with the United States, European Union, and United Kingdom to stop doing business with Russia. Wrote Farrell and Newman, "It was as if some alternative-world Switzerland had not simply abandoned its World War II neutrality but used miraculous technologies to transport Winston Churchill's government into an impregnable castle in the sky, where it could manage the war and conduct its ordinary affairs above the altitudes traversed by Nazi bombers." Google, Meta, Cloudflare, Amazon, Apple, Salesforce, and scores of other tech firms followed suit. They sent engineers into Kyiv to shore up Ukraine's cyber defenses. They donated servers, satellite terminals, routers, humanitarian aid, and millions in cash. Uber began offering unlimited free rides from the Ukrainian border to Polish cities for refugees and their families. Airbnb announced it would provide free short-term housing for up to one hundred thousand Ukrainians fleeing the war.[14]

What prompted tech companies to switch positions, jettison their traditional neutrality, and openly side with Ukraine?

A Delicate Balance of Interests

In the postwar era of the 1950s and 1960s, companies were generally subordinate to state authority. Companies relied on governments for security, to enforce property rights and uphold the rule of law. They only exercised limited agency when it came to pursuing independent market-driven objectives. Then, in 1971, President Richard Nixon ended the Bretton Woods system of fixed exchange rates, freeing capital controls. This decision cleared the way for cross-border financial flows and allowed corporations to globalize. It precipitated a shift in power—in favor of private actors and away from nation-states, elevating businesses as "key actors in contemporary geopolitics."[15] Today, businesses are crucial to geopolitics and economic statecraft. Yet private companies do not necessarily conform to state authority; indeed, they often harbor interests that run contrary to the goals of their governments.

While companies and governments aren't in a locked state of competition, their relationship is increasingly "dynamic and reciprocal" rather than hierarchical. In some cases, corporations operate in tandem with the strategic interests of the state where they're based. Other times, their actions diverge. Decentralized systems like that of the United States face special challenges. Traditionally, the US government has maintained an "arms-length relationship" with its companies and fostered a business culture that values autonomy and limited state intervention. Consequently, the state only has limited sway over corporate decisions. While certain industries like defense and aerospace enjoy collaborative relations, cooperation is more tenuous in other fields. "In the core sectors most important to contemporary geoeconomic competition—namely advanced technology manufacturing, finance, and information and telecommunications services—the US government cannot act as quickly and decisively as it might like, write scholars Geoffrey Gertz and Miles M. Evers."[16] While Trump has started to shift this relationship, established patterns run deep.

The increased significance of technology in international affairs has further emboldened large tech corporations. They are in the driver's seat in many areas. Tech firms are leading research efforts to achieve AGI. They are building out key infrastructure—data centers, underwater cables, satellite networks—to facilitate faster communication and productivity. They are spearheading the development of more powerful and smarter weapons. Tech platforms dominate the global information space, laying out the rules for how individuals and even nation-states can communicate with one another. In short, private tech companies have become major players.

But their power isn't unlimited, either. Geopolitics in the digital age entails a delicate balance of interests between nation-states dependent on tech firms, companies that are reliant on states to safeguard their markets, and international institutions attempting to forge common rules of the road. The Ukraine war threw these issues into stark relief.

The conflict spurred a new level of recognition from policymakers about their dependence on tech companies for matters of national security. I spoke to a former senior Biden official about the burgeoning power of tech companies. "When you're talking about Apple or Microsoft, they have the resources of a G20, maybe even a G7 country. I mean, if you have a $3 trillion market cap, that puts you in the G7, doesn't it? So that's pretty fucking big."[17] That being said, market cap, which represents the total value of a company's shares, may not be the best measure of a company's actual worth given stock market volatility—gross revenues can better indicate a company's real value. As table 1 shows, the top revenue producer in the world, Walmart, still lags the smallest G7 economy (Canada) by a significant margin. Yet the gap between the richest countries and the highest-earning businesses is narrowing.

GDP of G7 Economies (2025)[18]		Leading companies by revenue (2025)[19]	
United States	$30.34 trillion	Walmart	$680.98 billion
Germany	$4.92 trillion	Amazon	$637.95 billion
Japan	$4.39 trillion	Saudi Aramco	$480.58 billion
United Kingdom	$3.73 trillion	Sinopec	$444.74 billion
France	$3.28 trillion	Apple	$395.76 billion
Italy	$2.46 trillion	Alphabet (Google)	$350.01 billion
Canada	$2.33 trillion	Microsoft	$261.8 billion

Table 1. G7 Economies vs. Largest Companies by Revenue

These companies are powerful not just because they possess sizable resources, but because of how they can use them. "I think that the centrality of technology companies to national security is higher than it's ever been," said the same Biden official, especially "when you factor in that a lot of these companies have near sovereign control over critical domains—internet service providers, cloud computing, social media platforms—and now more than half the satellites in orbit are owned by SpaceX."[20]

The implication is that in today's climate, nation-states have to consider more than their own strategic interests when responding to a crisis; they have to factor in how tech companies will react. As a result, some commentators posit that soon, "having the support of Google or Meta will mean as much for a country as having the support of the world's superpowers."[21]

But there is a caveat to tech companies' growing power. The bulk of them are publicly traded corporations led by CEOs with fiduciary responsibilities to their shareholders. Their primary goal is to maximize shareholder revenue. To the extent that companies harbor geopolitical ambitions, it is to leverage markets and relationships so as to grow their businesses and generate higher profits. As such, tech firms are acutely sensitive to public pressure. There is peril to going too far out on a limb to support a polarizing cause or controversial political figure. As Elon Musk has learned from Tesla's slumping sales, companies can pay a steep price for their CEOs' activities.

In the case of the Ukraine war, tech companies initially faced intense pressure to support Zelensky's administration. US senators, such as Intelligence Committee chairman Mark Warner, accused Alphabet CEO Sundar Pichai of "profiting from disinformation." He criticized the company's failure to crack down on Russian state media sites such as RT, Sputnik, and Tass, which were monetizing their posts via Google's ad service and YouTube. He also called on X and Meta to "assume a heightened posture" against information operations sponsored by Moscow. Prominent voices in the public sphere—Brookings scholar Robert Kagan, *The Atlantic*'s Anne Applebaum, conservative pundit William Kristol—decried the brutality of Russia's invasion and called on companies to help Ukraine in its time of need. On social media, #StandWithUkraine became a global rallying cry, bringing rock musicians, athletes, and even Disney memes out in force.[22]

It also didn't hurt that for most tech companies, the economic impact of severing ties with Russia was minor. Despite its geographic expanse, Russia's tech economy is modest. The total size of its IT market in 2023 was just $29 billion; US foreign direct investment in Russia prior to the war was a paltry $12.8 billion. Companies taking a stand against Russia could benefit from a national security "double bottom line"—what was good for Ukraine could also be good for business—without sacrificing much revenue. For example, one of the "lessons learned" that Microsoft shared with clients about the war was that governments that had sought to centralize their operations in "so-called sovereign data centers" would imperil their citizens. Microsoft argued that a government's ability to "move data outside the country" in case of conflict was vital. The subtext: in the name of sovereignty, Ukraine had kept its data on its territory and almost paid a terrible price. Other governments would be wise to store their information in one of Microsoft's hyper-scale data centers rather than risk losing everything.[23]

Likewise Ukraine represented something of a watershed moment for Starlink's brand. Prior to the invasion, it had a mixed reputation for being an expensive and not always reliable alternative to ground-based networks. But the war served as a massive proof of concept for the company, with officials like Fedorov commenting that "Starlink is indeed the blood of our entire communication infrastructure now." Ukraine's experience showed that satellite internet could be a major tool to maintain connectivity during war or to help beleaguered citizens circumvent censorship. The value of earned media Starlink accrued easily ran into the millions of dollars.[24] As a result, even though aligning with Ukraine broke with precedent, it wasn't a hard choice for companies to make.

Looking Beyond Ukraine

While the Ukraine war presented tech companies with a relatively clear set of options, future crises could be a tougher call. How would US companies react to a very different conflict—say, an invasion of Taiwan by China—where the cost to revenue would be in the billions of dollars? Regardless, trends suggest that large tech companies will keep getting more powerful. Take computer processing power (or "compute," in Silicon Valley lingo), a commodity OpenAI head Sam Altman compares in value to oil. In an interview with podcaster Lex Fridman in early 2024, Altman opined that "compute is going to be the currency of the future. I think it will be maybe the most precious commodity in the world, and I think we should be investing heavily to make a lot more compute." While governments dither, private companies are leading the drive to finance, build, and secure these assets. Their dominance of compute resources will allow private actors to continue playing a pivotal role. This represents quite a turnabout from the

Cold War era, where most innovations were government driven and publicly funded. Today, the situation has reversed. "Almost all of the technology that is of importance in the future," says former Deputy Defense Secretary Bob Work, "is coming from the commercial sector."[25]

This dynamic sets up a quandary. On the one hand, major technology companies have given the United States "a real national security lead" when it comes to the technologies of the future, from AI and robotics to quantum.[26] Private labs have generated groundbreaking innovations that have made the United States a tech superpower. The country has benefited from adopting a hands-off model that has freed up its vibrant private sector to generate new ideas and transformative products. On the other hand, such a strategy comes with risks. Democratic governments can cajole companies to channel their resources toward certain national security objectives, but companies may not listen. Short of using legal mandates like the Defense Production Act (which gives the president powers to direct private companies to meet orders from the federal government) or nationalization (directly taking over a company's operations), US presidents—at least until Trump's second term—have been reluctant to compel firms to do their bidding.

In China, by contrast, the Chinese Communist Party (CCP) can ensure that the interests, capabilities, and goals of the government stay closely aligned with its companies. When persuasion fails, authorities do not hesitate to use coercion to keep business leaders in check.

The fall from grace of Jack Ma, celebrity founder of e-commerce site Alibaba and Ant Group, is telling. Ma was known as "China's most outspoken billionaire" and had an insatiable zeal for the spotlight, making him an ideal front man for China's booming tech industry. He presided over an annual televised shopping spectacular known as Singles Day and drew standing-room-only crowds everywhere he spoke. He hosted celebrity-packed gatherings with heads of state, media icons, and pop stars sitting side by side. At one well-publicized dinner Ma organized at Davos in 2018, he was photographed seated next to Bill Gates on his left and Norway's prime minister, Erna Solberg, to his right. Pictures from the evening showed Ma performing magic tricks for Rwandan President Paul Kagame and mugging for the camera with Canadian Prime Minister Justin Trudeau. During the dinner, Alibaba's PR team "pushed pictures and prewritten copy to Chinese media, flooding the internet with lines about Ma's feast and its attendees," portraying Ma as the symbolic leader of China's global economic expansion. Predictably, this high-rolling lifestyle began rubbing Beijing the wrong way. CCP officials felt it was improper for a private businessman to speak to the world on Beijing's behalf. Said one of Ma's collaborators in a *Financial Times* profile, "*Gonggao gaizhu*—a subject's achievements make the king feel uneasy. They don't like him representing China on the world stage."[27]

And then, Ma suddenly went silent. In late 2020 he disappeared from public view. He stopped visiting the business school he had founded, canceled planned television appearances, and withdrew from international conferences. For months beforehand, he had faced increasing pressure from Chinese regulators who were bent on reining in his companies. Ma had waged a behind-the-scenes campaign to push back against impending rules intended to restrict how his companies operated. But his true offense was deciding to take the fight public. In his remarks onstage at a high-profile October event in Shanghai attended by a group of senior officials and business executives, Ma confronted China's regulators directly, saying that the country's "financial system must be reformed," and that state officials were harming the financial technology sector with their "pawnshop mentality." Reportedly, top officials in Beijing were stunned, seeing the episode as "Ma attempting to win the public over to his side and influence policy." CCP leadership acted swiftly. Within a week, regulators abruptly canceled Ant Group's long-planned $37 billion IPO, intended to be one of the largest stock debuts in the world. Ma then went into isolation. The larger message was clear: challenge the state and face the consequences.[*28]

This stood in marked contrast to the Biden administration's handwringing regarding Elon Musk. At various points, Pentagon officials pleaded with Musk to maintain Starlink connectivity to support Ukraine's operations, often to little avail. One US defense official admitted, "Living in the world we live in, in which Elon runs this company and it is a private business under his control, we are living off his good graces." While the US government was forced to accept Musk's volatility, China made it clear to Musk that there were firm boundaries about what they would permit. In an interview with the *Financial Times*, Musk conceded that Beijing "made

---

* In an unexpected postscript, Jack Ma's stay in exile appears to have ended. Following his banishment from public view at the end of 2020, Ma quietly oversaw a strategic turnaround at Alibaba. His focus on developing AI models and applications paid off; analysts view Alibaba as primed to take advantage of China's growing AI demand. In a symbolic gesture, he was selected to sit in the front row with other top business leaders at a major meeting with Xi in February 2025.

clear its disapproval" of his rollout of Starlink in Ukraine and that it sought assurances that Musk "would not sell Starlink in China." (Additional reporting also disclosed that Vladimir Putin had pressured Musk not to allow Starlink services in Taiwan "as a favor to Chinese leader Xi Jinping.") The CCP has spoken out when it perceives that Musk has crossed sensitive lines. For instance, in 2023, after he tweeted about a US government report that assessed that the COVID virus could have originated from a Wuhan lab, Chinese state-run media cautioned Musk that he could be "breaking the pot of China" (similar to the saying "don't bite the hand that feeds you"). Given the billions in subsidies and cheap land China's leaders have bestowed on Tesla, the warning carried weight.[29]

Prior to Trump's return to office, the US government generally relied on three strategies to increase its influence over private companies. First, officials used contractual arrangements to guarantee reliable access to private infrastructure. In response to concerns about Starlink's trustworthiness, the Pentagon initiated a review of all of its contracts with SpaceX. Former Air Force Secretary Frank Kendall said that he needed to determine "whether [he] was comfortable depending upon billionaires and their potential fickleness for military services." Military leaders decided that a different arrangement was needed. Soon afterward, the Pentagon created a new communications system it called Starshield. It would resemble the existing Starlink network and work off of SpaceX satellites but would be "owned by the US government and controlled by the Space Force." For Ukraine, the deal came as a relief. The Starshield contract brought three thousand terminals under the Pentagon's direct control (and out of Musk's hands), thereby assuring he couldn't interfere with their military operations.[30]

Second, government agencies created interdependencies that reduced the chances that a private actor would restrict access to its infrastructure or exploit its control in some other fashion. For instance, NASA opened its launch site infrastructure to private businesses that offered services for it. In return, those businesses became dependent on NASA to keep the launch site operational and faced diminished options for moving their services elsewhere.

A third strategy was for governments to diversify vendors, thereby reducing the ability of a single operator from wielding excessive influence. For a democratic government like Taiwan, that meant shifting away from Starlink in light of Musk's entanglements with China. In 2023, the government signed a contract with London-based OneWeb to put LEO satellite internet receivers in every village and county in the country. While the cost was high, Taipei believed that the national security benefits from the deal were worth the expense. (As it later turned out, Taiwanese officials were disappointed with OneWeb's performance and subsequently discussed forming a partnership with Amazon Kuiper.) US officials have similarly tried to break the private sector stranglehold over AI infrastructure held by companies like Amazon, Google, Microsoft, and OpenAI. In January 2024, the National Science Foundation created NAIRR (the National Artificial Intelligence Research Resource), designed to provide US researchers with computational infrastructure, data, and training resources for AI innovation. Its goal was to "limit the access divide" between flush private sector companies and under-resourced public organizations like universities and nonprofits. Initial results were promising; however, NAIRR remained a drop in the bucket compared to the billions that private companies are plowing into AI.[†31]

Trump, however, has upended the rules of the game. If prior presidents scrupulously avoided interfering with private business, he has dived headlong in the opposite direction. Take, for example, Nvidia's June 2025 agreement with the White House to resume selling H20 chips to China. Two months earlier, Trump officials said they would ban H20 sales due to national security risks. But after CEO Jensen Huang personally met with Trump, the president issued a reversal—with a catch. Nvidia and fellow semiconductor giant AMD would each give Washington a 15 percent cut of their revenues from chip sales to China in exchange for permission to resume those transactions. "I said, 'I want 20 percent if I'm going to approve this for you,'" Trump told reporters. "For the country, for our country. I don't want it myself. . . . And [Huang] said 'Would you make it 15?' So we negotiated a little deal." Such a quid pro quo arrangement was unheard of in the United States. No American company had ever agreed to pay a share of its revenue to acquire export licenses.[32]

But that was just the start. Next on Trump's radar was struggling Silicon Valley chipmaker Intel. Trouble began on August 7, when Trump posted on

† Overall, academia is falling further and further behind the commercial sector in scientific research, such as AI development. "Today, no university in the world can build a frontier AI system on par with industry," says a December 2024 report from Stanford. "Companies have over 1000x more compute than universities, and they produce AI models that are 50x larger." In all likelihood, the window has closed when it comes to bridging AI's public research/private sector divide.

social media that the company's new CEO, Lip-Bu Tan, was "highly CONFLICTED and must resign," adding, "There is no other solution to this problem." According to reporting, Trump became inflamed by a Fox Business segment in which host Maria Bartiromo raised concerns about Tan's business entanglements in China. Tan paid a visit to the White House soon afterward, and Trump changed his mind about Tan's resignation. Instead, he proposed a new arrangement: Intel would sell the US government a 10 percent stake in its business, worth $8.9 billion, making the government its largest shareholder. The deal represented one of the biggest government interventions in an American company since Barack Obama's rescue of the auto industry during the 2008 financial crisis. Said Trump, Mr. Tan "walked in wanting to keep his job, and he ended up giving us $10 billion for the United States."[33]

Of course, this approach risked generating perverse incentives. Rather than letting market forces dictate the best product, buyers could feel pressured to choose Intel's chips, lest they fall afoul of the Trump administration. Over time, this could shift incentives away from innovation and toward rent-seeking behavior. If Trump's dealings with Nvidia and Intel were based primarily around financial considerations, his run-in with AI startup Anthropic over its Pentagon contract was a far different matter. In July 2025, Anthropic signed a $200 million deal with the military to provide frontier AI technology. Under the agreement, its AI model Claude would be used in the government's classified systems. The deal was part of a push by the firm to cultivate national security clients ahead of a planned public offering. Executives were enthusiastic about the venture, proclaiming that it "opens a new chapter" for the company. It wasn't long, however, before disagreements spilled over. Executives became worried that the Pentagon could eventually deploy Claude to carry out fully autonomous lethal operations or to conduct mass surveillance. The Pentagon pushed back, arguing that the military should have the last word about how it would deploy Anthropic's technology. Unlike other companies, such as Google, Elon Musk's xAI, or OpenAI, which had "agreed in principle" for their models to be used for any purpose allowed by law, Anthropic took a different tack. The company had developed a reputation for promoting AI safety and it held firm to it red lines. After months of tension, matters came to a head in early 2026. Secretary of Defense Pete Hegseth had become increasingly livid about Anthropic's pushback and issued a memo directing AI companies to remove all limits on their technology. In response, Anthropic requested renegotiating its contract. This led to further backlash at the Pentagon. Talks between the two sides intensified but Trump had soured on the company and wanted to make an example of it. After a negotiation deadline passed on February 28, Hegseth announced that the government had designated Anthropic a supply chain risk, something historically reserved for adversaries, and that the firm would be banned from working with any federal agency: "America's warfighters will never be held hostage by the ideological whims of Big Tech," he posted on social media. His declaration sent shockwaves through Silicon Valley, making it clear to other tech firms that they risked serious repercussions if they failed to back Trump's policies (right away, Anthropic's rivals such as Elon Musk and Sam Altman pledged they would comply with the Pentagon's requests, "positioning themselves as loyal patriots").[34]

In less than a year, Trump remade himself into the biggest decision-maker for one of the world's most strategically critical sectors, responsible for everything from cutting-edge AI systems to military weapons. These actions revealed a dramatic break from Reagan-era free-market conservatism, echoing the most aggressive government interventions in American history. Problematically, Trump's intrusions lack a consistent strategy or a coherent framework. Trump himself has admitted as much. Following his Intel announcement, he indicated that he would pursue similar investments in other companies, describing his economic strategy as trying to "get as much as I can."[35]

Trump's willingness to wield government power against private companies shares similarities with Beijing's economic strategy. Like Xi, Trump is leading an unparalleled intervention into the commercial market, acquiring stakes in corporations, demanding shares of profits in exchange for market access, even taking "golden shares"—a special class of equity that gives the government outsized control with a small percentage of ownership—in companies like US Steel. Still, such comparisons have their limits. In China, officials have asserted full control over the country's tech sector. Beijing's largest companies, regardless of their ownership structure, "serve at the pleasure of President Xi Jinping." The CCP has made it crystal clear that the state is in charge of its digital future.[36]

In the United States, despite Trump's meddling, private companies remain the engines of innovation and retain vast discretion over what to build, how to innovate, who to sell to, and what to prioritize. Trump's interventions perhaps represent the start of a shift, but it is too early to say how far they will go and whether they will last. Today, America's future competitiveness and national security remain reliant on cooperation with private industry. And for better or worse, Silicon Valley has become a crucial player in the United States' fight to retain its technological edge against China and preserve its military advantages. Will this bet pay off?

5

From the Last Supper to the "Pro-Military Arm of Silicon Valley"

As the Cold War drew to a close, the world seemed poised for more peace and stability. The standoff between the free world and the Soviet bloc was over. And the Pentagon was ready to downsize. In 1993, Secretary of Defense Les Aspin summoned defense industry executives to a secret dinner hosted in his private Pentagon dining room—a gathering referred to as "the last supper." The assembled guests had little idea what was in store for them. Recalled Norm Augustine, the former CEO of defense contractor Martin Marietta in an interview, "I said to him [Les Aspin], these were very hard times, and it was nice to get a free meal. But why are we here? He responded that you'll know in a few minutes, and you probably won't like it." After the executives finished their dinner, Pentagon officials led them into an adjoining room. Deputy Defense Secretary William Perry strode to the front and projected a graph onto the screen. "It was a stunning graph, so much so that the following day I went over to the Pentagon and asked for a copy of it, which I still have," says Augustine. "And what was startling about it was that the Defense Department was saying there are way too many companies in the defense industrial base. That we can't afford them."[1]

The direction from the Pentagon was clear: budget cuts were coming, and most companies wouldn't survive. The industry needed to consolidate . . . fast. Perry's warning set off one of the swiftest transformations in modern US industrial history. Fifty-one separate companies, nearly all of which were considered prime contractors or major subcontractors, folded into just five giant defense firms: Lockheed Martin, Boeing, Raytheon, Northrop Grumman, and General Dynamics.

Subsequently, Congress slashed the defense budget, and it became nearly impossible for military startups to access meaningful pools of money, cratering the defense tech ecosystem. "Because the five were so big, they undermined one of America's greatest strengths—its seemingly inexhaustible bounty of bright entrepreneurs with new ideas," writes Mark Bowden for *The Atlantic*. "The Big Five spent a lot on research and development and had the capacity to rapidly expand if a product took hold, but the galaxy of small entrepreneurial players was diminished. It became harder for start-ups to compete and thus to remain alive."[2]

The Valley's defense economy didn't completely disappear. "The Cold Warriors still held court at the Hoover Institution; the quiet hulk of Lockheed still brooded alongside Highway 101," writes author Margaret O'Mara. But the entrepreneurs headlining the internet era of the early 2000s—Steve Jobs, Sergey Brin, Larry Page, Mark Zuckerberg—had little desire or need for defense work. Accordingly, America's military innovation stagnated as the top minds in Silicon Valley focused elsewhere. In the ensuing years, relations between the defense sector and the Valley continued to deteriorate. "So then we fast-forward to the mid-2000s, so think 2012, 2013, 2014. This is when the Snowden leaks come out and Valley companies start thinking that the US government and the NSA are just as bad as authoritarian regimes in China," Raj Shah, coauthor of the book *Unit X* and the previous head of the Pentagon's Defense Innovation Unit, told me. "And they put them on the same playing field. So that was probably the low point. Very few venture firms would invest in startups focused on government work."[3]

A culture of mistrust proliferated. Many engineers had come of age watching the United States blunder its way through the Iraq War and had little interest in designing software or hardware that would help the military kill suspects or allow intelligence agencies to collect bulk data on US citizens. Edward Snowden's leaks about the National Security Agency's vast surveillance program and its infiltration of tech companies' data (including breaking into the main communications links connecting Yahoo and Google data centers) exacerbated the ill will. Things got so bad that when Defense Secretary Ash Carter visited the Valley in 2015 (the first Pentagon chief to visit in twenty years), Google would not let him onto its campus.[4]

If that weren't enough, tech companies viewed the military "as a hidebound and unprofitable customer incompatible with the fast-moving industry." They were not convinced they could generate meaningful profits from government contracts, and they were deeply averse to navigating the bureaucratic maze of the Pentagon procurement process. One telling episode involved a contract Google signed with the Pentagon known as Project Maven.

Late in 2017, under the leadership of Eric Schmidt, Google agreed to use its machine-learning AI software to help the Pentagon analyze drone surveillance footage. The challenge for the Pentagon was that the volume of video data collected by its drones far exceeded human analytic capacity. They needed algorithms to sift through the imagery and identify relevant information quickly enough to provide an edge against ISIS and other terrorists. From the outset, Google executives were sensitive about how the contract would be construed. In an email to colleagues, Google Cloud's chief AI scientist, Fei-Fei Li—one of the field's "brightest stars"—cautioned: "Avoid at ALL COSTS any mention or implication of AI. Weaponized AI is probably one of the most sensitized topics of AI—if not THE most. This is red meat to the media to find all ways to damage Google." But the company couldn't hide the contract from its employees nor from the prying eyes of reporters.[5]

Maven became a lightning rod for Google. It split the workforce and led to contentious staff meetings and internal disagreements. Employees circulated a petition signed by more than four thousand staff demanding that Google forswear ever building "warfare technology." As tensions boiled over, company leaders had enough. In June 2018, Google announced it was dropping out of Maven and committed to "entirely ban" the use of AI software in weapons systems. Said Diane Greene, CEO of Google Cloud at the time, "We do believe that the uses of our cloud and AI will prove to be overwhelmingly positive for the world, and we also recognize that we cannot control all downstream uses of our technology." Soon thereafter, Google pulled out of competing for a $10 billion contract to build the Pentagon's cloud-computing infrastructure—known as JEDI—saying that that the initiative jeopardized its "corporate values" regarding the use of AI.[6] Superficially, at least, Google appeared to walk away from doing business with the Pentagon.

The "Pro-Military Arm of Silicon Valley"

Not everyone in Silicon Valley thought it was wise to spurn Pentagon contracts. A small but vocal group of investors argued that the Valley's rejection of the military was hasty and shortsighted. One such firm was a startup named Palantir Technologies. It was cofounded by Peter Thiel and Alex Karp (who were classmates at Stanford Law School), along with three others in 2003—and backed by nearly $2 million from the CIA's venture capital arm, In-Q-Tel. The company's name refers to the *palantíri*, the "seeing stones" in J. R. R. Tolkien's *The Lord of the Rings*. From the outset, Palantir's founders were clear about the market opportunity they saw for the company. Thiel and Karp would position it as a "pro-military arm of Silicon Valley" and promote the firm's software as a powerful surveillance weapon for the burgeoning national security state. They had read the 9/11 Commission report and concluded that a big reason for America's unpreparedness was a "failure to connect the dots." Said Trae Stephens, who worked for Palantir at the time, "We had the data, but it was never pulled into a single place." Palantir's salespeople boasted that the firm's products could ingest volumes of data collected by soldiers, spies, and diplomats, and use proprietary algorithms to detect unseen patterns to uncover criminal syndicates and terrorist organizations, as well as anticipate future strikes from rival nations.[7]

Palantir had its critics. Analysts quipped that the company overhyped its products and often claimed credit for accomplishments it had no role in. One story that made the rounds was that Palantir was responsible for helping catch 9/11 mastermind Osama bin Laden at his hideout in Pakistan. The company did little to refute the account, even posting a news story "repeating the rumor on its website." When Karp was asked directly about Palantir's role in an interview, he responded, "That's one of those stories we're not allowed to comment about." But as reporters dug into the claims, they found no evidence that Palantir played "any significant role in finding bin Laden." Nonetheless, the company became a Silicon Valley darling—one of its few national security "unicorns" (a privately held company that achieves a valuation of $1 billion or more). Its 2020 IPO valued the firm at $16.5 billion; as of January 2026, Palantir was worth twenty-five times as much, with a market cap of about $401 billion, placing it with blue-chips like Bank of America, Chevron, and Procter & Gamble (and leading *The Economist* to question whether Palantir "might be the most overvalued firm of all time").[8]

Thiel and Karp's vision proved prophetic. In the ensuing years more tech companies would join in and attempt to win Pentagon contracts. The path wasn't straightforward. At several junctures, Palantir (along with Elon Musk's SpaceX) had to file lawsuits to force open the government's procurement process. But their persistence paid off. By 2023, Palantir started winning major military bids. It obtained a US Army contract for AI systems worth $250 million in 2023; the following year, the Pentagon awarded it two contracts for $480 million and $100 million apiece to run the Maven AI platform. Interest in the tech defense sector soared; with Trump's return, even more money flowed from the government spigot.

Just eight months into Trump's second term, records show that Palantir had collected at least $300 million in new business, including contracts with the Federal Aviation Administration, the Centers for Disease Control and Prevention, and Fannie Mae. "They were positioned in the right way at the right time, and they had already built the technology that had the capability to do what the administration is trying to do," said Matt Pearl, former director of emerging technologies at the National Security Council.[9]

A slew of firms with names like Anduril, Epirus, Hermeus, ShieldAI, and Skydio, referred to as "neo-primes," took a page from Palantir's playbook and launched a range of new products—from hypersonics and autonomous systems to data-management software and biomanufacturing—with the hopes of striking it rich. Venture capital investors, such as Founders Fund, Andreesen Horowitz, and Y Combinator, plowed millions more into the sector.[10]

This momentum would not only sweep up defense tech startups; it soon extended to Silicon Valley's tech giants as well. Despite the nasty fallout from Maven and its pullout from JEDI, Google had quietly maintained a robust defense portfolio. It won part of a large CIA cloud contract in 2020 and was jointly awarded a $1.3 billion deal with Amazon in 2021 to provide cloud services to the Israeli government, including its security agencies. But its real breakthrough occurred the following year. The aforementioned JEDI contract that Google walked away from never got off the ground; it became mired in litigation and eventually the Pentagon pulled the plug. In its place, the Defense Department bid out a new $9 billion contract known as the Joint Warfighting Cloud Capability. After a competitive process, Google was selected as one of four providers. The deal signaled that the company had come full circle from eschewing national security projects to becoming a leader in the sector.[11]

The swift growth of the defense tech sector and the corresponding influx of venture capital dollars was not accidental. Rather, it reflected a growing view that the Pentagon had to modernize or risk future defeat. Take drone technology. While the United States has long held an edge in this sector, reports indicate that China has matched or even surpassed the United States. Beijing's WZ-9 "Divine Eagle" rivals America's flagship spy drone, the RQ-4 Global Hawk, in operational capability—and the United States has no answer for China's WZ-8, which can fly at hypersonic speeds to evade air defenses. In the Ukraine war, low-altitude, long-range Iranian Shahed drones have proven to be potent weapons. China's ASN 301 offers similar capabilities; meanwhile, America has only begun rolling out the low-cost LUCAS drone. As for FPV drones, China holds a huge advantage. Its products dominate the market—Chinese drone maker DJI is capable of producing over a million Mavics per year. Reportedly 85 percent of the components going into Ukraine's FPVs come from Chinese factories. The closest US competitor, the Skydio X10D, costs three times as much as a Mavic and can only be produced in limited quantities—in the thousands.[12]

Consequently, many argued that the Pentagon's current mode of operations—reliant on a handful of plodding prime contractors to produce ever more costly equipment—had run its course. That without dramatic changes, US military innovation would lag China's. But overcoming this inertia wouldn't be easy. Political leaders had shown little appetite for reform. "It's easy to get complacent with the repetitive warning signs coming from Russia and China and underestimate how dangerous [they are]," writes former Pentagon official Michael Brown. "Is America ready? Unfortunately not. And the core of the problem relates to the U.S. defense-industrial base, which suffers from too much concentration, too little commercial technology, and an insufficient ability to produce munitions."[13] But several factors finally nudged the Defense Department in a new direction. Undoubtedly, a major driver of change was what Pentagon strategists witnessed on the battlefields of Ukraine.

6

"A Feeling of a Thousand Snipers in the Sky"

In the early morning of May 3, 2023, two Ukrainian drones, spaced sixteen minutes apart, penetrated Russian airspace and flew toward Moscow. Their target was the Kremlin residence of President Vladimir Putin. The first drone approached from the west and exploded into the dome of the Kremlin's senate palace building at 2:27 a.m. The second came from the east and detonated near a flagpole on top of the same dome at 2:43 a.m. The explosions caused a brief fire that workers scrambled to put out. Videos released later showed the Kremlin lobbing antiaircraft fire over its golden-domed churches for hours afterward in fear of more drone attacks. Fortunately for Putin, he had escaped harm that evening; he was sleeping at a compound twenty miles to the west of Moscow. But despite Russian attempts to downplay the attack, state authorities were clearly rattled. The strike was so brazen that some experts initially doubted that Ukraine had pulled it off, speculating that it was a false flag attack by the Russians to justify a violent response.[1]

In time, it became apparent that not only was Ukraine behind the strike, but Kyiv had independently developed a full-fledged long-range drone program. "Because the West was not willing to give those weapons to Ukraine, Ukrainians were like, OK, we'll have to build them ourselves," said *New York Times* reporter Christiaan Triebert. "So they got together basically anyone that had some idea of how to build a drone." Over the course of a year, multiple groups joined forces—government officials, private companies, mechanical engineers, hobbyists, software developers—and raced to assemble long-range UAVs that could fly hundreds of miles. What started as a scrappy, experimental effort morphed into a billion-dollar enterprise, which ultimately became a crucial tool for expanding the war into Russian territory. Now, Ukraine's long-range drones regularly hit targets more than a thousand miles inside Russia. Packed with munitions, their fuselages come inscribed with a warning for Moscow: "Az vozdam," or "I will repay."[2]

Ukraine offered military strategists a glimpse of the future of war. There hadn't been a protracted ground war between major militaries for decades, so many cutting-edge developments in warfare were largely theoretical. The conflict allowed the Pentagon to test its assumptions about weaponry, technologies, and tactics in real time. They saw which platforms worked. They watched Russia adapt to new innovations and devise countermeasures. And they identified weak points in essential systems. Some of the findings were stunning.

For fifty years, the Pentagon prioritized using exquisite, expensive weapons with the ability to strike targets anywhere in the world over lower-tech options that were cheap to produce and could be deployed in mass quantities. According to University of Pennsylvania scholar Michael Horowitz, "Leaders chose to scale down the size of their forces and focus instead on honing their technological advantages." Thus, the United States invested $2 trillion into the F-35 fighter plane platform and billions on the M1 Abrams tank. It developed an array of exotic capabilities unmatched by any other military. The theory was that the United States could outspend any other country and develop such decisive arms that it would be futile for rivals to compete. But Ukraine exposed a key error—countries no longer needed to trade off mass production for precision. "Militaries find themselves in a new era in which more and more actors can muster uncrewed systems and missiles and gain access to inexpensive satellites and cutting-edge commercially available technology," writes Horowitz. "With these tools, they can more easily conduct surveillance and stage accurate and devastating attacks."[3]

What did this mean in concrete terms? In Ukraine, the implication was that volumes of one-way attack drones that could strike a target all at once would bring major advantages over conventional systems. They were inexpensive, lethal, and transformed the battlefield. Moreover, Ukraine's drone war evolved rapidly. The country began producing FPVs at a massive rate; by 2025, analysts watched production ramp up to 3 million UAVs. Russia wasn't standing still, either. It started deploying one thousand drones every twenty-four hours for surveillance or for strikes. The result was that the battlefield front lines became "practically no-go areas for humans." Drones were packed with more and more explosives and started flying greater distances at higher velocity (speeds began to exceed seventy miles per hour). When FPVs were first introduced in 2023, they had a range of only ten kilometers. Two years later, they routinely flew thirty or more kilometers due to "powerful antennae and signal-repeater drones that let them communicate with base stations from greater distances."[4]

As drones became a mainstay in the war, engineers rushed to create new models, largely sourced from Chinese components. (Ukrainian and Russian engineers found themselves in the unusual position of visiting the same Chinese factories in Guangdong and Shenzhen to purchase identical processors, sensors, and engines needed to keep their drones flying.) Heavy-duty Vampire drones, nicknamed "Baba Yaga" after a mythological witch from Slavic folklore, were designed to carry fifteen-kilogram bombs, as well as transport food, ammunition, water, and medicine. Mothership drones—fixed-wing UAVs akin to "flying aircraft carriers" that could carry up to six FPV drones—were also deployed, opening up new possibilities. They not only extended drone range but shortened the time between detecting a target and engaging it with multiple coordinated strikes. Ukrainian engineers also started incorporating unmanned ground vehicles, or UGVs, to shuttle supplies, deploy and clear land mines, carry out surveillance missions, observe enemy activities, and evacuate wounded soldiers. The UVGs could be equipped with machine guns or rocket launchers, functioning as "stealthy mobile weapons." Like their aerial counterparts, UGVs shifted the burden of risk from humans to machines. Overall, drones started accounting for the majority of losses on the battlefield, supplanting artillery, and instilling "a feeling of a thousand snipers in the sky."[5]

The increased reliance on drone warfare raised the urgency for both Russia and Ukraine to hone their electronic warfare and jamming techniques. Initially, their jamming efforts focused on quashing GPS signals from satellites. This approach particularly impacted US precision missiles, which started failing at high levels. Next, they sought to disrupt the communication link between operator and drone. To adapt, drone producers modified radio frequencies and started shortening radio interval lengths. They even began attaching wired spools to FPVs, making them impervious to electronic disruption. (Using wired links brings some downsides, such as reduced flight performance, flight range limited to thirty-five to forty kilometers, and the risk of entanglement with objects; on the flip side, this became one of the few reliable ways to fly drones in heavily jammed areas.)[6]

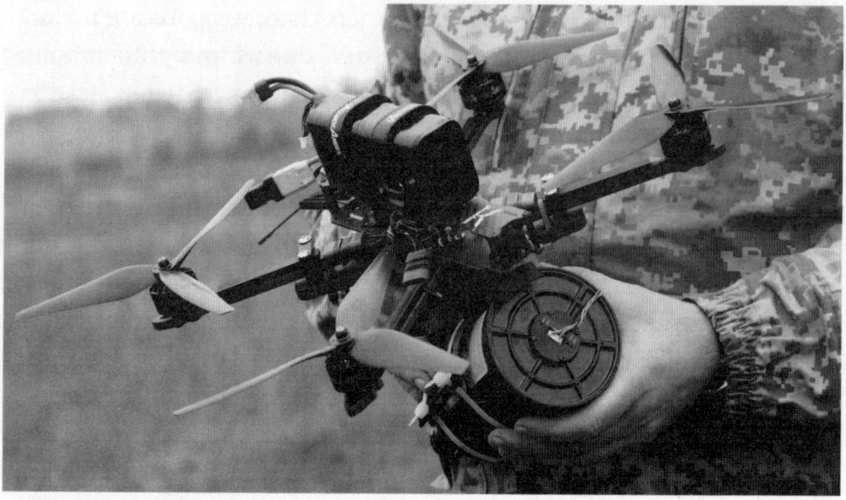

Ukrainian wired drone operator

Now, both sides are experimenting with AI software to pilot drones to their targets in the final stages of flight. Basic object-recognition software had already been in use since 2023, but the programs have significantly improved. US-based firm Auterion, for example, claims to have doubled the range at which a drone can autonomously engage a target, "from 500 meters to 1km or so" due to higher-resolution cameras. There are reports that Ukrainian units are experimenting with deploying swarm attacks—using up to twenty-five drones to autonomously target Russian soldiers, equipment, and infrastructure. And the strike accuracy of the top AI drones is high, sitting above 80 percent. But there remains a mismatch between cost and capabilities. FPVs cost about four hundred dollars each. Integrating a robust AI targeting system, such as Auterion's Skynode S autopilot system—which costs roughly as much as an Android phone—doubles the drone's overall cost, undermining its cost advantage. Instead, Ukrainian operators have turned to cheaper systems, such as the Raspberry Pi Zero, which is priced around one hundred dollars, and uses "pixel lock" technology to latch on to a target without a human operator. Unfortunately, lower-cost systems bring reduced accuracy. Said one UAV commander, "my pilots would never use pixel lock if they can maintain the video feed, because other than against specific classes of target, it leads to a massive reduction in effectiveness." Accordingly, the autonomous future has not quite arrived. Out of almost two million drones acquired by Ukraine in 2024, just ten thousand were AI enabled. Military analysts describe the battlefield as going through a "transitionary period"; weapons are no longer 100 percent controlled by humans, but they aren't fully autonomous, either.[7]

The impact of drone technology is not just felt on land. Ukraine's success using sea drones to disable Russia's Black Sea fleet also shows how quickly the character of war is changing. In August 2023, Ukraine faced a dire crisis. The previous year, Russia had signed an agreement allowing for the safe export of Ukrainian agricultural goods through the Black Sea. During that period, Ukraine was able to maintain its food export levels, sending 32.9 million metric tons of corn, wheat, sunflower oil, and other products to forty-five countries worldwide. But Russia decided to exit the deal a year later. Its intended goal: to inflict economic pain on Ukraine and further pressure Zelensky to end the war. Russian forces unleashed missile and drone strikes against ports in Odesa and Izmail, badly damaging shipping terminals and warehouses. In two months, they destroyed 280,000 metric tons of grains, and commercial activity largely came to a halt. While Ukraine had achieved scattered success against Russia's fleet, notably sinking the flagship cruiser, the *Moskva,* in 2022 using antiship missiles, it remained far outmatched. Kyiv possessed no large vessels to protect its shipping lanes and scant weaponry to threaten Russia's sea base in Sevastopol. But that didn't mean the Ukrainians couldn't develop options.[8]

In 2022, President Zelensky tasked the Security Service of Ukraine, the country's main security and intelligence agency, with finding a way to break Russia's stranglehold. The assignment fell to Brig. Gen. Ivan Lukashevych, a career officer who had a background in engineering and experience running clandestine operations. For months, he worked with engineers and weapons makers to come up with new designs that could sit low enough in the water to avoid detection yet pack enough explosives to incapacitate big ships. After one visit to a military training facility—where he watched ground operators pilot an aerial drone with help from Starlink—he commissioned a drone maker to create a similar sea prototype that would also be powered by Starlink communications. Several weeks later, Lukashevych's team tested a new model that was sixteen feet long, carried 220 pounds of explosives, and could be operated miles away from the shore. In the meantime, another Ukrainian agency also got into the game. The Main Directorate of Intelligence partnered with the same manufacturer as Lukashevych's team to develop the MAGURA drone. Those vessels were only a few meters long and powered by Jet Skis but had a range of nearly five hundred miles. As a result, they could patrol extensive stretches of Ukraine's coastline and ceaselessly hunt for Crimean targets.[9]

Both teams began unleashing their sea drones against Russian targets. The Ukrainians struck warships, shipbuilding plants, and the Sevastopol headquarters again and again. By early fall, Ukrainian authorities were ready to resume grain deliveries. Said Dmytro Barinov, the deputy head of its seaports authority, when the first grain vessel set sail in mid-September, "We were very nervous—we prayed that everything would go well." The ship passed through without incident, and by November trade began to return to prewar levels. In the following six months, Ukraine would ramp up its shipments, exporting 27.6 million metric tons of grain and oilseed through the Black Sea corridor.[10]

All the while, Ukraine continued to pummel Russia's fleet. By 2024, its forces had reportedly disabled some two dozen Russian ships and one submarine. Russia was so rattled that it moved the bulk of its fleet to a base in Novorossiysk, in the eastern part of the Black Sea. Even there, its vessels were not safe; in one reported incident, a Ukrainian sea drone crossed the water and damaged a Russian landing ship used to transport troops and military equipment. Russia's retreat came at a considerable cost; it was forced to divert precious resources—planes, drones, helicopters, electronic jammers—to the Black Sea, "easing the pressure on Ukraine's embattled ground forces" on the eastern front.[11]

Ukraine's remarkable victories in the Black Sea reinforced an emerging insight: small innovations could triumph over big platforms, and adaptive technologies could make a decisive difference in battle.* Ukraine showed that a country without an effective navy could build $250,000 Jet Ski–powered drones and use them to take down cutting-edge vessels, such as the *Sergey Kotov*—a patrol ship that cost $65 million to produce and was sunk by MAGURA drones in a predawn raid in March 2024. "Before, naval drones were used mostly for surveillance or logistics," said Lukashevych. "We are doing many things that no one in the world has done."[12] The math stopped adding up for the Russians; their fleet abruptly changed from dominating the Black Sea to becoming vulnerable prey.

The Ukraine demonstration effect was a big boost for the defense tech

* According to reports, the Ukrainians also received vital assistance from the United States to help them develop and produce maritime drones for use against Russia's fleet. Reportedly, the Americans gave Ukraine an early drone prototype "meant to counter a Chinese naval assault on Taiwan."

sector. "Russia invaded Ukraine and reminded us why defense technology is not merely something to debate in theory," said Katherine Boyle, a partner at VC firm Andreessen Horowitz.[13] For the Pentagon's planners, Ukraine's success on land and water exposed critical flaws in US strategy. If Russia's military crown jewels could be defeated by cheap Ukrainian drones, what were the implications for America's own billion-dollar platforms in a prospective war with China?

7
Defense Tech Gold Rush

In the foothills of Southern California, not far from the seaside town of San Clemente, military startup Anduril Industries is showing off its vision of the future of war. An assembled crowd watches as a faraway truck approaches a US military outpost. The firm's AI-powered Sentry observation tower identifies the vehicle as a potential threat, which is then highlighted in its software dashboard called Lattice. The program asks the operator whether to send the company's Ghost drone to track the truck more closely. The operator clicks their approval and the drone flies autonomously toward the truck, based on information collected by the observation tower. The truck disappears behind a bend, causing the tower to lose sight of it, but in the meantime, the surveillance drone has picked up the vehicle. The display shows an individual getting out of the truck and launching a drone, which Lattice promptly labels a threat. It queries the operator about whether to send an attack drone to intersect the UAV flying toward the base. The attack drone is also piloted autonomously, and the operator can direct it to accelerate its speed to neutralize the incoming attack. Wrote a reporter who watched the scenario play out, "The entire operation could have been managed by one person with a mouse and computer."[1]

Anduril's proposition is simple—in a future conflict between great powers, the winning side will not be the one with the most sophisticated weapons or the most firepower. Rather, the victor will be whichever military can "sort through and share information the fastest." Crucial decisions will not involve military headquarters in Washington or Beijing. They will instead occur "at the edge," right on the battlefield where threats rapidly emerge and have to be dealt with immediately. Said the company's CEO, Brian Schimpf, "You're going to need to really empower lower levels to make decisions, to understand what's going on, and to fight. That is a different paradigm than today."[2]

The Pentagon appears bought in. In 2022, it awarded Anduril a $1 billion contract from US Special Operations Command (helping the company to subsequently raise $1.5 billion in venture capital). In 2024, Anduril paired with General Atomics to win a nearly $9 billion contract from the air force to produce drones for the Collaborative Combat Aircraft program. But the largest deal arrived in March 2026, when the Army announced it had signed a $20 billion agreement with the firm to deliver AI-backed software for its systems. Said Garrett Smith, chief executive of Reveal Technologies, another defense tech company, "People are pointing to this moment as a proof point."[3]

Anduril's rapid rise belies its brief history as a company. The firm, whose name comes from the "Flame of the West" sword in *The Lord of the Rings* (an oblique reference to protecting Western allies against Eastern challengers like China), only started up in 2017. It was founded by Palmer Luckey, a controversial figure in Silicon Valley, who struck it rich when he sold off his virtual reality headset company, Oculus, to Meta for $2 billion. Luckey then went on to work for Facebook, but his tenure ended abruptly when reporters revealed that he had donated ten thousand dollars to an anti–Hillary Clinton group. (The organization he donated to, Nimble America, paid for ads mocking Clinton ahead of the 2016 election, including putting up at least one billboard that "featured a picture of Mrs. Clinton and the phrase 'Too Big to Jail.'") From the start, Luckey was an ill fit for Facebook's "liberal culture." He was a fan of large cars and military gear and drove a massive tan Humvee replete with "machine-gun mounts and orange toy guns" to work. Reportedly, he was forced to remove his vehicle from Facebook's parking lot after an employee called the police to investigate. With Luckey's tenure at Facebook cut short, he was ready for his next business venture.[4]

Earlier in his career, he had attended a 2014 retreat in British Columbia hosted by Founders Fund, a venture capital firm run by Peter Thiel. (Before cofounding Palantir, Thiel made his initial fortune helping to found PayPal, becoming part of the "PayPal Mafia," a group of influential Valley figures that include Elon Musk, LinkedIn cofounder Reid Hoffman, and Trump's AI and crypto czar, David Sacks.) At the retreat, Luckey connected with Trae Stephens, a former Palantir executive and intelligence analyst, who had joined Founders Fund to focus on the defense sector. Over lunch, Luckey mentioned to Stephens his interest in starting a weapons company, telling him, "I'm actually building a ramjet in my swimming pool"—an engine system developed for high-speed missiles and jet planes.[5]

The two hit it off and "decided to do something in the tech space." Stephens convinced Luckey of the need (and market opportunity) for starting up a modern defense contractor. As Stephens pointed out, the US military's technology was woefully out of date, lagging behind basic consumer electronics goods. "There is more AI in a Tesla than in any U.S. military vehicle," he wrote. "Better computer vision in your Snapchat app than in any system the Department of Defense owns; and, until 2019, the United States' nuclear arsenal operated off floppy disks." After extensive consultations and with backing from Founders Fund, they decided to launch Anduril. Thus far, their efforts have been richly rewarded. The company has experienced explosive growth in its short existence. In 2026, the firm raised $4 billion in its latest funding round—valuing it at $60 billion.[6]

Anduril's ascendance represented a recognition in the Pentagon that it needed to take a hard look at how it was operating and shake things up. The F-35 fighter jet was an instructive case. It is one of the Defense Department's most ambitious weapons programs and crucial to meeting America's security goals. Its price tag is also enormous. Over multiple decades, the Pentagon has sunk over $2 trillion into the aircraft. Throughout its lifespan, the F-35 has been plagued by billions in cost overruns, production delays, and critical deficiencies. The aircraft's gun, for example, has "suffered from the inability" to hit targets due to design problems. Its vaunted stealth capabilities can fail, and the readiness level of the aircraft is low; at last count, only 62 percent of tested fighters were deemed mission capable (the Pentagon threshold for all other aircraft is 80 percent).[7] The F-35 was symptomatic of larger problems dogging the US military. Its premier platforms were aging and underperforming expectations. Meanwhile, the Pentagon had failed to invest in next-generation technologies, such as AI, cyber tools, autonomous systems, and satellite networks. In short, analysts feared that the United States' military edge was eroding.

While the Ukraine war opened people's eyes in the Pentagon, China's emerging capabilities proved a greater catalyst for change. Beijing's leadership had made significant strides building out its technological base and enhancing its manufacturing. Its military-civil fusion efforts and prioritization of strategic technologies—from quantum computing and nuclear-capable hypersonic weapons to LEO satellites—had US strategists on edge. Defense tech companies saw where the trends were going and made countering China a central part of their marketing.

Palmer Luckey, for example, declared that Anduril was so focused on a potential war with China that many of his teams were only building weapons "that can be completed by 2027" (the year Xi Jinping has suggested that China should be prepared to invade Taiwan). SpaceX President Gwynne Shotwell said that "one of the only things that's preventing my grandchildren from speaking Mandarin is SpaceX."* Tech leaders organized high-level events on Capitol Hill and sponsored conferences to convince lawmakers about the risks of falling behind China in the AI race and the need to authorize billions to counteract this prospect. Trump's second term saw many defense tech allies—such as David Sacks, Scale AI's Michael Kratsios, and Palantir's Jacob Helberg—assume prominent government roles.[8]

The defense tech industry was ready for acceleration. At least three dozen funds were active, from Andreessen Horowitz's American Dynamism fund, Founders Fund, and General Catalyst's Global Resilience fund, to Shield Capital's frontier technologies fund. Even prominent startup incubator Y Combinator joined the fray and announced a new fund focused on defense, space, and robotics in early 2024. Six months later, it had backed its first startup, Ares Industries, a producer of low-cost cruise missiles. This flurry of activity reflected serious investment. In the last five years, over $180 billion in venture capital has flowed into defense tech startups. Already, over a dozen privately held companies are considered to be unicorns. Said Raj Shah, "I've never seen more entrepreneurs focused on building national security-oriented companies."[9] Figure 1 shows the explosive growth of VC deals in the defense tech sector since 2016.

Figure 1. Growth of VC Investments in the Defense Tech Sector (2019–June 2025)[10]

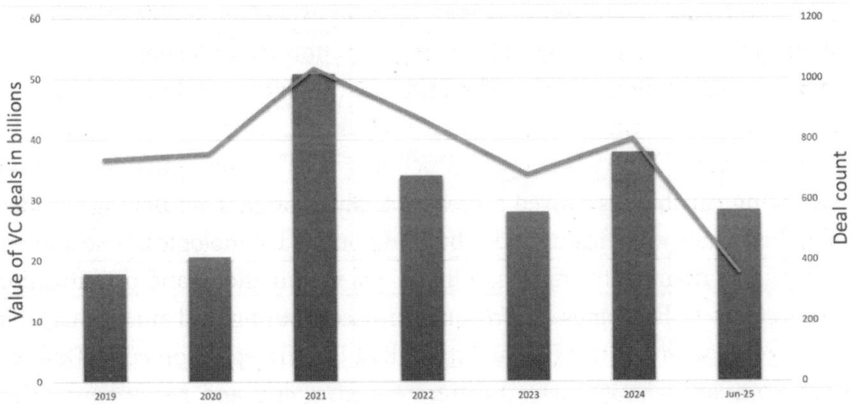

* Shotwell's remark came from an article published in *The Washington Post*. The full passage reads: "At an investor gathering last month, one of SpaceX's early investors, Ron Baron, recalled SpaceX President Gwynne Shotwell once telling him that 'one of the only things that's preventing my grandchildren from speaking Mandarin is SpaceX.' Shotwell backpedaled the remark on camera—'No, that was very bold of me'—but confirmed she believed SpaceX is 'very important' for U.S. national security."

As tech startups vied for a bigger share of the military's budget, former Pentagon officials moved into venture roles. Traditionally, retired defense officials sought positions with established defense contractors like Raytheon and Lockheed Martin. But many flocked to the Valley and joined VC firms, private equity outfits, or startups. Reporters identified at least fifty ex-Pentagon and national security officials who had transitioned to defense-tech-related ventures. One of the most prominent was Trump's former defense secretary Mark Esper. He joined the VC firm Red Cell in 2022 and has actively pitched Pentagon officials on anti-drone technology.[11] Table 1 shows top defense tech startups that have attracted significant attention and resources.

Company	Industry	CEO	Headquarters	Valuation
Anduril	Drones, software	Brian Schimpf	Costa Mesa, California	$60 billion
Scale AI	AI technology for defense	Alexandr Wang	San Francisco	$29 billion (Meta invested $14.3B in 2025)
Applied Intuition	Technology for unmanned systems	Qasar Younis	Mountain View, California	$15 billion
Shield AI	Drones, software	Ryan Tseng	San Diego	$5.3 billion (in talks for new funding round with $12 billion valuation)
Firefly Aerospace	Military space technology	Jason Kim	Cedar Park, Texas	$4.3 billion
Skydio	Autonomous drones	Adam Bry	San Mateo, California	$2.2 billion
Astranis	High-orbit satellites	John Gedmark	San Francisco	$1.6 billion
Epirus	Electronic weapons for air defense	Andy Lowery	Torrance, California	$1 billion
Hermeus	Hypersonic aircraft	A. J. Piplica	Atlanta	$702 million
Re:Build Manufacturing	Defense manufacturing and services	Miles Arnone	Framingham, Massachusetts	Undisclosed

Table 1. Leading Defense Tech Startups (2025–2026)[12]

While American defense tech firms led the charge, European startups also stepped into the arena. One of the better-known companies was a

German outfit named Helsing. It was founded in 2021 and initially focused on developing AI software for defense applications—similar to Palantir's business model. In the summer of 2024, Helsing raised $468 million at a valuation of $5.4 billion, "making it one of Europe's most valuable start-ups." More recently, it has moved into hardware. Its big bet is building autonomous weapons that can strike extended distances. Already, its HF-1 strike drones are being made in Ukraine by local producers. Helsing is now developing more sophisticated HX-2 drones at its own factory in southern Germany. The HX-2s have a range of one hundred kilometers and an expected cost of thirty thousand dollars per unit, offering "20 times the range of a Javelin anti-tank missile at a sixth of the price." The company hopes to set up multiple factories on the continent, making them resilient to potential Russian attacks and appealing to Europe's push for self-reliance. As cofounder Gundbert Scherf told *The Economist*, "One thing we've learned from Ukraine is that, when push comes to shove, you don't want to depend on anyone for weapons."[13] (That being said, Helsing's products may not yet be ready for prime time. According to *Bloomberg* reporting, Ukraine suspended its HX-2 drone purchases after they proved susceptible to electronic jamming and experienced high failure rates when taking off).[14]

Silicon Valley's AI giants also got into the act. Microsoft's involvement with Israel's intelligence sector is an instructive case. In late 2021, Microsoft CEO Satya Nadella met with Israeli spymaster Yossi Sariel, the head of Israel's prized signals intelligence directorate, Unit 8200. Sariel had an unusual request. His group was collecting a vast trove of top-secret intelligence material but needed somewhere to store it. The IDF didn't have sufficient storage nor computing power to handle this volume of data. Sariel asked, could Microsoft host this data on its cloud servers? He wanted Microsoft to provide a customized area within its Azure cloud platform for Unit 8200 to store, access, and process specific data. Nadella agreed to assist Sariel, paving the way for him to build a powerful mass surveillance platform that could capture and store millions of Palestinian mobile phone call records made daily in Gaza and the West Bank. Several intelligence officers from the group soon adopted a mantra that exemplified the project's ambition: "a million calls an hour." The amount of data they collected was stunning. By July 2024, Microsoft servers were hosting over 13.6 petabytes—roughly the equivalent of 14 billion printed books.[15]

How did the IDF use the data? After the October 7, 2023, surprise attacks by Hamas, Israel got on a war footing and began targeting suspects for retaliation. But it lacked information about who to strike and where they were located. This intelligence trove was crucial in helping Israeli forces identify bombing targets in Gaza. They used Azure to compile and transcribe phone calls, text messages, and audio. An intelligence officer described how Azure allowed him to rapidly search for specific "terms and patterns"—such as exchanges between individuals—which he then cross-referenced with the IDF's own systems to isolate locations.[16]

Microsoft denied any knowledge of how the IDF used its system or what data Unit 8200 planned to store on its servers, claiming that at no time was it "aware of the surveillance of civilians or collection of their cellphone conversations using Microsoft's services." But leaked documents from a joint investigation conducted by *The Guardian* and *+972 Call* in August 2025 suggested otherwise. "You don't have to be a genius to figure it out," said one source. "You tell [Microsoft] we don't have any more space on the servers, that it's audio files. It's pretty clear what it is."[17] Why then did Microsoft sign off on this partnership? Two reasons come to mind.

First, the company saw a lucrative commercial opportunity in front of it. The deal with Sariel had the potential to draw in "hundreds in millions of dollars in revenue" over a long period of time. Second, Microsoft executives viewed the partnership as "an incredibly powerful brand moment" for Azure. They figured that forging a successful relationship could help them win business from other governments requiring similar services.[18]

While Microsoft initially rejected the allegations, claiming that it "did not build or consult with Unit 8200" on a cloud-based surveillance system, nonetheless, the company commissioned an external investigation to investigate into the claims. One month later in September 2025, Microsoft announced it was terminating its partnership with Unit 8200, asserting that the IDF had violated its terms of service by storing surveillance data on Azure (the termination only affects its Unit 8200 relationship; other business with the IDF will continue).[19] The story's significance extends beyond merely exposing a major tech company's hypocrisy. As conflict surges, Silicon Valley has learned that war is a business opportunity. Microsoft's partnership with the IDF exemplifies a broader trend; multiple intelligence agencies and militaries are collaborating with major US tech companies on similar initiatives.

Take OpenAI, for example. In December 2024, OpenAI head Sam Altman announced that his company had teamed up with Anduril to build software to counter drone attacks. His announcement represented OpenAI's most significant involvement to date with the Defense Department and its first direct tie-in with a commercial weapons manufacturer. It also was a departure from the company's pledge to ban its AI products from being used in war. Employee reactions were not kind. In internal messages, staff asked how OpenAI could ensure its systems wouldn't assist lethal attacks. One person accused the company of downplaying the implications of doing business with a defense contractor. Another worried about reputational damage to the firm. Altman pushed back, arguing that it was important for OpenAI to "provide the best technology available to militaries run by democratically-elected governments" while noting that authoritarian governments would not hesitate to use AI for military purposes. That may be true, but there likely was a simpler explanation for OpenAI's change of heart: tech companies had figured out that working with the Pentagon could be immensely profitable, particularly under Trump. OpenAI's internal projections showed that it wouldn't turn a profit until 2029. In the meantime, it faced staggering losses—$5 billion in 2024 and as much as $14 billion by 2026. It came as little surprise that Altman was seeking new revenue streams to offset OpenAI's shortfalls.[20]

In the same timeframe, Anthropic announced a partnership with Palantir and Amazon Web Services to make its AI model Claude available to US national security and intelligence agencies—a decision it would later regret, as discussed previously. Anthropic's exit provided a window of opportunity for Sam Altman. Just hours after Hegseth labeled Anthropic a supply chain risk, the Pentagon announced it had reached a deal with OpenAI to access its models for its classified networks "for any lawful purpose." (Following an intense backlash, Altman went back to the table and demanded additional safeguards against mass surveillance use.)

Google's announcement also provoked a big reaction. On February 4, 2025, the company updated its ethical guidelines and dropped its commitments not to use AI technology for weapons, surveillance, or products that would potentially cause harm. Its executives argued that AI had become widespread and that companies based in democracies ought to use their influence to spread freedom and human rights in their government and defense sector engagements. What was the real reason for Google's turnabout? Money and deals. "It's clear that the company's principles were keeping it out of conversations and deals that it wants to be a part of," wrote tech reporter Casey Newton. "Updating the principles will give Google a seat at the table." Google's policy change had reverberations as far away as Beijing. The official WeChat account of the People's Liberation Army (PLA) posted a blunt critique, warning that Google's decision introduced big risks, and cautioned the United States against "opening Pandora's box and becoming a culprit endangering the world."[21]

All told, the defense tech sector had gotten frothy. Trump's return to office threw more fuel on the fire. He not only brought over a slew of political appointees with deep tech experience, but he followed up by issuing an April 2025 executive order directing all federal agencies to procure commercial products and services by default. "For defense tech start-ups backed by venture capital firms like Founders Fund, Andreessen Horowitz and Cerberus Capital Management, it was a signal to go all in," wrote *The Washington Post*'s Josh Rogin.[22]

Yet, for all of the hype, only a few startups had achieved breakthrough success. Aside from Palantir, SpaceX, and Anduril, many other venture-backed firms struggled to secure large-scale contracts. A key part of the problem was the Pentagon itself. Despite the Department of Defense's growing embrace of defense tech, it had only allocated a small amount of money to this sector. In fiscal year 2023, for example, venture-backed companies received less than 1 percent of Pentagon contracts worth $411 billion, representing only a small increase from 2010, when venture-backed firms received .5 percent of the department's $373 billion worth of contracts. "If there is not a shift in the concentration of contracts still going to the top defense primes toward these new vendors, this VC investment will dry up," warned Michael Brown, the former head of DIU.[23]

Part of the issue was risk aversion by the Pentagon's defense contract managers, who were unwilling to gamble on untested designs tied to companies with thin track records. After prior scandals involving fourteen-thousand-dollar toilet seats and malfunctioning naval ships, procurement officers were reluctant to venture beyond tried-and-true programs. And their assessment wasn't entirely wrong. New systems are risky. Some products will falter because their technology is developing more slowly than anticipated or they fail to perform in real-world conditions. Other products work well in civilian applications but are not rugged enough for military operations. American drone maker Skydio's experience in Ukraine illustrated these challenges. The firm sent hundreds of its products to bolster Ukraine's forces, but its equipment vastly underperformed. "Skydio's drones flew off course and were lost, victims of Russia's electronic warfare," reported *The Wall Street Journal*. The company had to go back to the drawing board and build a whole new fleet. But the retooled Skydio units encountered another issue—they were too expensive for an attritable conflict. "It's like choosing between BMWs and Skoda Octavias," says

Eduard Lysenko, an officer with Ukraine's defense-tech arm. "A BMW is faster and more comfortable, but it really doesn't help you if your task is to give everyone cars."[24]

Skydio's experience wasn't unusual. American drones were expensive, error prone, and difficult to repair, forcing the Ukrainians to seek alternatives. Not only was this bad news for startups hoping to generate sales and attention for their products, but it also threw a wrench in the Pentagon's plans to scale up its drone program.

Even when products performed well, defense firms still had to navigate countless bureaucratic hoops to obtain steady and predictable funding. The goal for most startups is to get their technology built into the Pentagon's five-year budget plan. Until that point, their best hope is to hang on for years through small-budget pilots or research grants. The unfortunate reality is that these agreements run out quickly and are seldom large enough for young companies to sustain their payrolls (or to generate the type of growth expected by VC investors). If a startup's product isn't included in the budget plan, Trae Stephens told *The Wall Street Journal*, "it is like the 'Field of Dreams' at that point. That is probably one too many miracles to bet on." Industry executives refer to this fallow period—between receiving initial government funding and reaching commercialization—as the "valley of death," where the glacial pace of federal contracting leads companies to hemorrhage their funding while they await decisions from the Pentagon.[25]

Is Defense Tech Worth the Hype?

In 2022, William LaPlante, the Pentagon's acquisition chief, delivered a stern rebuke to defense tech startups at an industry conference. "The tech bros aren't helping us too much in Ukraine," he cautioned. "Ukraine is not holding their own against Russia with quantum. They're not . . . holding their own with AI. . . . It's hardcore production of really serious weaponry. That's what matters." He added, "We're not fighting [in] Ukraine with Silicon Valley right now, even though they're going to try to take credit for it." LaPlante's words gained attention and sparked disagreement. He later clarified that he was criticizing "aspirational, often elusive technology" rather than companies such as Palantir, Anduril, and SpaceX that had delivered products for Ukraine's military.[26] But his critique ran true for many people. While few would dispute that new technologies have proven useful on the battlefield, many questioned whether they were bringing transformational effects, contrary to the claims of Silicon Valley's boosters.

For one thing, it is evident that new technology did not decisively turn the tide of the war in favor of Ukraine. Even when outpaced by drone technologies, Russia showed a remarkable capacity to withstand devastating combat losses. Analysts estimated that heavy fighting in early 2025 along Ukraine's eastern front lines resulted in more than fifteen hundred Russian troops being killed or wounded daily. Still, the Russians persevered and even increased the scope of their attacks. While the story of the war's first year was Ukraine's ingenious innovations—using Turkish TB2 drones to hammer sitting-duck tanks or leveraging Starlink connectivity to conduct daring raids behind enemy lines—the Russians soon adapted. They began using electronic warfare to misdirect Ukraine's drones. They started deploying their own one-way attack drones, inflicting major casualties on Ukrainian soldiers. Reports showed that Russian troops incorporated clever innovations—using electric scooters, motorcycles, and ATVs to disperse swiftly across the battlefield. "Hitting just one piece of equipment carrying 15 people, well, that's possible, it can be done quite easily," as one Ukrainian colonel put it. "But when those 15 people are riding electric scooters, then that's a very big problem."[27]

Russian troops posing with electric scooters near the frontlines

The fact that advanced technology did not prove more determinative in Ukraine was not surprising. For decades, US military leaders had been preoccupied with fighting a cleaner war that would reduce conflict "to a set of rules and a system of procedures—and therefore to make orderly and rational what is essentially chaotic and instinctive."[28] In the 1960s, Defense Secretary Robert McNamara and his "whiz kids" (a group of Ivy League–educated young men in their thirties and forties recruited to help run the Pentagon) sought to quantify all aspects of war into digital problems that could be solved through mathematics and calculations. General William Westmoreland, who commanded US forces in Vietnam, envisioned a more sanitized version of conflict that would occur on a surveilled, automated, and networked battleground, where the right technology would help commanders make sound decisions and permit them to be "continually aware of the entire battlefield panorama."[29] Technology would bring logic and rationality to war and allow American forces to exploit their superior capabilities and wield a decisive advantage over their adversaries. While it appeared that a cleaner version of war was just around the corner, it never quite materialized. Even when US scientific-precision tactics seemed ascendant, such as during the Iraq War, complicated realities on the ground—in that case an escalating Baathist insurgency that mired American troops in a decade of bloody fighting—undermined the notion that the United States could engineer its way to battlefield victory.

Still, the US military maintained its commitment to pursuing a techno-scientific approach to war, with Ukraine representing the most recent example of this thinking. Would this war prove any different? For military scholars like Columbia's Stephen Biddle, advanced technology in Ukraine once again failed to bring about transformational results. "Casualty rates in Ukraine have not been unusually high by historical standards. Attackers in Ukraine have sometimes been able to advance and sometimes not," wrote Biddle. In part, this was because Russia responded to Ukraine's innovations "just as their predecessors did: by adapting with a combination of technical countermeasures and further extensions of centurylong trends." When Ukraine deployed expensive drone fighters, the Russians countered with antiaircraft missiles. When Kyiv deployed simpler and cheaper UAVs, the Russians used antiaircraft batteries and jammers to negate Ukraine's advantage. Rather than represent a revolution, Biddle argued that Ukraine resembled an "incremental extension of longstanding trends." He wasn't alone in his criticism. Other analysts were equally skeptical about technology's effect in the war. Wrote Carnegie Endowment for International Peace scholar Gavin Wilde, while tactical systems like drones could deny battlefield mobility and keep Russian forces at bay in certain areas, "no amount of technological wizardry" could substitute for "recruiting, training, and equipping a capable fighting force over the longer term."[30]

That being said, one could make a reasonable argument that if not for Ukraine's innovations, Russia's dominant position in mass—its superior number of troops, tanks, and aircraft—would have collapsed Kyiv's defenses long ago. In this regard, T. X. Hammes, a scholar at National Defense University, observed that 90 percent of the weapons systems used in the war were developed and manufactured in the twentieth century: "the bulk of each force's formations are armed with decades-old equipment and tactics." As the conflict progressed, it was the other 10 percent of equipment that brought decisive effects. He highlighted three categories of new technologies that were "changing the game of land warfare" by dramatically improving Ukraine's defenses: (1) networked and high-speed command and control, (2) "near-persistent" surveillance of the battlefield, and (3) the skilled deployment of precision artillery, drones, and loitering munitions. Hammes argued that the combination of these tactics provided a pathway for smaller, lighter, and less heavily armed forces to defeat better-resourced militaries.[31]

Did all of this add up to a revolution in military affairs? Maybe not. But it was hard to deny the impact of these tools in checking Russia's military ambitions.

Present debates about military AI show signs of exaggeration and hype. A high-profile group of boosters, many with close ties to Silicon Valley, are steering public discussion and advocating for massive Pentagon investments in technologies they portray as warfare's next frontier. Many of these figures have little operational experience in the military but instead draw from their "impressive business successes" to make their case. "Through opinion pieces in high-profile journals, public addresses at acclaimed security conferences, and presentations at top-tier universities," they have spun a vision for a new model of fighting.[32] Certain persons, such as Palantir's Alex Karp, characterize the geopolitical stakes in existential terms. In a notable *New York Times* op-ed in 2023, Karp compared the development of AI weapons to the dawning of the nuclear age, calling this "our Oppenheimer moment," and arguing that if the United States failed to act then it would pay a heavy price. (Karp has also undergone a personal transformation, moving from being a self-identified "proud progressive socialist" to joining Peter Thiel as one of MAGA's "most zealous supporters from the industry.")[33]

It's difficult to trust such predictions. Yet, the strategic value of AI weapons has growing empirical support: from the thousands of US strikes against Iranian targets in March 2026 using Palantir's Maven Smart System and Anthropic's Claude, to Ukraine's locally built Delta system that identifies up to twelve thousand Russian military targets weekly, the evidence mounts.[34] How can we reconcile these competing viewpoints—between skeptics who doubt military AI's transformational impact and advocates who evangelize its benefits?

I posed this question to retired Air Force General Jack Shanahan, who was the inaugural director of the Pentagon's Joint Artificial Intelligence Center and oversaw the Maven program. He acknowledged that the military can only take things so far with old platforms and that eventually the Pentagon will have to "cut the cord" and build new systems from scratch. But he cautions that "we're nowhere close to doing that." For now, the Pentagon must learn to live with both aspects—optimize legacy platforms as much as possible while embracing emerging systems and weapons. "So in my very kind of bumper sticker way of talking about this, it won't be stealth bombers or FPV drones. It will be stealth bombers and FPV drones. Before long it will be AI-enhanced stealth bombers and AI-enhanced FPV drones," said Shanahan. "My point is that we've got to figure out the legacy plus the new together. It's a bridging solution."[35]

Others have less patience for the status quo. "I think it's a mistake if we sit here in DC and say, oh, that tech stuff is toys and doesn't work," Michael Brown told me. "A third of the Russian fleet is disabled or at the bottom of the Black Sea because the Ukrainians were pretty clever in saying, I'm going to use unmanned systems for an asymmetric advantage." There is technology working in Ukraine that the Pentagon and lawmakers can embrace. Unfortunately, he said, that's not how things normally work. "There's a built-in bias for big platforms built by big defense primes that translates to jobs in Congressional districts."[36]

When it comes to future military technologies, the path ahead is uncertain. Ukraine demonstrated how new modes of fighting and creative innovations can humble a much larger army beholden to traditional models of operation. A crop of energetic startups is eager to challenge entrenched defense primes for resources and funding. While the Pentagon's purse strings remain tight for companies sitting outside regular procurement channels, there are signs of a shift.

Overall, billions in VC money is flowing into the sector, reflecting confidence that the Defense Department will continue to embrace future technologies. Yet the Pentagon famously moves at a glacial pace. The amount of lead time required to plan, develop, and manufacture weapons systems is multiple years, due to an archaic process dating back to the Cold War. Even in a pro-business administration like Trump's, skeptics abound. Deputy Secretary of War Steve Feinberg, for example, has wondered aloud whether Anduril "is a real company" or whether it only makes "toys." Reuters reported on a tense meeting in July 2025 in which Feinberg appeared "unimpressed" by unmanned navy drones presented by two venture-backed startups, Saronic and BlackSea Technologies. "He views them as software companies—he's not sure how they will work in a mission environment," said one defense executive who speaks to Feinberg regularly.[37] In other words, change will not come swiftly or easily. But American policymakers can ill afford to keep pursuing business as usual. The United States won't be able to maintain its advantages over China and other rivals without bringing Silicon Valley along. Conversely, Valley firms need a sufficient incentive to keep vying for Pentagon contracts. Otherwise, they will put their talents into other sectors and America's strategic interests will suffer.

The next chapters describe how the rise of private tech forms the backdrop for a larger tech rivalry between Washington and Beijing. Both countries are squaring up for an extended fight—restricting the exports of key products, investing heavily in the industries of the future, and enlisting allies to support their agenda. They're wielding technology as both a carrot and a stick—promising benefits to nations that cooperate and threatening punishment for those that don't. Which strategy will prevail?

PART III

America's Tech War

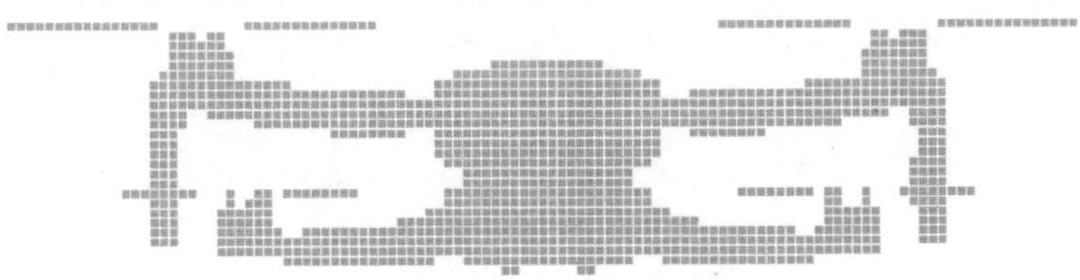

8

Rise of Economic Warfare

Late in the afternoon on April 2, 2025, Donald Trump stepped up to the Rose Garden lectern and delivered an announcement that shocked the world. "For decades, our country has been looted, pillaged, raped, and plundered by nations near and far, both friend and foe alike," he declared. "American steelworkers, autoworkers, farmers, and skilled craftsmen—we have a lot of them here with us today—they really suffered gravely." He unveiled 10 percent across-the-board tariffs on virtually any good imported into the United States and said that certain trading partners would face higher "reciprocal tariffs" aimed at penalizing them for their trade barriers. Among the hardest-hit countries were America's closest allies. The European Union was assessed with a 20 percent penalty, Japan received a 24 percent tax, and South Korea was hit with a 25 percent tariff. Trump reserved his greatest ire for China. Initially, Beijing was assessed with 34 percent tariffs, but this soon escalated once China retaliated with its own punitive measures against US exports.[1]

In one fell swoop, Trump upended decades of international trade orthodoxy and threw the global economic system into disarray. Unquestionably, his moves represented an unprecedented escalation of economic warfare, but the groundwork for these policies had been laid by prior presidents.

Under Joe Biden, American policymakers grappled with a dilemma. Coming into office, his administration had decided that the old free trade model with China was dead and that the United States would instead impose costs against Beijing's high-tech economy. The White House tightened restrictions against Chinese conglomerates like Huawei, but was dissatisfied with the results. Chinese industry continued to expand into strategically vital areas like artificial intelligence and supercomputing. Beijing's major cloud computing companies—Alibaba, Baidu, and Huawei—were competing ferociously with America's big three hyper-scalers: Google, Microsoft, and Amazon. And China's military was benefiting from its access to cutting-edge innovations. "America is not prepared to defend or compete in the AI era," warned ex–Google chief Eric Schmidt in 2022. "A second technological superpower, China, has emerged. It happened with such astonishing speed that we're all still straining to understand the implications."[2] All of this presented a threat to America's dominance that couldn't go unanswered.

Up until this point, US controls had focused on stopping individual Chinese firms from acquiring specific technologies, mostly by adding companies to a trade blacklist known as the "entity list." Starting with Obama and picking up speed during the first Trump presidency, scores of Chinese tech firms were placed on the entity list on the basis of national security concerns and prohibited from receiving American products. But policymakers realized that a more comprehensive approach was needed and pivoted to another strategy—rather than just sanction companies, they would target an entire industry.

Biden's secretary of commerce, Gina Raimondo, played a pivotal role in shaping the new approach. As the United States ramped up tech export controls against China and surged investment into domestic chip production, Raimondo became the public face of this policy and emerged as one of his most trusted advisers.

Raimondo's ascent was rapid. Before taking over Commerce, she served as the governor of Rhode Island. She was best known for signing into law free community college and kindergarten for state residents, raising the minimum wage, and cutting commercial taxes. She once stirred debate by picking a fight with commercial truck operators—proposing new highway tolls to repair Rhode Island's dilapidated bridges. Truck drivers were so incensed by Raimondo's plan that they circled the State House for months "blasting their horns in protest and rattling the nerves of Ms. Raimondo's staff." (Biden rallied to her defense as vice president, traveling to Providence to inspect a local bridge he described as "being held up by Lincoln Logs" and advising Raimondo to "let the horns blow" and to fix the bridges.)[3]

When Raimondo came to Washington in 2020, she had little China experience and was an outsider in the fights spilling over in the corridors of the State Department and West Wing about whether to continue Trump's hardline policies against China. Few people considered Commerce central or relevant to these discussions. The previous secretary under Trump, Wilbur Ross, hardly inspired confidence in the department. Ross had a habit of falling asleep in the middle of meetings, leaving one former administration official to comment that "Wilbur is good until about 11 a.m." It didn't take long for Trump to sour on Ross, once remarking to an aide that "Wilbur has lost his step. Actually, he's probably lost a lot of steps."[4] Expectations were low that Raimondo would play much of a role in shaping Biden's China policy, but those predictions were soon proven wrong.

Washington underestimated the personal stakes Raimondo felt about the loss of American industry to China. She grew up in a close-knit Italian-American family and often mentioned how her father lost his job at a Bulova watch factory, after twenty-eight years of employment, when the company shifted production to China. "The truth of it is what happened to my dad isn't unique, it's happened to millions of Americans over the last 40 years. And it's part of a much bigger story about America's declining investments in our workers and our economy."[5] Raimondo was intent on giving the department a greater voice in these discussions. She was also fortunate to come to Commerce at a time when the rules were changing. If the main focus of Commerce was to promote American businesses overseas, rising tensions between the United States and China—as well as Russia—elevated the department's "protect" function—its mandate to control the export of sensitive US products to foreign markets. Commerce became an important actor under Raimondo, thanks in part to its unique role in the federal government. While the National Security Council or State Department lead big-picture thinking about how to confront China and Russia, Commerce has specific tools that can turn policy ideas into practice; it's "where the rubber meets the road."[6] As successive White Houses leaned into sanctions and economic statecraft to constrain China's rise and deter Russian aggression, Commerce moved further into the spotlight.

When it came to developing a comprehensive package of tech sanctions against China, Raimondo and her team worked through a series of questions. How restrictive should the regulations be? How would China respond? What would be the impact on American manufacturers? She brought an "investment banking, Rhodes Scholar–type demanding direction" to the task, asking tough questions about the effect of the controls and whether Commerce officials had engaged sufficiently with industry.[7]

Finally, on October 7, 2022, Commerce was ready to make its move. The Bureau of Industry and Security (BIS)—an obscure office hidden away in the department—issued a rule designed to cripple China's chip manufacturing capacity. It stated that American companies could no longer provide advanced computing chips or chip-making equipment to Chinese firms unless they received a special license, imposing broad restrictions banning companies anywhere in the world from selling advanced chips to China if they were made with American technology, software, or machinery—using an authority known as the foreign direct product rule. The logic behind the regulation was straightforward: leading-edge microchips are foundational to supercomputers, AI systems, sophisticated weapons, and surveillance apparatuses. To stay ahead of rivals and maintain technological superiority, America needed to close off access to these tools. Biden's national security adviser, Jake Sullivan, stressed that export controls could be a "new strategic asset in the U.S. and allied toolkit to impose costs on adversaries, and even over time degrade their battlefield capabilities."[8] *The New York Times* described the move as tantamount to a "declaration of economic war on China."[9]

Using Economic Tools to Punish Rivals

Deploying economic tools to "punish, constrain or, more generally, to influence the behavior" of states or private actors is not a new trend. This strategy has been used throughout history, from Athenian sanctions against Sparta in 432 BCE and Spanish embargoes imposed against Dutch traders in the sixteenth century to Napoleon's blockade against Britain in the 1800s. Economic sanctions are a way for countries to materially exclude rivals from the global economy to protect specific objectives. They are "offensive in nature" and seek to "exert force over other actors" from a distance.[10]

While states often use sanctions to force a change in behavior from their rivals, they can also serve other purposes, operating as instruments of "denial, containment, or political symbolism." When US policymakers imposed their October 2022 chip restrictions, they did not make any associated demands against China. Rather, the controls had a simple purpose: to block a major rival from getting easy access to a critical technology. One of the original thinkers about trade as a tool of national power was scholar Albert Hirschman. Back in 1945 he described how trade lent itself to geopolitical manipulation, analyzing how Nazi Germany had leveraged economic dependencies to control weaker states. Rather than thinking about the exchange of goods simply in supply-demand terms, he argued that one should consider the "influence effect" of trade—how it fosters "relationships of dependence" between nations and serves as an instrument of power. Hirschman's theories laid the groundwork for contemporary research analyzing economic coercion and weaponized interdependence in the international system.[11]

Today, sanctions have become a preferred tool for policymakers, although it wasn't until the second half of the twentieth century that states' use of sanctions really took off. During the Cold War, the world experienced an average of 13.5 cases of sanctions per year. Subsequently, from 1990 to 2005, annual episodes of sanctions increased to 53.5 per year. The numbers are now far higher. The United States currently has active sanctions in place against more than 15,373 individuals, firms, and entire industrial sectors, targeting one third of all states worldwide with some kind of penalty—whether against individuals, financial institutions, governments, or entire industrial sectors.[12]

Figure 1. US Sanctions Trends, 1950–2022[13]

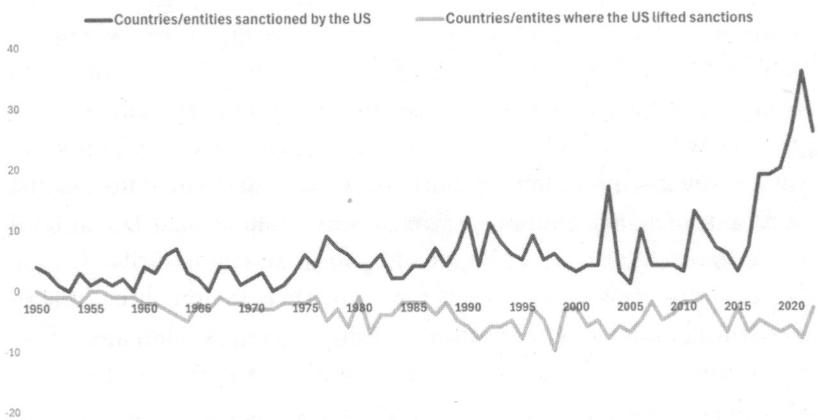

What explains this surge? US policymakers were drawn to these tools for two key reasons. First, along with military dominance, America's unipolar moment offered the United States economic supremacy. The US dollar reigned supreme. Governments and financial institutions worldwide became dependent on the dollar to conduct their financial transactions, giving US officials significant leverage. In practical terms, this meant that global banks had to access dollars through their US counterparts, forcing them to comply with the rules set by the US government. That made the Treasury Department "the gateway to the world's banking operations." This unique position allowed Treasury officials to impose sanctions on virtually any foreign person, company, or state deemed a threat to US interests. The penalty for violators was steep—getting placed on a sanctions list would freeze them out of the international financial system and prevent them from conducting basic transactions. American policymakers now possessed a powerful financial weapon and were increasingly willing to use it.[14]

Second, following the disastrous experience of the "Forever Wars" in Afghanistan and Iraq, US presidents grew reluctant to deploy troops to faraway places, but they still needed to protect national interests. Their solution was to use economic tools in place of military ones. When a coup took place in the Democratic Republic of the Congo, or the Syrian government instigated a brutal crackdown against civilian protesters, instead of using military force, the White House favored using sanctions to pressure regimes to change their behavior (unfortunately, leaders proved remarkably resilient against these economic measures, and sanctions rarely achieved their goals even if they caused pain).

As sanctions use increased, experts expressed doubts. Agathe Demarais, a senior analyst at the European Council on Foreign Relations, critiqued these measures in her book *Backfire*, arguing that they rarely accomplished their aims and that their overuse was "fueling resentment against Washington around the world." US policymakers also expressed reservations about overreliance on sanctions. Former Treasury Secretary Jack Lew, who served under Obama, cautioned that America should "guard against the impulse to reach for sanctions too lightly or in situations where they will have negligible impact," and that leaders should be "conscious of the risk that overuse of sanctions" could undermine American leadership.[15]

Nonetheless, Washington's appetite for sanctions remained unchecked. With China's emergence as the United States' main rival, policymakers increasingly deployed economic restrictions to contain Beijing. Starting in late 2017—with Trump's backing—Congress passed successive legislation, regulations, and licensing regimes intended to curb Chinese investment in critical American companies and to stop the transfer of foundational and emerging technologies to Beijing. When Biden took office, observers wondered if he would shift course. Instead, his policies proved remarkably consistent with Trump's.

I sat down with a former senior Biden official who told me that "you had a bunch of alumni of the Obama administration who I think in retrospect understood that we probably needed to have pivoted around 2015 or 2014 to a much sharper-elbowed approach" toward China. But intervening events kept getting in the way, from Syria's civil war and the wreckage of Libya to Iran's nuclear program and the threat of ISIS. As he recollected, "I think many who came into the Biden administration, looking back, essentially said that we had a couple of years at the end there where we should have had more of a correction."[16]

While the US strategy against China extended beyond tech controls, technological innovation emerged as the central priority. The main reason was that American policymakers saw a close link between technological innovation, military dominance, and state power. They worried that if China and Russia continued to erode America's technological edge, it would endanger US defense commitments and projection of power. China already held advantages in advanced manufacturing, commercial drones, robotics, and advanced batteries; permitting Beijing to gain more control over global digital networks and the prospective ability to build new weapons systems that could "overmatch U.S. capabilities" could reduce confidence in the US military and undercut its deterrence.[17]

Technological innovation also translates into economic power. Countries whose technologies dominate the future economy will capture trillions of dollars' worth of value. From biotechnology and quantum to clean energy and AI, the companies and products that lead the next wave of innovation will bring unmatched prosperity to their nations. Yet certain technologies are far more critical than others. Which ones deserve priority, and what should the US do about them?

9

What Is a Strategic Technology?

If you were to ask the average person to name a critical technology with clear strategic benefits, there's a decent chance they would mention nuclear weapons. Nuclear technology is difficult to harness and incredibly powerful, which incentivizes governments that possess it to keep it from their rivals. Its potential to reap massive destruction makes it an especially fearsome tool. Eighty years after atomic energy was first harnessed, only a small number of states—nine at last count—possess the know-how to build nuclear weapons.

But few technologies offer such a clear set of facts. Take drones, which are a good example of a "dual-use" technology, offering a clear military purpose as well as civilian applications. Off-the-shelf drones, readily purchasable online, have proven to be a game changer in the Ukraine war and other conflicts. They have transformed the battlefield, allowing individual soldiers to surveil and target tanks, armored carriers, and artillery batteries with ease. At first glance, it would appear sensible for governments to curb access to drones. After all, why would countries allow their adversaries to purchase such disruptive weapons simply by handing over a credit card? But drones also provide many nonmilitary uses. They can deliver lifesaving water and medicine to hikers trapped in remote areas, scan for endangered species, or alert emergency responders to the presence of wildfires. Moreover, drone technology is commonplace. Even if the United States tried to restrict drone access, such an effort would surely fail. According to the Federal Aviation Administration, there are nearly eight hundred thousand registered drone users in the United States alone.[1] So where should policymakers draw the line between high restrictions and minimal restraints?

Policymakers have struggled to answer this question and clearly specify what qualifies as a strategic technology. Historically, unless a particular technology was closely aligned with military necessity and had a direct application in war, it was not considered strategic. But this perspective had limitations, as it excluded relevant technologies (such as AI-enabled surveillance) that lacked a singular military purpose, while also presuming that all military technologies deserved to be called strategic (when many of them, such as high-capacity rifles, are ubiquitous). Today, views have shifted. Most experts agree that strategic technologies must include a wider array of capabilities and even incorporate some general-purpose technologies. But defining the precise contours is tricky.

In this regard, a good starting point is the Pentagon's list of six critical technologies (reduced from fourteen priorities), updated in November 2025. Named technologies include applied artificial intelligence, biomanufacturing, contested logistics technologies, quantum and battlefield information dominance, scaled directed energy, and scaled hypersonics. Unfortunately, this list contains few details about which specific innovations will be prioritized or what steps the Pentagon will take to develop useful applications in support of its military goals.[2] The pared down list also raises questions about how we should treat the technologies that were cut from prior lists, such as quantum, space, microelectronics, and renewable energy. Are these technologies no longer strategic priorities? Should we interpret this list narrowly as reflecting only military objectives, with civilian agencies maintaining their own priorities? If so, what are they?

These questions matter because they fundamentally impact what resources the government will channel, what innovations it will prioritize, and the scope of regulations it will pass. For example, it is one thing if America's goal is to deny China access to a narrow band of military technologies. But if the objective is to curb China's capacity in entire technological domains (such as for frontier AI applications), the implications are larger—and would mean curtailing most forms of technological cooperation between the two countries.

Overall, the main deficiency of such lists is that they lack a consistent rationale for why certain technologies are deemed strategic and how the government should handle these innovations. To crack this problem, I propose applying a simple three-part test to determine whether an innovation is strategic:

- ***How important is the innovation? Does the innovation bring significant economic value or national security utility?***
- ***Is the technology likely to be fought over between rival states?*** **(Certain technologies, such as military aircraft manufacturing, are less likely to diffuse to other states, making them more strategically important. Other sectors, such**

as general medical research, easily diffuse across borders, limiting their strategic value.)[3]

- ***Are there barriers to developing the technology?*** **(Such as steep R&D requirements or high fixed costs)**

If the answer to these three questions is yes, then the technology can be considered strategic. But that still doesn't tell us what, if anything, the government should do about it. After all, many strategic technologies—such as cloud computing—are better left untouched by bureaucrats. What circumstances *would* justify a government intervention? Three situations come to mind.

The first situation involves innovations that provide wide benefits but face funding challenges because companies do not believe they will capture the full returns from the project—good ideas that are poor financial investments. Infrastructure overhauls like high-speed railroads fall into this category. Building high-speed rail is capital intensive; companies need to assemble guideway and track elements, build stations, terminals, and intermodal facilities, and construct support facilities, such as train yards, shops, and administrative buildings.[4] Put together, these elements can run into the billions of dollars, while revenue returns can spread out well into the future (meaning that it could take decades before companies recoup their investment costs). Moreover, businesses that benefit from the presence of high-speed rail can free ride off its construction—restaurants, shops, office buildings, and apartment complexes adjacent to high-speed rail hubs do not contribute money toward building the rail transport network but profit financially from its presence. As a result, rail companies are reluctant to assume the full expense of these projects absent government support.

The second situation occurs when the barriers to entry for a specific technology are high and the technology has limited commercial appeal, such as space exploration or satellite technology. Companies in this sector face high fixed costs and steep learning curves for production—making government subsidies a prerequisite for entry. Elon Musk's rocket company SpaceX is a good illustration. Despite perceptions of Musk as a self-made entrepreneur, his companies depend on subsidies; SpaceX and Tesla have received at least $38 billion in government contracts, loans, subsidies, and tax breaks since 2003 for launching astronauts for NASA, building satellites for the Pentagon, or producing EVs for consumers. In 2024 alone, SpaceX and Tesla received at least $6.3 billion from federal and local governments.[5]

The third situation occurs when the technology has high barriers to entry yet maintains substantial commercial value. High fixed costs limit participation to a few companies, making these products vulnerable to supply disruptions, especially if providers are based in a foreign country. Civilian aircraft are a good example. The market for large commercial jets is small, dominated by just two companies: Airbus and Boeing. They require billions in upfront funding to develop the engineering, design criteria, and technical blueprints for a new aircraft platform. Accordingly, both companies have benefited from major government subsidies, without which their success would be uncertain.[6]

I would argue that if any of these three situations exist, then there are compelling reasons for a government intervention—either providing commercial subsidies, or as a fallback, penalizing foreign competitors to protect domestic industries.

Some readers may question whether these strategies remain important, or even relevant, given Trump's turbulent approach to trade negotiations and policy. In truth, Trump has discarded any coherent framework in favor of an approach predicated on short-termism, transactionalism, and crony capitalism. His treatment of TikTok is a useful illustration.[7]

There was a growing consensus in Washington that allowing China to retain control of the platform was an acute national security risk. To that end, Congress passed a law in 2024 requiring TikTok's parent company, ByteDance, to divest control of the platform. After several delays, Trump finally acted. Following a phone call with President Xi in September 2025, he announced he had reached a deal for an American investor group, led by Oracle, to take an 80 percent stake in TikTok's US operations.

Almost immediately, the agreement raised concerns. For one, it failed to resolve a core issue—who would control TikTok's algorithm, its "secret sauce" determining what users see on the app and serving as a potential weapon for Chinese influence operations. Under the agreement, the algorithm would remain in ByteDance's hands, and a copy would be licensed to the US investor group. While the group could then retrain it using data from American users, ByteDance would still be responsible for issuing updates to the algorithm and would potentially oversee other aspects of its operation. Rather than solving a national security dilemma, the deal allowed Beijing to preserve significant influence. But the agreement brought other benefits for Trump World that carried more weight. For one, it was disclosed that US investors were set to pay $10 billion to the US Treasury as a "transaction fee" for the government's role in brokering the deal.[8]

Further, the main beneficiaries of the deal would be Trump's handpicked allies; Larry Ellison and Rupert Murdoch headlined the investor group. Many observers cried foul. "The TikTok deal doesn't look open to all market participants," observed Rush Doshi, a former national security official under President Joe Biden. "It seems like it's open mostly to those close to the administration already, including its political supporters." The deal would not only enrich Trump's allies, but would also pave the way for partisan content-moderation decisions, mirroring concerns about Chinese control. In this instance, and many others, Trump appeared indifferent about whether his decisions would advance US national security priorities. Rather, he put his personal interests first: How would he profit? To what extent could he enrich his friends and allies? Could the policy be weaponized against his enemies?[9]

As I'll describe in the next chapters, US policy—at least through the end of the Biden administration—coalesced around two principal aims: pursuing strategies of denial (curbing a rival's access to innovations), and implementing policies of subsidy (such as the CHIPS Act, a $100 billion piece of legislation to reinvigorate US semiconductor production). Trump pursued a similar agenda—penalizing rivals and subsidizing certain domestic industries—albeit in a much more slapdash fashion.

None of these interventions were risk-free. Fundamentally, production subsidies cause a misalignment between price and production costs. They can help correct "market imperfections"—circumstances where private markets fail to deliver socially desirable or strategic outcomes. They can also incentivize businesses to invest in R&D that benefits the company as well as the public, and help startups survive an early loss period until they turn a profit. But they can also generate significant drawbacks.

Take oil and gas subsidies—a big priority for Trump. Propping up gas prices not only keeps the valuation of energy firms artificially high, but it can dampen investments in renewable energy sources—undercutting funding for more efficient and environmentally friendly products. Subsidies can also diminish innovation, as they undercut market motivation for investing in more efficient production processes. (Why should a company invest in renewables if the government will continue paying it a generous fee to continue using carbon-intensive methods?) Over time, resources are diverted to less-productive uses—forcing more efficient firms that have not received a subsidy to exit the market. The end result undercuts productivity and promotes rent-seeking behavior (a condition where an individual benefits from the allocation of resources at the expense of society).[10]

Still, forgoing economic efficiency for security interests can be a beneficial trade-off. The United States learned the hard way about the pitfalls of relying on markets alone to support strategic industries. In the aftermath of World War II, the US and the UK possessed large naval and commercial fleets along with major shipbuilding industries. Eight naval shipyards and at least sixty-four commercial shipyards were "actively building large naval or merchant ships" in these countries. In fact, America's real problem was excess capacity: it had far more ships than it needed while maintaining shipyards built for wartime production. By the 1950s, global trends started to shift. European countries as well as Japan rebuilt their facilities, replacing them with efficient modern ones. Several newer entrants also came into the market. In contrast, America's industry was headed in the opposite direction. Naval orders plummeted, and by 1950, the US shipbuilding industry "had almost no work."[11]

In the ensuing decades, high labor costs and fierce competition from foreign companies eroded America's shipbuilding advantage. But it was President Ronald Reagan's deregulation push in 1981 that sounded the death knell for US manufacturers; he slashed government subsidies and left shipbuilders to compete on their own. Globalization took care of the rest. It no longer was cost-effective for American factories to compete with foreign shipbuilders, particularly newer Chinese entrants, and the country's net shipbuilding capacity fell accordingly, from producing 1 million gross tons of merchant ships in 1977 to just three hundred thousand gross tons in 2005. Today, the numbers show a complete collapse. In an average year, America builds around five naval vessels. In contrast, China produces over a thousand comparable ships. In 2023, over half of the world's commercial shipbuilding output came from China—one Chinese shipyard today "has more capacity than all US shipyards combined." (At present, China has twenty large shipyards building military and commercial ships, along with 140 dry docks for rapid expansion and upkeep.) European manufacturers account for approximately 5 percent of global output. Meanwhile, American shipbuilders contribute "next to nothing."[12]

The demise of American shipbuilding, an industry closely linked to the United States' geopolitical power, illustrates how depending on markets to sustain a strategic sector can be a bad bet. Applying a more thoughtful approach for selecting and safeguarding strategic technologies could help

America avoid these mistakes. Biden officials seemed to grasp this lesson, attempting to break this pattern by using government funds to restore US competitiveness in key sectors and keep China's technological innovation at bay. Trump's team? Not so much.

10

"We Cannot Let China Get These Chips. Period."

When Joe Biden came into office in 2020, he dismantled most of Trump's agenda. But one area held firm between the two presidents—their assertive approach to China. Not only did Biden continue Trump's policies, but he added heft to them. His team adopted a "small yard, high fence" strategy when it came to tech controls—designed to stop a small number of countries from accessing a "narrow slice of technology."[1] Initially, they framed these export controls as a way to preserve relative technology advantages over China—to stay "a couple generations ahead." Over time, the strategy shifted. Senior officials increasingly coalesced around a more expansive interpretation, arguing that current technological developments were so fundamental that export controls should be leveraged to build as large a lead as possible. By 2024, a concrete aim had emerged: stopping China from obtaining artificial general intelligence (AGI) before the United States.

As we've established, the concept of AGI is blurry, even among the experts. Mustafa Suleyman, cofounder of DeepMind, posited that a good test for AGI would be whether a model could take $100,000 and turn that sum into $1 million without direction. Steve Wozniak, a cofounder of Apple, suggested an even simpler milestone: "a machine that can enter an average home and make a cup of coffee."[2] Despite the uncertainty about what AGI could accomplish, US officials viewed it as essential to beat China to the AGI finish line.

China's AI research has generally lagged behind the United States, although it is quickly making up ground. A big issue has been the disparity in funding. Over an eleven-year period (2013 to 2024), private investment in US AI labs totaled $470.9 billion, while Chinese AI labs received $119.3 billion in support. In 2024 alone, private AI funding in the United States totaled $109.1 billion—almost twelve times higher than China's $9.3 billion. These resources have allowed America to remain the leading producer of frontier

AI models, but Beijing remains close behind. In early 2026, frontier models from Anthropic, OpenAI, and Google Gemini continued to set the pace, but open-weight competitors, such as DeepSeek, Alibaba's Qwen, and Kimi, had narrowed the performance gap to about seven months. "While US lab leadership persists, China is the clear #2, and open models now provide a credible fast-follower floor," wrote Nathan Benaich for the State of AI Report.[3]

Beijing's mounting success came despite bipartisan American efforts to curb its tech sector. Starting with Biden, US officials targeted China's semiconductor industry, vowing they would do everything in their power to keep leading-edge chips out of its hands. Said Raimondo at a defense gathering in December 2023, "We cannot let China get these chips. Period. We're going to deny them our most cutting-edge technology."[4] Biden also blacklisted over two hundred Chinese corporations, from leading commercial firms to defense tech companies, and tightened inbound investment screening to stop foreign investors from taking stakes in critical American startups.[5]

More significantly, the United States transformed *the ways it wielded* export controls and sanctions. I spoke to Kevin Wolf, who oversaw export controls under President Obama. For Wolf, it was emphatically clear that this paradigm shift "broke the thinking on what a strategic vision would be on China." Past efforts focused on restricting specific items needed to develop missiles or other weapons systems; this was the first time since the Cold War that the United States had gone deep into the ecosystem of a particular country, affecting "billions and billions of dollars of trade." The new controls were characterized by the fact that they were unilateral (at least until Japan and the Netherlands joined with the United States), they focused on a single country, and they targeted largely commercial items found several stages earlier in the supply chain than products traditionally subject to export controls. Ultimately, Wolf said, "it's a big deal both from a policy and export control philosophy point of view, and in terms of economic impact."[6]

America's growing reliance on coercive economic tools is a function of its dominant position in international trade and financial networks. As the scholars Henry Farrell and Abe Newman have theorized, the United States is in a powerful position to exercise "weaponized interdependence" against its rivals. While convergence and globalization offered countries the promise of greater prosperity through integration, weaponized interdependence is the other side of that coin—a lesson in what happens when one country controls most of the world's networks. The modern global economy is a system of networks through which money, goods, and information travel; there are crucial hubs or nodes that can choke off economic or information flows, compel policy change, and deter undesirable actions, and the United States wields disproportionate influence over them. This position gives American officials the power to unilaterally cut off countries like Iran or North Korea from the international banking system or to deny them access to essential goods.[7] As such, America's potential retreat from globalization risks undercutting the very leverage it now weaponizes for export controls and for prosecuting its tech war. The supply chain concentration of AI chips illustrates choke point dynamics at work: a handful of companies dominate the design, manufacturing, and fabrication of high-end chips, illustrating America's unique leverage.

ASML, for example, based in the Netherlands, is the world's lone producer of extreme ultraviolet (EUV) lithography machines, tools that etch intricate circuitry patterns on pieces of silicon wafer "tens of thousands of times thinner than a human hair."[8] Each of ASML's smooth white machines, which are about the size of a large SUV, costs hundreds of millions of dollars. Without access to ASML's tools, it is impossible to manufacture cutting-edge AI chips; control over the company brings de facto control over the entire AI semiconductor supply chain. Unsurprisingly, the United States pressured the Dutch to restrict ASML's exports to China. After months of diplomatic maneuvering, US officials finally convinced the Netherlands (as well as Japan) to bar shipments of semiconductor manufacturing equipment to Beijing. ASML had little choice but to play ball, but that didn't stop its CEO at the time, Peter Wennink, from taking a shot at America's strategy: "The laws of physics in China are the same as here. The more you put them under pressure, the more likely it is that they will double up their efforts" to build equivalent machines.[9]

Deploying economic coercion tools can be a slippery slope; America's dominance of weaponized interdependence is not a persistent condition. Over time, there is a risk that countries will retaliate by weaponizing their own trade networks, constructing parallel networks shielded from US interference, or accelerating their industrial self-sufficiency. Russia's resilience against Western sanctions—discussed in detail in chapter 13—is a good case in point. Initially, these measures took a heavy toll on Russia's economy and derailed its ability to procure microchips and other components for its missiles, tanks, and drones. But Moscow found alternatives. It formed new trade routes with China to tap into Beijing's semiconductor supply chain. And it partnered with Iran to import and then jointly produce thousands of Shahed one-way attack drones in a new factory in Russia's Tatarstan region. The Kremlin also relied on "webs of intermediaries" in places ranging from

Türkiye to Dubai to obtain prohibited telecommunication equipment, microelectronics, and surveillance gear.[10]

In general, weaponized interdependence can be an ill fit for technological innovation. At its heart, the concept is predicated on a linear mode of innovation—that technologies develop along a narrow pathway and if one route is cut off (e.g., stopping the export of leading-edge chips), then targeted countries will have limited pathways to advance. The reality is more complicated. "There is more than one way of achieving one's goals; hence, there is no 'secret sauce' that can protect anyone for long," writes the Stimson Center's Ansgar Baums. "Restricting access to technology first and foremost incentivizes the replacement of it."[11]

Take, for example, Chinese startup DeepSeek's remarkable AI advances. DeepSeek's origins lie in finance, not technology. Its parent company is a Chinese quantitative hedge fund named High-Flyer that made its money using AI to place bets on the Chinese stock market. High-Flyer was cofounded in 2015 by Liang Wenfeng, a thin, bespectacled engineer who had studied at the prestigious Zhejiang University in the eastern city of Hangzhou (a tech hub that is home to Alibaba). Liang is a "hands-on leader" who sometimes slept in his office as he pushed to get projects completed, and colleagues have said he fits the "typical profile" of a Chinese engineer—he puts little thought into his clothes and hairstyle, uses logic and data to make personal decisions, and loves playing soccer.[12] Liang was long interested in automated stock trading—even as a student he had experimented with building algorithms to pick stocks—and High-Flyer soon became a pioneer in applying newer AI techniques.

In 2019, Liang began using Nvidia GPUs to build computing systems for his firm. By 2022, they had acquired more than ten thousand A100 chips and spent roughly $180 million to construct two large data centers, capabilities only a few Chinese companies possessed. "It is like buying a piano," Liang told a Chinese tech publication in 2023, regarding the chip purchases. "Firstly, it's because you can afford it. And secondly, it's because you have a group of people who are eager to play music on it." In 2023, he decided to take his interest in AI a step further and established DeepSeek. Rather than focus on financial trading, the new venture was centered around research and dedicated to building large language models. In interviews, Liang emphasized that he wasn't in it for the money. Noting that the original backers of OpenAI weren't focused on profits but wanted to "pursue the mission" of advancing artificial intelligence, he shared a similar motivation to "develop AI for the public good." One of his former employees confirmed that the company rarely discussed commercializing the applications they were building. "During my time there, we did not talk much about how we make money. They just focused on making a great foundation model."[13]

Liang pledged that DeepSeek would make most of its training results available to the public (a commitment the firm has so far upheld), and he expressed concern about the potential for a small group of individuals or companies to dominate AI technology. A key advantage for DeepSeek was that it didn't need to secure funding to cover the costs of training large models. Rather, it could draw upon High-Flyer's "vast reserves of computing power" to advance its research.[14]

In January 2025, DeepSeek released R1, a specialized AI model designed to solve complex reasoning problems. It stunned the world, performing equivalently to products from leading competitors like OpenAI, despite seemingly being developed at a fraction of their cost. DeepSeek claimed that its total training expenses were less than $6 million. (Said Andrej Karpathy, the former head of AI at Tesla, DeepSeek made it "look easy" to train a frontier model "on a joke of a budget.")[15] While the $6 million figure failed to account for the massive AI infrastructure costs already incurred by High-Flyer, DeepSeek's emergence still shook the industry. The market effect was immediate; Nvidia shed nearly $600 billion, or 17 percent of its value, representing the largest single-day drop in stock market history. Tech figures like Marc Andreesen lauded DeepSeek's model as "one of the most amazing and impressive breakthroughs I've ever seen."[16]

What was particularly galling to US officials was that DeepSeek's innovation was precisely the scenario export controls were designed to prevent. How did the company do it? DeepSeek pushed its engineering "to its absolute limits." It incorporated a range of optimization techniques—key-value caching, mixture of experts, reinforcement learning, and chain-of-thought reasoning—to squeeze as much processing power as possible out of its chips. DeepSeek is emblematic of China's "good-enough" innovation strategy. As *Wall Street Journal* reporter Lingling Wei explains, DeepSeek and companies like Huawei, BYD (China's leading EV producer), and Alibaba have "excelled in taking existing technologies—often developed by the U.S. and other Western powers—and rapidly scaling them for mass production or consumption." While their products or services may not be on par with top products from OpenAI, Apple, or Tesla, they are good enough to make China a global leader in those sectors.[17] (US companies are not convinced that DeepSeek's achievements are simply due to engineering optimization. OpenAI, for example, claims that DeepSeek is free riding on American innovation by exploiting a technique known as "distillation" that involves using an existing AI model to transfer learning and improve a newer model. In this case, OpenAI alleges that DeepSeek is closing the capability gap by illicitly accessing its frontier models and harvesting outputs for distillation purposes.)[18]

Regardless, DeepSeek's accomplishment startled policymakers and sparked concerns about the relevance of export controls. Analysts like Anthropic's Jack Clark even declared that AI diffusion had become irreversible and attempts to control AI capabilities were becoming moot.[19] Time will tell, but DeepSeek's case clearly illustrates a broader conundrum—there are many paths for Chinese companies to enhance their computer processing and power their AI models. When one door closes, others will open.

The Sovereign AI Push

America's tech containment agenda brought other implications. As global interest in generative AI technologies soared, more and more countries sought to develop "sovereign AI." If AI is critical to national interests, then governments are best served harnessing AI to suit their own needs, rather than ceding this power to US-based corporations (which are naturally aligned with Washington's agenda). In concrete terms, many countries sought to control their own data centers and retain their own supply of advanced chips rather than lease these capabilities from cloud providers like Google or Microsoft. "Each country wants to have some sort of control over our own destiny on AI," said Michael Kratsios, Trump's director of science and technology policy, to *Politico*. "The easiest thing that they can do is try to create a model that is fine-tuned to the language, culture, tradition or specifics of that country."[20]

Countries like Singapore announced plans to build their own national large language models (LLMs), noting that their "cultures, values and norms differ from those of Western countries, where most large language models originate." The Dutch government released a development plan that prioritized building its national AI model, GPT-NL, as well as investing in large-scale AI infrastructure like supercomputers. Taiwan also started building a government-funded LLM called Taide as an alternative to Chinese models. In capitals worldwide, governments began treating AI like a utility, an essential service that states felt obliged to fund or "risk falling behind rivals economically and militarily."[21]

These efforts aligned nicely with Nvidia's interests; the company was coming to terms with losing access to China's market and looking for new ways to boost its revenues. The sovereign AI trend offered potential. At trade shows and conventions—in public and private conversations—Nvidia advocated that all countries should "own the production of their own intelligence." To turn this concept into reality, Nvidia set up a dedicated division to work with various governments and sent out company officials to meet with leaders around the world and convince them to build their own data centers "filled with Nvidia's chips." These efforts gained traction. Many countries considered forgoing the cloud computing rental model (where governments lease computing power from large American firms and are spared the expense of updating their software or troubleshooting problems) in favor of directly controlling their infrastructure.[22]

Biden officials reacted cooly to the sovereign AI proposition; among their concerns was that more countries acquiring advanced chips and building AI infrastructure would make it easier for China to siphon Nvidia's GPUs. While they weren't terribly worried that allies in Europe would allow their GPUs to circulate beyond their borders, they had less faith that countries in the Middle East or Southeast Asia would be vigilant about guarding this technology.

Trump, however, saw things differently. Officials such as David Sacks and senior science and technology advisor Michael Kratsios took a dim view of export controls. In their judgment, the best way to compete with China was to sell as many chips as possible to the widest range of countries. "American chip companies must be allowed to sell the American technology stack abroad, albeit with security requirements, else we forfeit the AI race to China," Sacks posted on X. "If we refuse to do business with a country, we push it into China's arms." Sacks, in fact, made a case for loosening export restrictions to China, arguing that fostering Beijing's dependency on US products was the surest way to promote America's "global technological dominance."[23] Unsurprisingly, industry welcomed the change in tone from Biden to Trump.

Industry Gains an Audience with Trump

Silicon Valley sees itself as a center of innovation beyond the reach of government red tape and bureaucratic politics. Yet its business is profoundly affected by decisions made thousands of miles away in Washington. During Biden's presidency, companies bristled over his export control policies. To better understand their perspective, I traveled to Sunnyvale, California, to visit Nvidia's headquarters. It was founded in 1993 by Jensen Huang, who dreamed up the company while sitting with two friends at a Denny's in San Jose. Nvidia's campus features two gigantic buildings clad in black glass and concrete, resembling spaceships, facing one another on a large parcel of land adjacent to the busy San Tomas Expressway. The interior emits a serene calm, with studious-looking engineers darting from glass meeting rooms and pop-up coffee dispensaries into a warren of labs. Nvidia is the world's largest chipmaker, with a market cap of about $4.5 trillion. The company's flagship product is GPUs that are essential for AI computing, smartphones, gaming systems, and the world's most advanced computer clusters.

Nvidia's chips are in high demand everywhere. When the Commerce Department enacted its 2022 semiconductor controls, the company immediately had to stop selling its top-of-the-line H100 and A100 chips to China. Nvidia executives argued they were paying too heavy a price for these policies. Jensen Huang said that forcing the firm to curb its sales to China had left Nvidia with its "hands tied behind our back," causing enormous damage to the tech industry.[24] Historically, 20 to 25 percent of the company's revenue derived from China. This revenue stream had experienced a significant drop, directly affecting the company's R&D investments (Nvidia's sales to China now only account for one eighth of its revenue). How was the company supposed to make up this revenue? As its profits from China fell, Nvidia contended that non-US firms were reaping the benefit. "What you risk is spurring the development of an ecosystem that's led by competitors," said Tim Teter, Nvidia's general counsel. "That can have a very negative effect on the US leadership in semiconductors, advanced technology and A.I."[25]

A 2024 report from the Federal Reserve Bank of New York reinforced this view. The authors found that export controls incentivized US companies to decouple from the target country (from all sectors, not just targeted areas), and these actions caused affected companies to lose an aggregate of $130 billion in market capitalization. The drop in stock market valuation was accompanied by "a decline in revenues, profitability, bank credit, and employment." Conversely, Chinese competitors attained advantages from US controls; targeted firms established new relationships with alternative Chinese producers, signifying Chinese "reshoring." US firms failed to recover lost profits in other markets; there was no evidence that American firms found new customers to counteract the loss of business from China. Instead, the authors concluded that "the inability of affected suppliers to quickly find alternative customers may therefore harm the very same firms whose technology US export controls are trying to protect."[26]

Still, Biden's team argued that preserving the United States' technological superiority against China was a price worth paying. As a former Biden White House official told me, "There is no balancing national security. National security comes first. It always comes first. Full stop." She explained that "we protect what we must and we promote what we can. . . . So there is no tension between economics and national security."[27]

But things changed considerably for Nvidia under Trump's presidency. About two weeks after Trump's inauguration, Huang paid a visit to the White House to make the case for looser restrictions on the company's chips. Huang had gone increasingly farther out on a limb against Biden's export controls, calling them "a failure" and arguing that if America stopped providing AI technology, other countries would step into the void. He even pushed back on calling China an adversary, insisting Beijing was a "competitor" but "not our enemy."[28]

Huang hoped that he would get a more favorable hearing from Trump, but when the two men met in the Oval Office, they "struggled to connect." A follow-up meeting at Mar-a-Lago also went poorly, with Huang failing to persuade Trump from curbing Nvidia's sales to China. By the spring, however, their relationship began to warm after Huang committed to big investments in American factories. Trump saw this deal as a success and began calling Huang "my friend." In a speech in Saudi Arabia in May, he publicly referenced Huang—"What a job you've done," he said. "We are proud to have you in our country." As Trump recognized the global demand for Nvidia's chips and the leverage that brought, Huang became a regular travel companion of the president. In advance of a state dinner in Britain in the fall, Trump remarked that AI "is taking over the world," turning to Huang. "You're taking over the world, Jensen," he said, while signing a tech agreement with UK Prime Minister Keir Starmer. "We both hope you're right."[29]

Yet there were still limits to Huang's influence. In October 2025, shortly before Trump sat down with President Xi in South Korea, he contemplated whether to greenlight approval for Nvidia's exports of its newest generation of Blackwell AI chips. These would be worth billions in sales and Huang had "lobbied relentlessly" to remove the curb. But Trump's senior aides pushed back. Secretary of State Marco Rubio, supported by US Trade Representative Jamieson Greer and Commerce Secretary Howard Lutnick, warned that the chip sales would compromise national security and enhance China's AI capabilities. Confronted with nearly unified opposition from his advisers, Trump chose not to raise the issue with Xi, representing a rare win for national security over Nvidia's interests. That being said, just a month later, Beijing and Nvidia received a nice consolation prize—Trump lifted restrictions on the export of Nvidia's second-best AI chip, the H200, prompting outrage from national security experts.[30]

While export controls have grabbed most of the headlines, US policymakers had other levers they could pull, including carrots to change incentives. American officials understood that they could no longer afford to let private corporations dictate the nation's technological direction. Instead, Washington would have to take a heavier hand.

11

Supercharging American Innovation

Using federal money to bankroll research and critical industries was hardly a new idea for the United States. Washington had been doing this for decades, dating back to the Second World War. Before that, most scientific research was carried out by government labs, which conducted experiments, developed innovations, and tested applications in-house. In the pre-nuclear age, this was a sensible way to advance research for military application. But in the context of a global war, America needed a better system. To that end, in 1941, President Franklin Delano Roosevelt created a new outfit, the Office of Scientific Research and Development (OSRD), and put Vannevar Bush, an MIT scientist, in charge. The goal was to channel hundreds of millions of research dollars and thousands of PhDs to support the war effort.

Bush immediately faced a dilemma. Every hour of every day was crucial—scientists who were not researching new ways to defeat the Axis risked giving Germany and Japan critical windows in which to inflict harm upon Allied forces—but how could he mobilize so many scientists as quickly as possible without sacrificing productivity? Bush decided to use an outsourcing model to harness America's best and brightest minds right where they were—to contract out basic and applied research to university labs. Almost overnight, the government funneled millions of dollars into academia, a considerable sea change for most departments, which had previously considered themselves lucky to obtain "a private donation of a few thousand dollars for an industrial research project or a modest amount of foundation work." Once Bush turned on the spigot, academic labs started receiving resources "many orders of magnitude larger." As OSRD got into full swing, it assembled a "great army of scientific men and university research laboratories" to win the war.[1]

This strategy delivered spectacular results during wartime, but the end of hostilities abruptly halted the expansion. Research funding plummeted and factories and labs were mothballed as policymakers hoped to reap the dividends of a postwar peace. This downswing, however, proved temporary. The Soviets successfully tested their own atomic weapon in 1949, barely four years after the conclusion of World War II. Soon thereafter, the United States found itself embroiled in a hot war on the Korean Peninsula. As more and more troops deployed to the Korean front lines, a new anxiety gripped Washington about whether it had invested sufficient resources to meet the Cold War threat. Recognizing how swiftly the landscape had changed, in 1950, Congress authorized a public agency to support basic scientific research—the National Science Foundation (NSF)—and backed it with generous funding. Like OSRD before it, the NSF was not set up to conduct basic research itself; it parceled out grants to university researchers across the country in a highly competitive process. The army and navy followed suit, forking over billions of dollars to companies to design high-tech weapons.

Frederick Terman, the dean of Stanford's engineering school, saw which way the winds were blowing. Ten years earlier, he had been recruited to run Harvard's Radio Research Lab, which focused on developing radar countermeasures and represented the "pinnacle" of US research on jamming techniques.[2] At the time, universities were just getting into the game of securing large government research contracts, and the money was going to only a handful of universities—Harvard, MIT, Berkeley. Most other institutions, including Stanford, lagged behind. As federal funding surged into university labs once again in the 1950s, Terman didn't want Stanford to lose out on the bonanza; tapping into it would not only elevate Stanford's status, but could "unlock a wave of new economic growth for an entire region."[3]

California was already home to a flourishing defense contractor industry. Facilities producing ships, missiles, radar systems, aircraft, and aerospace components were scattered throughout the state. Terman recognized that California's vast open spaces, dry sunny climate, abundant energy, and highly skilled labor force offered numerous advantages over the East Coast, especially when it came to providing testing facilities for nuclear weapons, advanced missiles, and supersonic aircraft. He urged Stanford to seize the opportunity to transform into a global powerhouse. At the time, the jump seemed improbable. As Margaret O'Mara wrote in *The Code*, "Who could have imagined that this university would become a hub of cutting-edge electronics research? Who would have thought a small Northern California college town would become the capital of the high-tech world?" Yet, thanks to the urgency of Cold War competition, the willingness of Congress to dedicate billions toward scientific research, and the vision of individuals like Terman, the groundwork was laid for Silicon Valley's ascendancy.[4]

Public funding of research labs was one thing, but federal dollars on their own wouldn't be enough to unleash Silicon Valley's full entrepreneurial potential. Once again, Terman's guidance made a difference. Seeing an opportunity to tie Stanford's research into the wider business community, he encouraged many of his students to forgo obtaining PhDs in favor of founding startups. He urged professors to consult for these companies and set up a licensing system where faculty could take innovations from Stanford's labs and swiftly commercialize them in the outside world. Terman changed the career incentive structure so that going outside the university walls could benefit a faculty member's academic trajectory. As Stanford lecturer and entrepreneur Steve Blank says, Terman was not interested in building production systems for the military. "I want to do research. I want other people to start companies and have the military fund them, and these are the companies we'll consult for and sit on boards."[5] And so the Valley began to flourish.

While the region was able to cement its status as a center of tech innovation, its budding entrepreneurs still faced difficulties accessing capital. Old-money institutions such as family endowments, the in-house investment arms of big corporations, and hidebound Wall Street firms were loath to support "go-it-alone entrepreneurship," which required patient capital staked on individuals "who'd never worked outside an academic lab, and who often were building products for which there wasn't yet a market."[6] When startups succeeded, they could yield astronomical returns, but the hit rate was low—just one in ten firms made it through. The Valley required a new risk capital structure to support these long shots.

In September 1959, "two military generals and the former chair of RAND and the Ford Foundation" launched Draper, Gaither, and Anderson, Silicon Valley's first venture capital firm. The idea was simple: marry East Coast money with West Coast innovation and watch profits blossom. The Draper firm would last for only seven years, but it spawned a model of venture investing that became the de facto means for financing the tech industry's explosive growth. This VC infusion ushered in a second wave of expansion that would exceed the military-led growth of the 1950s. While Stanford and the government were responsible for establishing the entrepreneurial culture, and the military "primed the pump as a customer for key technologies," it was venture capital that turned the region "to volume" by funding consumer applications.[7] Thus, the Valley steadily moved away from military technology and government research and repositioned itself toward other sectors—largely for financial reasons.

The conclusion of the Cold War brought another rupture. With the Soviet Union in disarray, US lawmakers saw a rare chance to reap a peace dividend. They slashed defense spending and reduced government support for science and innovation. As figure 1 shows, federal funding for R&D peaked in 1964, equivalent to 1.86 percent of GDP. In the ensuing decades, its share steadily declined; by 2022, it had fallen to 0.63 percent of GDP. Private industry largely filled this gap. While private R&D funding sat below 1 percent of GDP through the 1960s, by 2022 it had reached 2.59 percent of GDP.

Figure 1. R&D Funding Trends in the United States as a Percentage of GDP, 1953–2023[8]

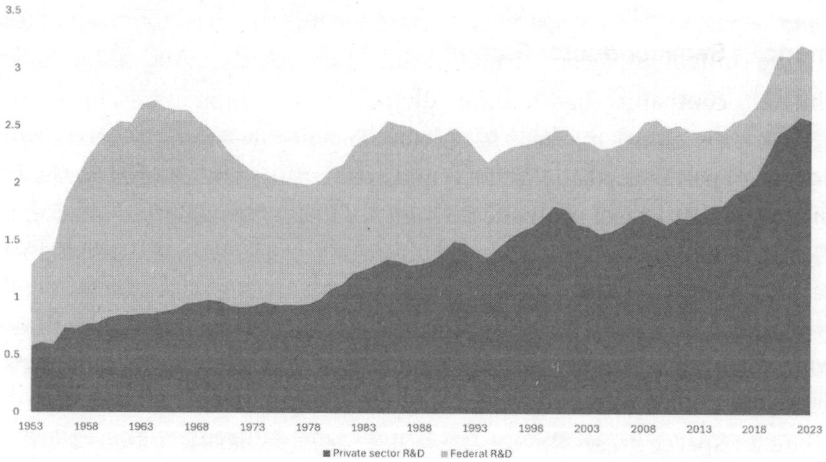

The transition to private sources of funding changed research incentives. The argument was simple: America had won the Cold War and democracy was ascendant. Rather than keep a tight lid on its technologies, national interests would be better served by commercializing innovations and spreading them across the globe, thereby binding countries closer to the United States. This open technology approach made possible the emergence of a whole new suite of technologies that remade the world. "Open technological innovation has yielded clusters of technologies including smartphones, unmanned aerial vehicles, robotics, gene-editing tools, additive manufacturing, machine learning, and even accessible forms of AI," notes scholar Audrey Kurth Cronin.[9] Yet it also meant that Silicon Valley would become increasingly detached from government priorities, whether building futuristic military systems or manufacturing advanced semiconductors.

Another consequence was that private companies began driving technological breakthroughs. New technologies were much more likely to be invented in the private sector, financed largely through private sources of capital, and marketed to global clients. When it came to semiconductors, this theory of capitalism worked well up to a point. In the aggregate, allowing private companies to shape the chip market generated huge efficiencies and facilitated breathtaking innovations. But letting the market dictate terms brought two problems. First, the chipmaking supply chain became concentrated among a tiny assortment of companies—creating major vulnerabilities if anything were to happen to one of the choke points. Second, America ceded its chip manufacturing capacity to other countries, namely Korea and Taiwan.

America's Semiconductor Gambit

While US companies had traditionally dominated chipmaking, in recent decades their global market share had plummeted—from 37 percent in 1990 to 12 percent today. "We rely on a very small number of factories in Asia for all of our most sophisticated chips. That's untenable and unacceptable," said Raimondo. "It's an economic security problem. It's a national security problem. And we're going to change that." Semiconductors and computational power are foundational for AI, and reinvigorating domestic production became a top priority for Biden. Congress passed the CHIPS Act in 2022, a $100 billion program to restore US leadership in the semiconductor space. Its secondary aim was to create fallback options for the industry in case one of its key nodes—say, TSMC's fabrication plants in Taiwan—was destroyed in a military conflict with China. CHIPS legislation authorized billions in public subsidies and loan guarantees to solve this dilemma. It also incentivized billions more in private sector investments, to the tune of nearly $400 billion.[10]

After CHIPS became law, it took months for the program to get off the ground. Commerce had to build an office from scratch, hire a new team to administer the funds, and come up with criteria for how it would disburse the money. It was not until the spring of 2024 that Raimondo announced a string of deals: $8.5 billion for Intel to build chip plants in Arizona, Ohio, New Mexico, and Oregon; $6.6 billion to TSMC for a leading-edge fabrication plant in Arizona; and $6.4 billion to Samsung for a fabrication facility in Texas. There were many outstanding questions about America's chip-making effort. How long would it take to build these new facilities? Would the new plants create enough redundancies to alleviate America's security concerns?

Undoubtedly, there were bumps in the road. Take TSMC's investments. The Taiwanese company is the world's leading fabricator of semiconductor chips, with an estimated market cap of $1.75 trillion. Almost every chip design firm in the world relies on TSMC to actually manufacture sophisticated silicon wafers. For example, Nvidia's Blackwell B200 GPU "superchip," the most powerful AI chip in the world, is not actually made by Nvidia. Instead, Nvidia owns the IP behind the Blackwell B200, but it is TSMC that manufactures the physical item.[11]

This specialization of tasks is intentional. TSMC's founder, Morris Chang, decided that it would be better to focus on a single aspect of chipmaking—fabricating the chips—than handling anything else. TSMC would be a "neutral player around which other companies would design their products." To that end, he formed an alliance with dozens of companies that "design chips, sell intellectual property, produce materials, or manufacture machinery." While these firms competed with one another, none were in the chip fabrication business, and they were willing to coordinate with TSMC to set up compatible standards and feed directly into its manufacturing process. TSMC poured billions into refining its facilities, investing in R&D, and staying on the cutting edge of semiconductor manufacturing. Chang's prescience paid off: today, TSMC has no real rivals to its business and controls 90 percent of the market.[12]

But as chips took on greater geopolitical importance, policymakers grew increasingly concerned that TSMC's Taiwan facilities were vulnerable to disruption from China. Starting in 2019, Trump officials reached out about building a plant in the United States. In 2020, the Commerce Department announced a deal to open a $12 billion plant in Arizona to diversify TSMC's production line. After passage of the CHIPS Act, TSMC announced it would construct a second plant on the same Arizona site, increasing its total investment to $40 billion.[13]

TSMC's transition to Phoenix was not smooth. Construction of the new plant was beset by delays, in part due to labor shortages and disagreements. The cultural clash between Taiwanese and American engineers became problematic. TSMC brought a famously rigorous work culture. "If [a machine] breaks down at one in the morning, in the U.S. it will be fixed in the next morning. But in Taiwan, it will be fixed at 2 a.m.," said Morris

Chang, adding that the wife of a Taiwanese engineer would "go back to sleep without saying another word." TSMC veterans described their US counterparts as lacking in the commitment and obedience they believed were fundamental to their company's achievements. In turn, Americans resented TSMC's "long hours, high stress, unrealistic deadlines, and 'Asian culture.'"[14]

Yet for TSMC, the American investment was as much about politics as anything else. While TSMC would make fewer profits from its US facilities, "the only thing they need Phoenix for is to make sure but the US government doesn't turn against them," says Chang-Tai Hsieh, a University of Chicago professor. Delays plagued the first two sites, but TSMC nonetheless agreed to construct a third plant in Phoenix subsidized by $6.6 billion from CHIPS. Based purely on market considerations, it made little sense for TSMC to expand its US plants; locations closer to Taiwan have worked out much better. For example, TSMC opened a chip factory in Japan in 2024 that took just three years to complete and came in under budget "thanks to strong government support, local partnerships, and a low-cost labor force that worked 24/7." But increasing TSMC's investments in Japan, even if it addressed the security resilience problem, didn't satisfy America's political needs. And so TSMC has plowed ahead with its American investments.[15]

With Trump back in office, America's innovation push continued, albeit in a distorted form. On the day of his inauguration, Trump announced a $500 billion AI infrastructure called Stargate. OpenAI, Japanese investment firm SoftBank, Oracle, and MGX (an Emirati state-owned equity investor) committed $100 billion up front to the venture, with plans to spend up to $500 billion over the next four years. Even for Silicon Valley, this was an eye-popping amount that showed how heavily people were willing to bet on AI as the future of American power.

One of the biggest beneficiaries was OpenAI's Sam Altman. He had undergone a transformation since he first came onto the tech scene as the brash, idealistic founder of a disruptive tech startup intended to advance the public good. In time, he pivoted the company into a for-profit model, built up a massive lobbying operation, and made a big play for federal contracts. Now, shifting to a new narrative, Altman framed his company as an engine of growth for the entire US economy. This pivot found a warm reception in Washington. "The U.S. government—leadership of both the Democratic and Republican parties—is intensely focused on preserving America's lead in AI," says Center for Strategic and International Studies analyst Gregory Allen. "As long as OpenAI can demonstrate that they are strongly contributing to America's overall leadership, they will have open ears in Washington for their ideas."[16]

Looking beyond Stargate, Altman took his road show abroad to drum up interest in a whopping $7 trillion proposal to build a string of semiconductor factories and data centers. Overseas, the response was tepid. Reportedly, when Altman visited TSMC's headquarters in Taiwan and briefed company officials on his trillion-dollar vision, they found his idea "so absurd that they took to calling Mr. Altman a 'podcasting bro.'" In a subsequent visit to Tokyo, Altman explored building data centers powered by shuttered nuclear plants decommissioned after the 2011 Fukushima disaster. He told Japanese officials that he needed five gigawatts of power, "about a thousand times the power that an average data center consumes." The pitch was met with open laughter.[17] So while the global competition for AI tech was heating up, there were still limits to how far the most ambitious governments were willing to go to fulfill the aspirations of dreamers like Altman.

In the last few years, a growing number of countries have been rolling out their own industrial subsidies. The EU committed to spending 43 billion euros on its version of the CHIPS Act, Japan allocated $13 billion for chip manufacturing, and India announced it would spend $15 billion on chip plants. China, of course, remained the world leader. It had invested major sums in semiconductors for over a decade; by 2030 its funding commitments would likely exceed $150 billion. Motivated by security concerns, governments worldwide have ramped up their involvement across economic sectors, actively picking winners and losers in strategic industries.[18]

As a former CHIPS Act official told me, "we're going to make a lot of companies unhappy because we had to make strategic choices about what is the best investment to further our national and economic security."[19] These are trade-offs that US officials willingly accept to prevent vital industries from falling to China and to protect core national interests.

12

The Pitfalls of "Damming Half the River"

It is one thing to subsidize critical industries, but it is quite another to rally together sovereign governments to impose punishing tech restrictions against a particular country. Even under Biden, America's allies remained on the fence about how far they would commit to the US tech war. Washington knew that going it alone against China was not tenable. As former Commerce official Eric Hirschhorn notably said, working unilaterally to impose sanctions is akin to "damming half of the river—it makes you feel as if you did something, but in reality, it doesn't have much effect." Sanctions denial strategies require multilateral coordination to work; if countries fail to act in unison, targeted states can easily find other sources to replace US expertise and capital. This can lead to "policy disaster," where the targeted country doesn't suffer any material loss of capabilities but American companies are cut off from a key source of revenue, allowing foreign competitors to take advantage of the market opening.[1]

By definition, choke points involve only a handful of players. To leverage a choke point doesn't mean that the United States has to harness tens of countries to support a particular action; all that is required is alignment between those countries and companies controlling the choke points. In the case of semiconductor controls against China, American officials only had to enlist American, Dutch, and Japanese chip lithography manufacturers to have an impact.

This dynamic gets more complicated when export controls are expanded to include additional technologies and capabilities. America's full tech containment strategy requires that other allies join the cause. Getting the Europeans on board—as well as Japanese, Korean, and Taiwanese manufacturers—is critical, but they all have unique interests to consider. Meanwhile, Trump's saber-rattling has caused its allies to second-guess the wisdom of following Washington's path. It's a conundrum for America's partners. How much should national security interests supersede economic considerations? Are their governments willing to take hard steps even if it costs their companies dearly? And do they still trust the United States?

Europe Weighs the Pros and Cons of Confronting China

The United States bought into the idea of convergence in the 1990s—that economic integration would liberalize China's political system and incentivize it to compete over business rather than fight wars. European leaders followed suit and focused on building deep commercial ties with China. Beijing's growth model relied on a simple formula: large-scale capital investment (financed by a combination of domestic savings and foreign investment), high-productivity growth, and the surplus production of cheap manufacturing exports. It paid off spectacularly, with China's share of global manufacturing climbing from 2.3 percent in 1991 to 19 percent in 2013. While they acknowledged growing inequities in the relationship, many countries in Europe, such as Germany, were insulated from the initial shock of this export boom. At the time, China's exports centered on consumer electronics, furniture, apparel, and household appliances, "not the automotive and engineering sectors at the heart of the German economy." Instead, it was American manufacturing that bore the brunt of these losses. Between 2001 and 2011, the United States lost an estimated one million jobs—concentrated in Midwestern manufacturing hubs like Detroit, Akron, and Pittsburgh—that left behind shuttered factories, hollowed-out cities, and a devastated workforce.[2]

As long as America shouldered the costs, the EU could accept the status quo. But Europe would face its own consequences before long. Many analysts viewed 2019 as a turning point. In March of that year, the European Commission released a groundbreaking strategy paper labeling China as a partner, competitor, and systemic rival. It held that while the country remained a cooperative partner in certain areas, it had become an economic competitor "in pursuit of technological leadership," and had also emerged as a systemic rival "promoting alternative models of governance."[3] Soon enough, this triptych of "partner, competitor, rival" found its way into countless speeches, frameworks, and policy documents. The phrasing "had a kick. It introduced a term, 'systemic rival,' that has become part of the European lexicon ever since, to China's deep disquiet," wrote author Andrew Small.[4]

Two events would shift Europe further away from China: the COVID-19 pandemic and Russia's war in Ukraine. The pandemic put Europe in an unenviable position. The bloc relied heavily on globalized supply chains for critical items from medical gear to semiconductors. Overnight, the pandemic broke these links and Europe found itself on the short end of the stick. Basic items such as personal protective equipment were in short supply. Europeans perceived China to be hoarding essential products, or worse, engaging in price gouging. Chinese media responded by going on the offensive. In France, the Chinese embassy posted a fabricated story that French retirement homes were allowing old people to die. In Germany, Chinese diplomats pressured government officials to "heap public praise" on Beijing. In Italy, bloggers distributed manipulated video clips purportedly showing Romans playing China's national anthem in gratitude for medical supplies.[5]

This campaign backfired spectacularly. Across the continent, negative views of China surged, leading to reputational collapse in countries that were formerly close partners. As senior EU tech envoy Gerard de Graff told me, "From the perspective of Europe, there's been a rude awakening during COVID and subsequently where our leaders realized that we've become overly dependent on geographies that aren't always friendly towards us."[6]

China's support for Putin following the invasion of Ukraine was another jolt to EU-Chinese relations. When Russian tanks steamrolled toward Kyiv in 2022, Europeans harbored few illusions that China would condemn the aggression, but some leaders hoped that China could serve as a voice Putin "must listen to," and "a possible restraint on Russia." Instead, Xi doubled down on his relationship with thc Russian leader, welcoming him to Beijing, where the two proclaimed a "no-limits friendship."[7] China refrained from providing Russia with lethal aid, but most other items were fair game.

One glaring example involved exports of construction excavators in September 2022. The previous month, Russian forces had faced a series of devastating defeats around the northeastern city of Kharkiv. The Russian military had been focused on other regions, allowing Ukrainian troops to surprise them and reclaim over three thousand square miles of territory, about half the size of New York's metropolitan area. Still, the defeat could have been much worse. Facing disaster across the front lines, Russian forces quickly fell back and began erecting the Surovikin Line, a network of trenches and fortifications designed to stop the Ukrainians from progressing farther. Excavators and front-end shovel loaders were essential to this defensive effort, and records show that Chinese exports of both types of machines surged during the construction of the Surovikin Line, tripling over the previous year. Without this equipment, it is doubtful that Russia's "defense in depth" strategy could have stopped Ukraine's attacks.[8]

As China tilted toward Russia, European leaders developed a "greater awareness of the risks of interdependence with China, and a greater awareness of where China was going," Mark Leonard, head of the European Council on Foreign Relations, told me. EU officials began laying out the case against China in public. In a forceful speech in March 2023, European Commission President Ursula von der Leyen warned that Xi's visit to Moscow had conveyed in "a thousand words" China's reimagined vision for the international order, and that it was imperative for Europe to "derisk" its relationship with Beijing.[9] But what that would look like in practice was not clear.

Von der Leyen, who lives in a 269-square-foot windowless converted apartment next to her office on the thirteenth floor of the EC's Berlaymont headquarters (she says it's to avoid Brussels' notorious traffic and extend her workday), has been instrumental in making the EC more politically relevant. She has moved the body into uncharted territory, such as pursuing an 800-billion-euro joint borrowing package in response to the COVID pandemic, enacting an ambitious green energy agenda, and pushing through multiple sanctions packages against Russia. (Said one official approvingly: "she has a very structured and analytical mind, as you might expect from a doctor."[10])

While Europe's attitude toward China hardened, its leaders struggled to balance two competing objectives—reducing overdependence while maintaining strong economic ties. Toeing this line was not easy, but von der Leyen's persistence paid off. The EC released a new package of export controls in 2023 and committed 43 billion euros' worth of investments to increase Europe's share of global microchip production from 10 percent to 20 percent by 2030. Put together, these efforts represented a major turnabout.[11]

China didn't take kindly to these developments. Beijing blasted von der Leyen's derisking approach, arguing that "there is a sense that 'de-risking' might be 'decoupling' in disguise," and accused Brussels of caving to Washington's demands. Her uncompromising views won her praise from Biden, but the feeling in European capitals was mixed.[12]

Germany in particular sat in a tight spot. Under former Chancellor Olaf Scholz, it had stayed relatively silent on China. Noah Barkin, a senior fellow with the German Marshall Fund, compared its leadership to an ostrich with its head stuck in the sand. "US pressure on the neighboring Dutch? Not our problem. Restrict Huawei? Take two years to introduce a baffling new law instead, and then refuse to implement it. Shield the European car industry from an onslaught of subsidized competitors? Not Germany, where faith in free trade remains unbroken, even in the face of seismic Chinese distortions."[13] Barkin and others argued that eventually Germany would have to take a stand—one that could provoke Chinese retaliation. Yet, Germany continued to punt.

The case of Zeiss, a specialized lens and glass maker in Germany, illustrated the conundrum. Under American pressure, the Netherlands, Japan, and Taiwan imposed severe chipmaking restrictions against China. Zeiss supplies ASML with optical lenses that are crucial for its machines. It also manufactures devices that help chipmakers identify defects in silicon wafers. For Chinese chipmakers aiming to independently produce advanced semiconductors at scale, Zeiss's components were invaluable. Reportedly, US officials visited Berlin ahead of Scholz's 2024 trip to China and pressed Germany to curb Zeiss's exports—but Chancellery officials were hesitant to pick a fight and rebuffed the Americans. Zeiss CEO Karl Lamprecht accompanied Scholz on his visit to Beijing the following week where there was scant talk about export controls.[14]

Germany's largest businesses remained eager to access China's market and were unwilling to risk confrontation. "Global value chains have been building up over the last 50 years. How naive do you need to be to believe that this can be changed within six or 12 months? This is about decades," said a senior officer from Siemens, one of the nation's leading tech companies.[15]

Still, Germany's views weren't fixed. A polarized debate began to take place about how to deal with China, reflecting partisan splits. According to Leonard, Germans are "trying to reconcile the fact that on the one hand, they are much more invested in China, and a lot of it is also quite a concentrated relationship." But they are also changing their policies "based on an understanding of what's happening in China and of their own interests. And they're much more clear-eyed about some of the challenges of excessive dependence."[16]

So, where did this leave things for Europe? By the end of Biden's term, the EU had become increasingly wary about its relations with Beijing. There was a growing consensus that the United States and Europe would stand together against its unfair trade practices, rising tech capabilities, and geopolitical agenda. But Trump's reelection upended the bargain.

From the outset, Trump took a hard line against Europe. He described the bloc's trade policies as "an atrocity." In public remarks and internal communications, his team evinced contempt for the transatlantic alliance and lambasted Europeans for failing to uphold their security commitments. (In a leaked Signal chat regarding an imminent strike against Houthi forces in Yemen, Secretary Hegseth quipped: "I fully share your loathing of European free-loading. It's PATHETIC.")[17]

Soon after taking office, Trump released a raft of "Liberation Day" tariffs, assessing 20 percent penalties against EU member states—a shocking tally for treaty allies with an eighty-year history of cooperation. Europeans were furious. French Senator Claude Malhuret remarked, "Trump's message is that being his ally serves no purpose, because he will not defend you, he will impose more tariffs on you than on his enemies, and he will threaten to seize your territories, while supporting the dictators who invade you."[18]

Even worse was Trump's mad dash to pry Greenland from Denmark. From the beginning, it wasn't clear why he was so focused on controlling the territory. A popular theory was that Greenland's outsized portrayal on a Mercator map convinced Trump that it must become part of the United States. *The New Yorker*'s Susan Glasser recounted a visit she paid to Trump at Mar-a-Lago in 2021, when he pointed to Greenland on a map and said, "Why don't we have that? You take a look at a map. So I'm in real estate. I look at a corner, I say, 'I gotta get that store for the building that I'm building,' et cetera. You know, it's not that different. I love maps. And I always said, 'Look at the size of this, it's massive, and that should be part of the United States.'"[19]

In January 2026, his rhetoric hardened and he began talking in earnest about seizing Greenland outright. In response, seven European leaders, including from Germany, France, and the UK, issued a joint statement saying that existing borders were nonnegotiable. Danish Prime Minister Mette Frederiksen went further, declaring that "If the United States chooses to attack another NATO country militarily, then everything stops." After a week of tension and private text entreaties from worldwide leaders (many of which Trump posted publicly on Truth Social), he flew to Davos and gave a fiery speech where he delivered a "verbal assault" against the Western alliance. Then, hours later, he de-escalated, announcing he had "formed the framework of a future deal with respect to Greenland" during a "very productive meeting" with NATO Secretary General Mark Rutte. While European leaders breathed a sigh of relief, their anger was palpable. Said one EU diplomat, "Our American Dream is dead. Donald Trump murdered it."[20]

In the whiplash of these events, the EU began to earnestly reassess its options. European leaders recognized they were facing a grim new reality. On the one hand, they were forced to navigate a hostile US administration that was freelancing policy without any constraints. ("He's at the peak of just not giving a f—anymore," said one White House aide. "Bad news stories? Doesn't give a f—. He's going to do what he's going to do. He's going to do what he promised to do on the campaign trail.") On the other hand, China showed little interest in reaching an accommodation with Europe. "One might view the breakdown in the transatlantic relationship as an opportunity for China, but I think China views it as a sign that Europe is increasingly weak," said Barkin. "Its best friend is gone. And so, from China's perspective, its leverage has risen with Europe and that has made it even less willing to compromise."[21]

Even prior to Trump's polarizing actions, Europe and the United States were far apart on several core issues. Fundamentally, Europe believed that the United States had overemphasized the "protect" side of its agenda at the expense of the "promote" angle, maintaining that Americans were focused too much on export controls and not enough on advancing innovation. Europeans lamented that the United States continued to grow its small yard—adding new measures all the time, from greater chip controls to new data and cybersecurity restrictions. Many were concerned that the United States was suffering from "groupthink" driven by a "relentless search for new chokepoints." Under Trump, the United States showed scant interest in collaboration, preferring to push protectionist policies that risked cleaving the alliance apart entirely. In short, Europe feared it was caught in the middle—pressured to cut economic links with China while being denied access to the US market.

In light of these developments European intellectuals argued there was an urgent need to chart an alternate course—keeping Beijing at a distance while limiting Europe's embrace of Washington. Columbia Law School's Anu Bradford, rejecting the assumption that the EU must align with either the United States or China, contended that in the battle of digital empires, Europe could show a "third way" that checked corporate power while protecting fundamental rights and democratic institutions. By forging its own path, Europe could attract other governments—like Australia, Brazil, Canada, and South Korea—to adopt its special mix of regulations and rights.[22]

Her ideas paralleled a similar effort gaining traction: proposals to develop a "EuroStack," digital infrastructure exclusively produced within the twenty-seven-member bloc. The concept was based on a simple argument: despite the continent's strengths as a regulatory power, it was heavily dependent on outside technologies, importing more than 80 percent of its digital products and services, from cloud storage and smartphones to telecom equipment and AI applications. EuroStack promoters contended that it was vital for Europe to develop and use its own technologies. An Italian economist named Cristina Caffarra headlined the push. She had spent years involved in the digital competition policy space, advising companies like Apple, Microsoft, and Amazon, as well as government regulators. She didn't mince words about the need for Europe to break free from its dependence. "It's a massive disgrace that when I have a video conference with the Commission, I use [Microsoft] Teams," she told an audience at a January 2025 gathering in Brussels. "The reality today is that we are a colony."[23]

EuroStack proponents worried that if the continent failed to innovate, it would be beholden to a set of unpalatable choices—either reliance on America's oligopolist tech firms or dependence on China's providers. For them, the solution was for Europe to build its own infrastructure, based on open-source principles and values. Certainly, there was merit to their contentions. Reducing its dependence on Big Tech and diversifying its tech ecosystem could safeguard Europe's digital sovereignty. But success wasn't assured. "We've already seen the bloc try to create its own version of Google. That failed miserably," writes the Atlantic Council's Mark Scott.[24] A larger question was whether Europe could pull this off logistically. Did it have the innovation backbone and the capital required to make such a venture a success? And realistically, how long would it take to get things off the ground?

While these ideas were gaining appeal, the reality was sobering. The EU's tech companies are not globally competitive, and its innovation ecosystem remains weak. While the EU's regulatory capacity and large consumer market can serve as partial checks on American or Chinese power, its leverage remains limited. Consider Mario Draghi's critique of Europe's competitiveness (he is a former prime minister of Italy and ex-president of the European Central Bank). In a widely anticipated 2024 report outlining Europe's future growth challenges, Draghi wrote that the continent had "largely missed out on the digital revolution led by the internet and the productivity gains it brought" and that it is weak "in the emerging technologies that will drive future growth," harboring only four of the world's top fifty tech companies. He warned that if Europe could not become more productive, member states would be forced "to scale back some, if not all, of our ambitions."[25] Given the bloc's track record, there was little reason to think these changes would happen quickly or easily.

Taiwan's Dilemma

Europe wasn't alone in struggling with how to navigate long-term relations with both the United States and China. Middle powers around the world confronted a similar dilemma; nowhere were the stakes higher than in Taiwan. Their leaders face a grave strategic threat from China, and they have few illusions about Beijing's designs on their territory. Logically, this makes Taiwan a natural ally for the United States in its tech war and a willing proponent of export controls. To a large degree, this bears out. Just one day after the Commerce Department announced its semiconductor restrictions against China in October 2022, Taiwan indicated it would follow suit.

A complicating factor for Taiwan, however, is its semiconductor industry, considered to be the most advanced in the world. In 1985, a powerful Taiwanese bureaucrat named K. T. Li reached out to Morris Chang, then an executive at electronics manufacturer General Instruments. Li convinced Chang, who was fifty-four years old at the time, to return to Taiwan and take a job at a technology research institute. Chang said Li also had another idea for him: "He felt I should start a semiconductor company in Taiwan. So that was the start of TSMC." Taiwan's government staked the firm by purchasing a 48 percent share, with Dutch company Philips and Taiwan's private sector providing the remainder of funding. Taiwan's bet paid off. As discussed, TSMC is a behemoth. It holds a 53 percent share of the global semiconductor market and supplies 92 percent of all cutting-edge chips. Taiwan recognizes how crucial semiconductors are to the country; its leaders refer to the industry as the "sacred mountain that protects the nation."[26]

TSMC's success gave rise to the "silicon shield" theory, the idea that global reliance on Taiwanese semiconductors was so significant that it would protect the country from military invasion. As former Taiwanese President Tsai Ing-wen wrote in a *Foreign Affairs* piece in 2021, the silicon shield "allows Taiwan to protect itself and others from aggressive attempts by authoritarian regimes to disrupt global supply chains." The concept was premised on two assumptions. The first was that Beijing's dependence on Taiwanese companies to supply chips would serve as a check on Chinese aggression. Indeed, China's semiconductor factories still only produce about 10 percent of the chips needed to sustain its consumer electronics industry; it relies on TSMC to "make up 70 percent of the deficit," importing over $47 billion worth of Taiwanese chips in 2023 alone.[27]

The second aspect was that Taiwan's silicon shield would incentivize third countries, specifically the United States, to protect the country from invasion. If TSMC and other Taiwanese companies stopped supplying chips, the global economy would quickly fall into a tailspin. By one estimate, a hypothetical Chinese blockade of Taiwan's chip industry would lead to a stunning $2.5 trillion in annual losses to the world economy. Thus, the global appetite for Taiwan's chips would ensure that protecting its foundries remained an international priority.[28]

However, while Taiwan's semiconductor industry is indeed crucial, it is hard to imagine that chips alone would stop China from invading under certain circumstances. One senior Biden administration official noted that the silicon shield theory is "reflective of an older school versus a more updated mentality. I actually don't think it's a major driver for key individuals in Taiwan." She told me that Beijing has a "whole host of calculations" regarding Taiwan and what would compel Beijing to use force. "I do not think that the question of Taiwan's semiconductor capacity is on that list, or if it's on the list, it's very far down the list."[29]

Even so, Taiwan's leaders do not want to see their semiconductor industry weakened, either from export controls or chip diversification. TSMC founder Chang, for instance, has expressed concerns that America's push to on-shore chip manufacturing could "hollow out" Taiwan's industry as a result. He notes that the United States already possesses 39 percent of global semiconductor equipment, manufacturing, design, and intellectual property capacity, alongside 11 percent of the world's actual manufacturing. "So what is their goal with the CHIPS act?" Chang asks. "To go back to 30, 40 percent? Or is their goal just to maintain the supply essential for national security?" To protect against these risks, TSMC's current boss, C. C. Wei, has made it clear that he intends to keep 80 to 90 percent of the company's capacity in Taiwan. While TSMC's Arizona plants are being set up to produce cutting-edge chips of 4 or 5 nanometers, by the time they are up and running, the technological frontier will have shifted to 1.4- and 2-nanometer chips. It is unclear whether TSMC will move fabrication of these next-generation chips off the island.[30]

What does all of this mean for Taiwan? While Taipei has strong incentives to go along with US policies and downgrade its ties with China, its links to the mainland should not be underestimated. The two countries' economies remain deeply intertwined. China is still Taiwan's largest trading partner and favored destination for investment; an estimated 242,000 Taiwanese citizens work on the mainland. Taiwan's exports of microchips to China have stayed robust, averaging 56 percent of its total semiconductor exports over the last decade. "My hunch is that the large manufacturing sectors will try to stay in the Chinese market, even with harsh conditions,"

Alexander Huang, from Taiwan's opposition Kuomintang Party, has said. "If you talk to those business owners, they say, 'Nah, no way will I give it to my competitors.'"[31]

In all likelihood, Taiwan will continue to hedge for the foreseeable future. In an optimal scenario, the country would have it both ways: adherence to US semiconductor restrictions while maintaining chip exports to China with some boundaries (e.g., blocking sales of certain advanced GPUs and preventing Chinese companies from stealing tech IP). However, as tensions between the United States and China persist—possibly leading to a larger standoff—Taiwan faces a difficult reckoning: Can it withstand the consequences of withdrawing from China's massive market, and would such a move provoke further hostility from Beijing?

▪ ▪ ▪

The previous chapters have largely focused on US strategies to counter China's technological rise. China represents the predominant challenge to American power. It is the sole country that possesses both the motives and capabilities (economic, military, technological) to reshape the international order and supplant the United States. But that doesn't mean American officials have overlooked Russia.

Moscow's 2022 invasion of Ukraine underscored the risk Russia posed to US interests. Its open defiance of global norms directly challenged America's leadership. As such, Washington and Brussels imposed a raft of sanctions against Russia. Initially, their aim was to deter Putin from invading Ukraine. When those measures failed, they sought to compel Russia to halt the invasion, or at least hinder Moscow's ability to prosecute the war. The world watched as Western leaders made an extraordinary gamble—wielding sanctions to technologically isolate and strangle a nuclear power. The United States and its allies bet that modern warfare's dependence on advanced technology created unprecedented vulnerabilities. As Russian factories ground to a halt and military production lines scrambled for alternatives, both sides asked: Had Washington found Moscow's critical weakness—or awakened the Russian bear?

13

Punishing Russia's War Machine

American actions against China were largely preemptive, trying to block a rising power before it could reach equal footing. But the United States' strategy against Russia followed a completely different playbook. Its aim was to shut down Russia's war machine and weaken the Kremlin's ability to continue its campaign against Ukraine. Following Moscow's February 2022 invasion, the United States and its allies enacted a massive sanctions program designed to implode Russia's economy.* They targeted the country's largest banks—freezing their assets, blacklisting them globally, and cutting them off from the US financial system. They banned Russian energy products and imposed export controls on critical technologies—microchips, computers, lasers, and telecommunications equipment. The sanctions totaled up to more than 18,381 designations against Russian individuals and companies. US officials were blunt about what they wanted to achieve. Said former Deputy Treasury Secretary Wally Adeyemo, "We're going to cut [Putin] off from Western technology that's critical to advancing his military, cut him off from Western financial resources that will be critical for feeding his economy and also enriching himself."[1]

They initially hoped that the sanctions threat would deter Putin from invading Ukraine in the first place. Their goal was to convince him that the costs of war would far outweigh any benefits. The White House signaled it would go after the assets of Russian oligarchs in the Kremlin's inner circle and would ban the children of Russia's elite "from attending prestigious universities." When Biden was asked whether he would target Putin

* US sanctions against Russia for its military activities in Ukraine predated the 2022 invasion. Beginning in 2014, when Russia forcibly annexed the Crimea region of Ukraine, US and European officials imposed a raft of penalties targeting Russian leaders, defense companies, energy firms, and banks. While these were significant steps, they were limited in scope. American officials were optimistic that these measures would be enough to deter Putin from expanding his military operations into the eastern Donbas region. Ultimately, their deterrence value proved marginal.

directly, he responded, "Yes. I would see that." Other leaders, such as UK Prime Minister Boris Johnson, warned Putin that war would be an "utter catastrophe." (Reportedly, Putin then threatened Johnson with a missile strike, telling him it "would only take a minute.") None of these efforts worked. Early in the morning on February 24, Putin announced the start of a "special military operation." Minutes later, deafening explosions reverberated in Kyiv and urban centers across Ukraine. The largest military invasion in Europe since World War II had begun.[2]

Why did these warnings fail to deter Putin? Sanctions threats typically fall short for three reasons:

- **The target doesn't find the threat to be credible and thinks the sender is bluffing.**
- **The sanctions package is not sufficiently coercive; the target views sanctions as a lesser evil than complying with the sender's demands.**
- **The target expects sanctions to be imposed whether or not it yields to the sender's requests.[3]**

In Russia's case, the first and third conditions were unlikely. The West had imposed an escalating set of sanctions against Russia since 2014, so there was no reason for the Kremlin to think that the United States wouldn't follow through on its latest threats. And Biden took pains to emphasize that the Kremlin faced a choice: Russia could proceed with the invasion and face extraordinary economic punishment, or it could refrain from attacking and avoid any such penalties. Thus, the second explanation was the most credible: Putin's resolve to invade Ukraine outweighed any sanctions threat from the West. For Putin, conquering Ukraine was an essential part of his vision for Russkiy Mir ("Russian World") and critical for restoring Russia's greatness. "With a waning window of opportunity to influence events in Ukraine," writes Tufts scholar Daniel Drezner, "it is unlikely that any purely economic threat, no matter how accurately it was signaled, would have deterred Russian aggression."[4]

Once deterrence failed, policymakers switched focus to a second objective: coerce Putin to halt the invasion.

Western leaders imposed a slew of measures intended to bring Russia's economy to heel and compel Putin to change course. They kicked seven Russian banks out of the SWIFT financial payments system, froze billions in Russian assets, and implemented a range of export controls. Russian officials were taken aback by the West's sharp response. Said Russian Foreign Minister Sergey Lavrov, "When the reserves of the Central Bank were frozen, no one would think, out of those who made predictions, what sanctions the West might apply," while also accusing the West of committing "thievery." The Kremlin was especially caught off guard by the coordinated freezing of $300 billion worth of Russian Central Bank reserves held in the United States and Europe, as well as the flood of Western companies withdrawing or divesting from the Russian market, a number that increased from 250 to 1,000 in the course of several months.[5]

These measures had an immediate impact on the Russian economy. Almost overnight, the ruble plunged in value, crashing from an exchange rate of 75 rubles to the dollar to 143 rubles to the dollar—leading Biden to prematurely declare, "As a result of these unprecedented sanctions, the ruble almost is immediately reduced to rubble." In scenes reminiscent of the Cold War, Russian citizens waited in long lines in the freezing cold to withdraw their savings from ATMs, raising concerns about bank runs. To stave off financial collapse, the Central Bank increased its discount rate by 20 percent in a "desperate bid" to shore up its currency. Anti-war protesters even took to the streets to rail against the war. On the first night of the invasion, thousands rallied in Moscow, St. Petersburg, and other large Russian cities, chanting, "No to war!" Western policymakers privately mused about whether political pressure from demonstrators and aggrieved oligarchs would force a change of heart from Putin.[6]

Then reality set in. Russian security forces swiftly cracked down on the fledgling demonstrations, detaining nineteen hundred people the very next day and sending a ripple of fear throughout the country. After initial volatility, the economy stabilized as well, shrinking by only 2 percent in 2022 (far less than projections of a 10 to 15 percent contraction).[7] Life for Russia's middle and upper classes returned to normalcy—they continued packing bars and restaurants and attending music festivals, blithely ignoring the slaughter taking place five hundred miles away. It became apparent that Putin was prepared to weather the West's sanctions, and that asset freezes, tech controls, and visa bans would do little to change his calculus. By the spring of 2022, Western leaders had largely abandoned their coercive objectives and pivoted to a third strategy: use sanctions to deny critical components to Russia's military, thereby hindering the war effort.

Constructing a New Export Controls Regime Against Russia

If the West couldn't deter Russia from invading Ukraine or coerce Putin into stopping the invasion, Biden figured that Moscow should pay a heavy price for its actions. This was strategically vital. If Putin faced little resistance conquering Ukraine, what would stop him from continuing his aggression elsewhere—attacking Moldova, threatening the Baltic states, or even turning his sights on Poland? To preserve balance, it was essential to impose consequences on Russia, and the most straightforward way to accomplish this was to arm Ukraine and degrade Russia's military. From a normative perspective, Western leaders believed it was also critical to uphold a core tenet of international law: countries cannot acquire territory through the threat or use of armed force.[8] If Russia got away with attacking Ukraine without penalty, it would weaken the pillars upon which the modern international system rests.

The West's recalibrated strategy hinged on denying Russia critical technological inputs. US officials knew that simply imposing unilateral controls would be heavy on symbolism but light on results, so they pursued a novel approach. The United States needed to recruit as many countries as possible to punish Russia. But first it had to build a new sanctions regime from scratch. No existing body was suitable for this purpose. In the post–Cold War era, countries relied on four export control regimes to stop the proliferation of weapons: the Nuclear Suppliers Group, the Australia Group (for chemical and biological weapons), the Missile Technology Control Regime, and the Wassenaar Arrangement (for conventional arms and dual-use items). These groups operate by consensus, meaning that any changes to their mandate or control lists would require agreement from all parties. Russia is a member of three of these groups (it is not a participant in the Australia Group), allowing it to easily block new export controls. So what options did the United States have?

They would have to create an entirely new export control infrastructure. This would require participating states to amend their laws and agree to a "common licensing policy" to deny controlled items to Russia and its close ally Belarus. The new regime would work as follows: The United States would come up with a list of prohibited items (referred to as the Common High Priority List—CHPL). Other countries would then follow suit and also restrict these items. The list would include products directly used in Russia's weapons—such as semiconductors for hypersonic missiles and sensors for missile guidance systems—as well as commercial goods, such as milling machines or aircraft parts, that were important to Russia's military. US officials divided the CHPL into four tiers of priority, reflecting their military utility. Fifty products in total were included on the CHPL, each identified by a special six-digit code.[9]

At last count, thirty-nine other countries implemented parallel restrictions based on the CHPL. Such a complex system required close cooperation at a level unseen since the end of the Cold War. "That thirty-nine-country coalition, it's unprecedented. That just hasn't existed before and it's really amazing," one senior Commerce official told me. "Because if you don't act together, all you've done is hamstrung US industry."[10]

But would this effort bring results?

Initially, the outcomes were promising. Gina Raimondo testified in a May 2022 Senate hearing, "We have reports from Ukrainians that when they find Russian military equipment on the ground, it's filled with semiconductors that they took out of dishwashers and refrigerators." The White House claimed that Russia's two major tank factories—Uralvagonzavod Corporation and Chelyabinsk Tractor Plant—had stopped production due to lack of foreign components. US officials also noted that tech exports had fallen close to 70 percent since sanctions were imposed, and that shipments of semiconductors, lasers, maritime technology, avionics, and telecommunications equipment had "decreased 85 percent and their value has decreased 97 percent" compared to 2021."[11]

But contrary reports emerged as well. A Reuters investigation found that five months after the war had started, "thousands of shipments" of computer parts, such as microcontrollers, programmable chips, and signal processors, were still making their way to Moscow. Customs records tied these shipments to prominent Western companies such as AMD, Analog Devices, Infineon, Intel, and Texas Instruments. Part of the problem was system lag. Infineon and Texas Instruments claimed their products were "already in transit" when the invasion occurred and there was simply no way to prevent them from reaching Russia. Intel maintained that its shipped goods represented "internal company deliveries" that ceased after the company pulled its Russian operations in April.[12]

Despite these hiccups, Biden officials claimed the sanctions program was on track. They cited major shortages of semiconductors and avionics components that forced Russian firms to "cannibalize existing airline parts" (indeed, *The Wall Street Journal* reported on a surge of "midair fires and malfunctions" on Russian airlines due to a scarcity of spare parts). American officials asserted that Russia's hypersonic missile production had nearly come to a halt, that Moscow's production of its "next-generation airborne early warning and control military aircraft" had reached a standstill, and that factories producing surface-to-air missiles had likewise shut down due to shortages. Reportedly, during a meeting of Russian technology entrepreneurs in 2022, a research analyst working on hypersonic missiles directly confronted Putin about the growing shortages: "As a developer, I am very worried about our microelectronics." Putin replied, "Me too, honestly."[13]

As the sanctions program continued into its second and third year, however, the data showed a more complicated picture. In the first months of the war, Russia's battlefield imports hit a cliff and plunged by 50 percent or more. Yet by October 2022, just eight months after Russia attacked Kyiv, aggregate imports had rebounded to their prewar levels and continued to increase. By December 2022, imports had reached a twelve-month high and have remained at elevated levels ever since. Figure 1 illustrates the challenge.

Figure 1. Russian Imports of Critical Battlefield Components (Millions of Dollars)[14]

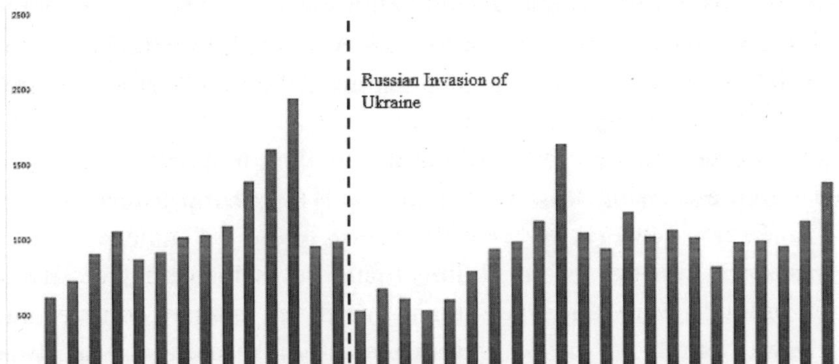

What explains this upswelling? As Western governments curtailed exports, Chinese firms plugged the gap. Trade data for 2023, for example, showed that 63 percent of battlefield goods came from Chinese manufacturers (coalition countries provided only 20 percent of these items).[15] Figure 2 breaks down Chinese exports of dual-use goods (the dark shaded area) and CHPL items (lighter shaded area) powering Russia's war machine.

Figure 2. Chinese Exports of Critical Technologies to Russia (Millions of Dollars)[16]

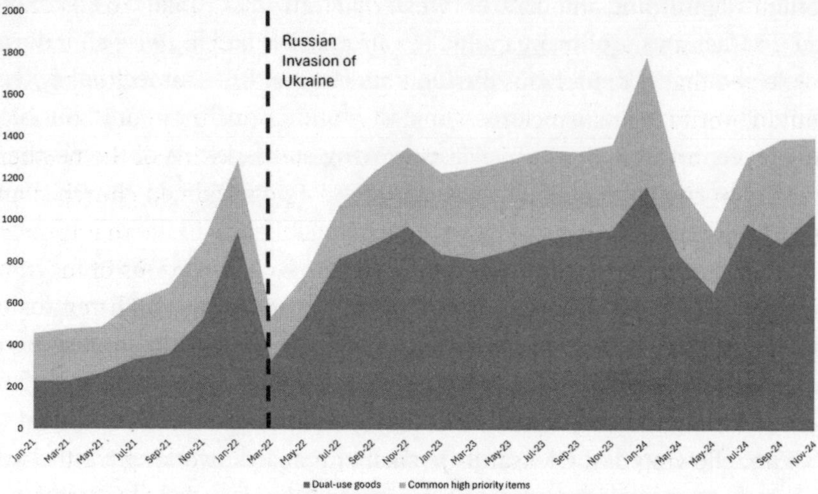

After an initial dip in exports following Russia's invasion, Chinese exports snapped back to prewar levels and climbed higher in the ensuing months. That said, European and US components also continued to play a meaningful role supporting Moscow's military-industrial complex. As of March 2026, the Ukrainian government's database of foreign-produced weapons components identified 5,564 components in 193 separate weapon units primarily originating from four countries: the United States, Switzerland, Japan, and Germany.[17]

The case of Russia's Orlan-10 drone was illuminating. The Orlan-10 is a workhorse for the Russian military—it is frequently deployed for intelligence, surveillance, and reconnaissance purposes, but also used for lethal attacks, responsible for "killing up to 100 soldiers per day," according to Ukrainian officials. They are inexpensive to produce, costing about $87,000 to $120,000 per unit (far cheaper than the cost of a cruise missile). They are also filled with Western components. Since the start of the war, the drone's manufacturer, St. Petersburg–based Special Technology Centre (STC), has "drastically increased imports of critical Western-manufactured components." Financial records reveal that the firm has given millions of dollars to a network of affiliates to procure microelectronics from foreign suppliers. Shell companies have shipped banned microelectronics to STC for the Orlan-10. Items include amplifiers made by American company Qorvo and computer-on-module units supplied by US firm Gumstix.[18]

But the Orlan-10 represents only the tip of the iceberg. Many other military systems used by Russian forces—from cruise missiles to tanks—contain significant amounts of Western electronics. Take, for example, one of Russia's flagship weapons, the Iskander-K cruise missile. Analysts suspected that Iskander missiles contained Western electronics, but they couldn't verify their suspicions. Then, in August 2023, they got a break. On August 19, an Iskander-K missile swept across the skyline of the northern Ukrainian city of Chernihiv—passing above golden-domed churches and blooming parks—before striking a white stone theater in the main square. The target of the strike was a nearby conference where dozens of individuals involved in military drone manufacturing had gathered. The explosion "unleashed a ball of fire that caused trees to bend and hurled thousands of half-inch cubes of metal through the air." It blew the roof off the theater and resulted in hundreds of civilian casualties.[19]

The following day, Oleksandr Vysikan, an analyst working for the Kyiv Scientific Research Institute of Forensic Expertise, traveled to Chernihiv to take a look at the damage. His job is to study munitions recovered from the battlefield to better understand the inner workings of Moscow's weapons. Often, there is little for him to examine beyond scraps and fragments. But not this time. In Chernihiv, he salvaged whole pieces of the Iskander, including its SN-99 navigation system, which contained components made by four Western companies. Vysikan surmised that a Russian brigade likely launched the missile from the nearby Kursk region and relied on the navigation system to guide it two hundred miles to Chernihiv. He noticed that markings on the Iskander's engine showed that it was assembled "no earlier than March 2023," over a year after the United States imposed export controls. Ukrainian officials were not surprised. They had been griping for months that the sanctions regime was full of leaks. Said Vysikan, "The missile was fresh. It had components that enabled it to communicate with a satellite, which meant one thing: It was more deadly." Vysikan identified several Western companies whose parts were found in the Iskander. They included Silicon Laboratories, an Austin-based firm; Analog Devices, another American chip manufacturer; Germany's Infineon Technologies; and Integrated Silicon Solution Inc., an American company with Chinese ownership.[20]

The greatest number of parts came from Analog Devices, headquartered in Wilmington, Massachusetts. Ukrainian investigators found that 21 percent of the four thousand Western components used in Russian weapons originated from the firm—more than from any other Western corporation. Analog Devices is one of the oldest chip companies in the world, occupying an important niche producing sensors, microelectronics, and radio frequency transceivers (used to convert audio and visual data into digital signals). While it is smaller than industry leaders like Nvidia and TSMC, it remains one of the top suppliers of basic semiconductors that power mobile phone networks and many electronic devices.[21]

According to Russian customs data, millions of Analog components have found their way to Russia since the start of the war. In 2023, Russia imported $326 million worth of these products. However, no direct sales were recorded by Analog to Russian companies; instead, its products flowed through third-party dealers in China and Hong Kong. Analog argued that it was following the law. Ukrainian officials maintained that "without those U.S. chips, Russian missiles and most Russian weapons would not work."[22] Whether or not Analog was aware that its chips were powering Russian missiles, two facts were clear: Russia's war machine heavily relied on Analog's devices, and this arrangement was lucrative for the company.

Have Tech Controls Worked Against Russia?

The impact of US and allied sanctions against Russia's military has proved underwhelming. Undoubtedly, export controls made it more difficult for Russia's armed forces to obtain key equipment and replenish their supplies, impeding the Kremlin's ability to prosecute the war in Ukraine. But Moscow compensated for the reduction in Western goods by bolstering imports from China. Moreover, while Western exports slowed down, they did not come to a halt. Critical technologies such as microelectronics and radio navigational aids continued to flow into Russia.

There are several explanations for the disappointing results. First, lacking China's cooperation, Western controls against Russia faced inherent limitations. In the globalized marketplace, Chinese companies matter a great deal—both in terms of what they can produce and their ability to redirect critical supplies to Russian end users. The United States simply lacked the tools to compel China to follow its sanction regime.

But the challenges extended beyond China's noncompliance. Western companies faced a steep learning curve when it came to ensuring their products did not end up in the hands of bad actors. As the Semiconductor Industry Association said in a statement, "We have rigorous protocols to remove bad actors from our supply chains, but with one trillion chips sold globally each year, it's not as simple as flipping a switch." Even prominent companies like Nvidia were not immune to supply chain diversions. Reports indicated that Russia's Lancet drone relied on Nvidia's Jetson TX2 AI computing module to autonomously identify and track specific objects on the battlefield. (The 7.5-watt Jetson TX2 is a new generation of processor designed to bring AI edge computing to mobile devices like drones; it has been described as "the fastest, most power-efficient embedded AI computing device.")[23] Over time, companies like Nvidia should get better at stopping their products from reaching unauthorized parties. (Conversely, one could argue that companies that made millions from the diversion of chips to Russia had perverse incentives to keep turning a blind eye to leakages.)

Finally, sanctioning great powers is hard. "Great powers are most likely to possess the strategic reserves and capacities necessary to withstand sustained economic pressure," writes Daniel Drezner. "Almost by definition, great powers will anticipate frequent conflicts with the rivals that are sanctioning them," reducing the chance they'll yield to punitive measures.[24] Russia is a big country with a substantial industrial base. Putin has put its economy on a wartime footing, retooling existing factories and building new facilities to augment its military. And because he anticipates ongoing confrontation with the United States—beyond the Ukraine war—he has little incentive to comply with US demands, reducing America's coercive leverage.

However, one shouldn't write off the impact of US tech controls, either. A big issue is misplaced expectations. From the outset, American officials overpromised about what sanctions could achieve. As Russian tanks rolled toward Kyiv, Biden stood at the White House lectern and vowed that new sanctions would "cut off more than half of Russia's high-tech imports" and scuttle Russia's military modernization. "It'll degrade their aerospace industry, including their space program. It will hurt their ability to build ships, reducing their ability to compete economically. And it will be a major hit to Putin's long-term strategic ambitions."[25]

Several years later, few of Biden's warnings came to pass. Russia's high-tech imports were not halved; rather, Moscow ably substituted Chinese tech for banned Western items. Sanctions may have degraded Russia's long-term military modernization, but in the near term, its armed forces held their own against Ukraine's NATO-supplied military. They refurbished aging equipment at a sufficient rate. They plugged weapons gaps and acquired new capabilities by leaning on partnerships with China, North Korea, and Iran. Russia's circumvention networks adequately offset Western tech controls across the board. As Kevin Wolf told me, "I think expectations were stated too high as to what export controls could do. . . . You have no common denominator with export controls. You've got every country, every location, every mom-and-pop shop, putting stuff in a box and sending it through the mail." Cracking down on these different leakage points takes time and resources. But as Wolf notes, "Export controls are never perfect. Even during the Cold War when there were these embargoes of Russia, there was always leakage. There's always escapes."[26]

Thus, assessing the effectiveness of export controls comes down to definitions. If the goal, as one US official put it, was to "prevent Russia from acquiring weapons and means to carry out this war," then America's efforts fell short. But if success is measured more modestly—how much harder did it become for Russia to procure critical items? How much more expensive? What was the cost to Russia in lost manpower and equipment by having to rely on Soviet-era tanks and older stocks of ammunition?—then it can be better argued that US export controls had a meaningful impact.[27]

▪ ▪ ▪

While the United States and its allies continued to churn out sanction after sanction, neither China nor Russia remained idle. China's industrial transformation had been underway for over a decade, giving it a competitive edge in many sectors. As America's tech controls ramped up, China was ready. It invested furiously in developing new innovations and capabilities. Together with Moscow, Beijing implemented adaptations and creative workarounds in response. Beijing enlisted a host of third countries to facilitate alternate trade networks. Its actions would sorely test the limits of American power.

PART IV

China Seizes the Future, Russia Adapts

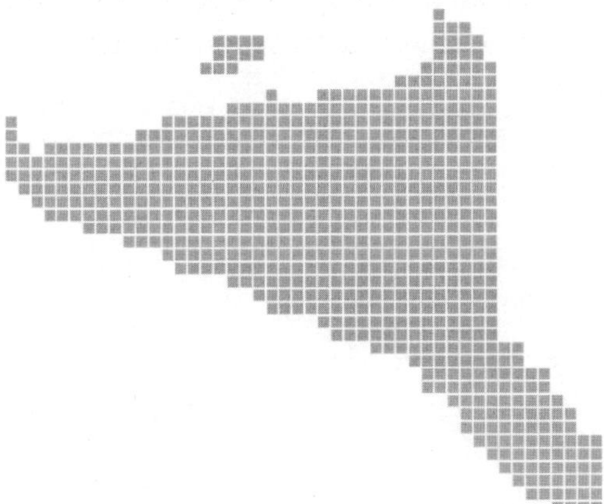

14

China's Quest to Become a Tech Superpower

One of Xi's foremost priorities is the "great revival or rejuvenation" of China—referred to as the "China Dream." The China Dream's core thesis is that for five thousand years, China shaped world civilization as a source of innovation and expansion. The narrative recalls an era when China, as the Middle Kingdom, demanded tribute from surrounding nations while pioneering transformative innovations, "creating paper, gunpowder, printing, and the compass" as well as commanding vast fleets of ships to all corners of the globe. Yet China's modern history was marked by foreign occupation of its territory and a departure from its former status as a great power. When Mao Zedong and the Chinese Communist Party (CCP) seized control of the country in 1949, this opened a new chapter for China, but its struggles would continue. In subsequent decades, it would experience more disastrous chapters, including the Great Leap Forward, which caused famine and the deaths of millions in the late 1950s, and the Cultural Revolution, which led to about 1.6 million deaths by 1976.[1]

China's modern turning point came in 1978, when Deng Xiaoping stepped into power. He initiated a sweeping program of reform, opening up the economy and loosening the state's control over society. His actions set in motion "forces that would in mere decades pull hundreds of millions of people out of absolute poverty, transform China into the workshop of the world, and set it up as a great power in the twenty-first century," writes NYU's Chen Jian. By 2022, China's GDP had nearly caught up to that of the United States (when accounting for purchasing power parity, its economy slightly exceeds the United States'). No other country in modern history had transformed as fast as China. Its economic miracle created the conditions to challenge the United States for geopolitical supremacy. Paradoxically, without sustained US investment, diplomatic encouragement, and technical support, it is unlikely that China would have reached a position where it threatened to "occupy the commanding heights" of global power.[2]

Multiple factors are responsible for China's economic miracle: government reforms that removed state interference from the economy, liberalized foreign trade and investment, strategic investments in industrial manufacturing, and educational gains in China's workforce. Unquestionably, however, the United States played a critical role in China's rise. Beginning with President Richard Nixon's opening to China, successive American policymakers bet that rapprochement would not only drive a wedge between Beijing and Moscow but would set the conditions for eventual political liberalization in China. As Kurt Campbell and Ely Ratner wrote in *Foreign Affairs*, cooperating with China offered something for everyone. "Free traders and financiers" foresaw inevitable and growing openness in China. Integrationists argued that "Beijing's ambitions would be tamed by greater interaction with the international community." And the hawks believed that "China's power would be abated by perpetual American primacy." So America threw open its market to Chinese goods, invested heavily in Chinese factory production, and turned a blind eye to licit and illicit technology transfers occurring in the country. US policymakers bestowed most-favored-nation trading status to China in the 1990s, supported its accession to the World Trade Organization in 2001, established a high-level economic dialogue under President George W. Bush, and negotiated a bilateral investment treaty under Barack Obama. What resulted was a deep network of "trade, investment, technology, education, and personal ties" between the two countries.[3]

Despite China's impressive economic gains and the reassertion of its military ambitions in the East and South China Seas, by the time Xi came to power in 2013, there was also a sense of drift in Beijing. As Stanford scholar Elizabeth Economy writes, Xi had a front-row seat to growing challenges facing the country: "The Communist Party had become corrupt and devoid of an ideological center, the provision of public goods had fallen dramatically behind society's needs, and even the economy needed a new infusion of reform." Xi saw this as a mandate to undertake dramatic reform in order to save the party and allow China to realize its potential as a great power.[4]

For Xi, restoring the China Dream meant building China's economic strength and revitalizing its global standing. He believed that the world had entered a new period of turbulence—of "great changes unseen in a century." The external arena had become more unpredictable and alarming than even during the pitched rivalries of the Cold War. At the root of this disorder was a shifting balance of power between the United States and China, fueled by breakthrough innovations in AI, quantum, space, and biotechnology—what China's leaders referred to as the "Fourth Industrial Revolution."*[5]

Winning the Fourth Industrial Revolution was crucial to Xi's theory of success.[6] He was convinced that the coming decade would be decisive for who would lead the next industrial age. Xi's aim was to turn China's domestic industries into global leaders and transform the country into a science and technology superpower. His ambitious technology agenda is detailed in table 1.

10 strategic technologies in "Made in China 2025" (released in 2015)[7]	Priority technologies in the 15th Five-Year Plan (2026–2030)[8]	Critical choke point technologies requiring investment†[9]
New generation IT	Artificial intelligence	High-end semiconductor components (e.g., photolithography machines)
Robotics	Integrated circuits	Underwater network connectors
Aerospace equipment	Industrial machine tools	High-end vacuum evaporators
High-tech ships	High-end instruments	Aircraft engines and aviation design software
Railway equipment	Basic software	Specialized steel alloys and bearings used in high-end mechanical products
New energy vehicles	Advanced materials	Heavy-duty gas turbines
Power equipment	Biomanufacturing	High-end radio frequency components
Agricultural machinery	Hydrogen energy and fusion energy	Microspheres
New materials	Quantum technology	Transmission electron microscopes
Biomedicine and medical devices	Brain-computer interfaces	Smartphone operating systems
	6G mobile communications	Lithium battery separators

Table 1. China's Major Tech Priorities

A key priority for Xi was to build up China's technological self-reliance. From the moment he took office, Xi viewed China's dependence on foreign technology and know-how as a strategic vulnerability. "Now even the import of relatively usual technology faces all kinds of restrictions," Xi warned. "When we were weak, everyone wanted to sell their technology

* The concept of the Fourth Industrial Revolution originated from Klaus Schwab and the World Economic Forum in 2015 but was enthusiastically embraced by China's leadership.

† In 2018, the state-run *Science and Technology Daily* newspaper detailed thirty-five dual-use technologies and components representing critical import dependencies for China. Georgetown's Center for Strategic and Emerging Technologies shortened the list to fourteen technologies based on a three-part criteria. I further condensed the list to include eleven technologies and components featured in the table.

to us. Now that we are developed, no one wants to sell us their technology. They are afraid that this will make us big and strong." The self-reliance goal took on added urgency in light of US tech controls—Chinese officials increasingly fretted about "technological strangulation."[10] Xi recognized that China's inability to manufacture certain "choke-neck technologies," such as semiconductors, was a big vulnerability, and he directed major funding to break America's stranglehold.

China's Military-Civil Fusion Push

One area that received considerable attention was China's military-civil fusion (MCF) initiative. The defense sector had long been dominated by plodding state-owned enterprises focused on preserving their monopolies rather than innovating. They remained isolated from the dynamic commercial economy, yielding inefficiencies and meager returns. As a result, the arms industry languished and the People's Liberation Army (PLA) was forced to import crucial systems from foreign suppliers. While the CCP started to reform the defense sector, even today China continues to import critical components, such as helicopter systems, gas turbines for destroyers, and aircraft engines (between 2011 and 2020, China ordered nearly a thousand aircraft engines from Russia).[11]

MCF was supposed to fix this problem by reducing barriers between the civilian and defense sectors, fostering joint projects, and developing militarily relevant technologies. The concept isn't new; starting in the 1980s, the CCP pursued civil-military integration, though with limited success. But under Xi, China made a renewed push for MCF. He established a powerful body, the Central Commission for Integrated Military and Civilian Development, to guide its strategy and execution. Without Xi's "herculean efforts," MCF would have received only passing support; instead, it was elevated to a national-level priority supported by meaningful commitments.[12]

Accordingly, MCF began to show impact. Analysts documented a growing number of Chinese companies that worked jointly with the PLA. One area of emphasis was military AI research. In early 2024, researchers from Shanghai Jiao Tong University published a study showing how AI could be used to deploy weapons systems in "automated 'kill webs'" that could adapt to real-time battlefield conditions at sea. A week later, the PLA announced that it had awarded the university a sizeable defense contract to turn the concept into reality. This wasn't the first defense contract Shanghai Jiao Tong had received. As *The Wall Street Journal* reports, since 2023, the university had won multiple awards in areas ranging from underwater drone designs, drone swarms, and using layered AI models to track military targets. As China pushed to adopt future technologies into its military, MCF proved its value, allowing the PLA to go outside its traditional network of state-owned contractors to tap a wider network of suppliers, private companies, and civilian universities. Data from Georgetown demonstrates the extent to which this trend took root. Researchers analyzed thousands of AI-related contract award notices from the PLA and found that "an emerging class of firms and universities" were playing a consequential role in China's AI military procurement. Many of these companies were relatively new and focused on developing dual-use technologies, "indicating that these firms see both the civilian and defense sectors as avenues for growth."[13]

Still, some experts argue that MCF's significance is exaggerated. Tai Ming Cheung of the University of California, San Diego, wrote that "MCF has conveniently emerged to represent all that is evil and threatening about the Chinese Communist party-state: a dark-of-the-night boogeyman seeking to fatally undermine the American way of life." He contended that "it is a very long way from posing an existential danger to the United States and its allies." Similarly, a report coauthored by think tank analysts Elsa Kania and Lorand Laskai observed that MCF was shrouded in myth, but its actual effect was more prosaic. Rather than give China a decisive security advantage, they argued that it remained in its early stages and had encountered significant headwinds exposing the "much messier reality of how MCF has taken shape."[14]

Nonetheless, US leaders deemed MCF a major risk to national security. Concerns initially took hold under the first Trump administration, when officials warned that China's commercial technologies were tipping the "scales of global power" toward America's rivals. In the waning days of Trump's first term, he issued a set of executive orders barring Chinese students with ties to PLA-affiliated universities from entering the country and prohibiting Americans from investing in companies tied to the PLA. "American capital should not be used to finance the construction of Chinese Communist weapons literally aimed at killing Americans and driving the U.S. military out of Asia," said White House aide Peter Navarro.[15]

Biden continued Trump's policies, maintaining a laser focus on China's MCF efforts. When Gina Raimondo was asked whether there was a point where Biden's export control yard would become too big, she responded, "I don't think we're there yet, I really don't. Xi Jinping's civil-military fusion strategy makes it hard, because everything's militarized." She added, "He could walk into any company at every minute and take whatever he wants."[16] For US policymakers, even if MCF had been riddled by false starts and underwhelming results, that hardly mattered. Its potential to tap into China's dynamic tech ecosystem and equip the PLA with state-of-the-art capabilities, especially for emerging technologies like AI, was sufficient evidence for a robust response.

Funding China's Techno-Security Ambitions

Xi's techno-security agenda requires lots and lots of money, which raised the question: How would China pay for all of this? One of Xi's main fund-raising vehicles are "government guidance funds." These are essentially venture capital funds administered by government authorities at different levels. The funds pool together public and private sources of capital to support major projects, initiatives, and strategic sectors. While government guidance funds have been around since the early 2000s, they took on new prominence under Xi. Hard figures are difficult to come by, but analysts estimated that as of 2020, Chinese officials had created 1,741 guidance funds with a target size of 11 trillion yuan ($1.55 trillion)—a figure exceeding Canada's GDP. Reportedly, however, the funds have raised less than half that amount, pulling in 4.76 trillion yuan ($672 billion).[17]

As a continuing source of capital, government guidance funds offer several advantages. They allow the Chinese state to leverage market efficiencies while offering long-term "patient capital." This funding can complement other industrial subsidies, bringing additional financial leverage. For example, one of Beijing's most prominent efforts is its national semiconductor fund, created in 2014 and commonly known as the "Big Fund." Its third and latest round of financing brought in $48 billion from government ministries, state-owned banks, and investment vehicles linked to municipal governments with the goal of weaning China from semiconductor imports.[18] When combined with other sources of financing—both state-backed and private—this represents a major inflow of resources into China's chip sector. (By comparison, the CHIPS Act totaled about $100 billion in funding, while the EU's version of the CHIPS Act committed 43 billion euros' worth of investments for its semiconductor industry.)

But China's funding model also brings risks. The Big Fund has repeatedly missed its targets. Its initial goal was for China to make 40 percent of its chips domestically by 2020. It fell short by a wide margin, barely reaching 17 percent. For 2025, analysts projected that China would manufacture 30 percent of consumed chips in its factories, remaining far short of its 70 percent aim.[19]. Meanwhile, several chip projects have faltered due to “frozen funding and mismanagement.” State-backed chip consortium Tsinghua Unigroup, for example, has warned of potential default on nearly $2.5 billion in international bonds.[20]

Corruption concerns are also rife. In 2022, authorities arrested four top executives of the Big Fund and charged them with financial graft. This included taking Ding Wenwu into custody, the head of the whole fund. By the time the probe wound down, dozens of executives and state officials had been arrested and charged with corruption. The net effect was to temporarily reduce the fund’s investment activities to nearly zero, causing the CCP to reassess its chip strategy. China’s government subsequently introduced reforms, but it is unclear whether these steps were enough. In sum, these funds have been beset by bureaucratic inefficiency and undermined by unrealistic goals and poor management. They often raise less money than anticipated (and experts accuse them of crowding out private sources of capital). Their rapid proliferation has led to criticisms about “redundancy and inefficiency.”[21]

Moreover, the scale of China’s investments should be put into context. Experts commonly assume that China is spending orders of magnitude more on its security and technology investments than the United States. In fact, the opposite is true: US expenditures continue to exceed China’s investments. Estimates show that China’s core institutions receive the equivalent of 3.5 percent of its GDP. When the definition expands to include funding for dual-use items and related activities, the total rises to between 5.5 percent and 5.7 percent of GDP. In comparison, core funding in the United States equals 5.9 percent of GDP, with expanded funding totaling 7.1 percent of GDP.[22]

The West’s overestimation of the size of China’s techno-security state could lead to policy distortions. As Tai Ming Cheung writes, some analysts assume that only full decoupling and imposing massive export controls will “prevent any assistance from ending up in the hands of techno-security-affiliated entities.” But given its more modest size, taking these steps is “akin to wielding a sledgehammer to fix an errant nail.” Rather than fully sanctioning China’s economy, Cheung argues that American policymakers should focus on narrowly targeted measures to curb its techno-security apparatus while preserving the broader economic relationship.[23]

Increased Tensions with the West

Before Biden imposed comprehensive export controls against China’s semiconductors and Jake Sullivan talked about the need for America to maintain "as large of a lead as possible" against Chinese technology, Donald Trump took a first shot at reining in Beijing's ascendance. Trump's early tech war, however, happened somewhat by accident. He had campaigned in 2016 on a pledge to "get tough on China," arguing that current trade terms had decimated American industry. Upon entering office, his team instituted hundreds of billions worth of tariffs against Chinese companies—eventually penalizing half of all Chinese exports. Trump's motivations were strictly economic. He viewed China as an economic rival, "bigger but not fundamentally different from other countries that ran large trade surpluses with the United States." In fact, Trump often said that Europe "treats us worse than China," criticizing, for example, Germany's large bilateral trade surplus.[24] When it came to the national security threat posed by China, he wasn't overly concerned—but he viewed national security as an opportunity to pressure China for a more favorable trade deal.

Not everyone on Trump's team saw things the same way. Many of his first-term aides, such as economic advisor Peter Navarro, national security adviser John Bolton, and National Security Council senior director for Asia Matthew Pottinger, were highly distrustful of Beijing. Said Pottinger, "In the United States, competition is not a four-letter word. We in the Trump administration have updated our China policy to bring the concept of competition to the forefront." To them, China represented an existential threat to US power and had to be tamed; Trump's trade war gave them the perfect opportunity to inflict harsh measures against the country's industry. In May 2018, they got their chance. ZTE, one of China's most successful technology companies, with seventy-five thousand employees and annual revenues of $17 billion, was accused of violating trade controls against Iran and North Korea and failing to punish its employees. The Commerce Department lowered the hammer against ZTE—no American companies would be allowed to provide any components to the firm. This meant that American microchips powering ZTE's wireless stations and routers would have to be pulled. US-made optical components powering its fiber-optic networks could no longer be supplied. And its smartphone partnership with Google's Android operating system would cease as well. These penalties were so severe that commentators described ZTE as "facing a death sentence."[25]

Unsurprisingly, Chinese leaders were alarmed. ZTE was a crown jewel company, and America's actions would potentially drive the firm out of business. Xi scheduled a special call with Trump to complain about America's treatment of the firm. To the chagrin of his aides, Trump relented and agreed to take ZTE off of Commerce's blacklist "as a personal favor to Xi." A week later, he tweeted: "President Xi of China, and I, are working together to give massive Chinese phone company, ZTE, a way to get back into business, fast. Too many jobs in China lost. Commerce Department has been instructed to get it done!" For Trump, using ZTE as a bargaining chip helped him maintain personal relations with Xi and presumably brought him closer to a trade deal with China.[26]

But for China's leaders, the episode served as a harsh wake-up call. In the course of a few months, ZTE's financial state had shifted from profitability to "imminent bankruptcy." During the crisis, Xi harkened back to Mao and declared that despite US sanctions, China would persevere. "By tightening our belts and gritting our teeth, we built 'two bombs and one satellite' [referring to a milestone weapons development program implemented by Mao]," he declared. "The next step is to do the same with science and technology." Many commentators likened the ZTE penalties as China's own "Sputnik moment," in which Beijing would fully mobilize for a tech arms race against the United States.[27]

Almost immediately, the CCP sought to identify its technological vulnerabilities. In speeches, Chinese officials suggested that the experience was a major turning point for the country. China shifted its semiconductor policy—accumulating a vast stockpile of microchips and chipmaking machinery to hedge against future sanctions, as well as starting a big effort to build out its domestic chips industry.[28] Thus, while some experts view America's October 2022 export controls as the opening salvo of the US-China tech war, the ZTE blacklist more accurately represents the beginning of the current round of tensions.

Today the two countries find themselves in a changed landscape. Xi has been clear that he considers the United States to be China's foremost adversary. In different venues, Xi has accused US leaders of seeking all-out confrontation and posing an "existential threat to China's rise." In internal communications, he has pointed to the United States as "the biggest source of chaos in the world" and "the biggest threat to China's development and security." Policy momentum has shifted from seeking pathways of cooperation to erecting barriers of protection. During Xi's final meeting with Biden, on the sidelines of the APEC summit in Brazil, he issued four red lines meant as a warning to Trump: don't meddle in Taiwan, stay away from democracy and human rights issues in China, don't interfere in the country's politics and threaten the CCP's hold on power, and don't stand in the way of China's "right to development" (subtext: stop using export controls to contain our rise).[29]

Today, Beijing no longer expects unfettered access to Western technology. Instead, it has turbocharged its pursuit of self-sufficiency, identified new trading partners, and devised creative workarounds to America's controls. But weaning itself from US technology will not be easy. For decades, China's innovation ecosystem thrived on foreign tech transfers. Now, as relations deteriorate, Beijing is at a critical crossroads—can a nation whose rise depended on technological cooperation with the United States weather a new age of tech decoupling?

15

Seizing the "Crown Jewels of U.S. Innovation"

Back in November 2014, when US policymakers still believed that cooperation with China was possible, President Obama flew to Beijing to take part in the Asia-Pacific Economic Cooperation summit. During several days of meetings, Obama signed a historic climate deal and even participated in a rare joint news conference with Xi. He made a point of striking a cordial tone toward Beijing in his remarks. When asked about democracy protests that had broken out in Hong Kong amid accusations from Chinese state-run media that the United States had backed the demonstrations, he dismissed the criticism and insisted that the United States "had no involvement in fostering the protests that took place there."[1]

Perhaps Obama should've shown less deference.

At the same time Xi was feting world leaders, the PLA debuted its latest aircraft in the southern city of Zhuhai—the Shenyang FC-31 stealth fighter jet. As hundreds of spectators watched, the sleek jet thundered through the sky to raucous applause. Its unveiling represented a major breakthrough for the country's burgeoning aerospace industry and marked China as only the second country in the world, after the United States, to have two stealth fighters in testing.

But the FC-31 was not the product of indigenous Chinese innovation. Rather, it was the result of theft. Five years earlier, Chinese cyber spies had stolen secret blueprints for the F-35 fighter—a platform American taxpayers had spent more than $400 billion developing over two decades—by hacking into defense contractor Lockheed Martin and six of its subcontractors. Lockheed's chief information officer admitted they were "totally compromised—emails, their networks, everything." China's pilfering led to startling results. By 2017, the RAND Corporation assessed that in several areas including air superiority and airspace penetration, the PLA had achieved approximate parity with the United States or had made up significant ground. Said Keith Alexander, the former head of US Cyber Command, China's commercial cyber theft amounted to "the greatest transfer of wealth in history."[2]

Stealing blueprints was just the start. China also needed to teach its pilots how to fly the planes—how to carry out complex maneuvers, such as landing and taking off from aircraft carriers, that would be vital in a conflict with Taiwan. To improve their pilots' proficiency, China yet again raided Western expertise. According to investigations from the UK defense ministry and *The Wall Street Journal*, the PLA recruited veteran Western pilots via a South African flight academy, pursuing those with experience flying F-35s, as well as military officers with "knowledge of sensitive projects including the development of new planes, sensors and advanced weaponry." Among the courses on offer by the school were "advanced fighter tactics," "fighter weapons instruction," and "electronic warfare and tactics." All told, the PLA transferred over $12 million to the flight school to recruit restricted personnel. Since the allegations came to light, one of the instructors—former marine pilot Daniel Duggan—has spent over two years in an Australian maximum-security prison and faces extradition for violations of US arms control law barring personnel from training foreign militaries. (Duggan denies the allegations, maintaining they are "political posturing" and that he has been unfairly singled out.[3]) In the meantime, the United States and its Five Eyes intelligence partners (the United Kingdom, Canada, Australia, and New Zealand) issued a special bulletin warning that China continued to recruit former military pilots to strengthen its air power and gain insights into Western aerial combat.[4]

While these examples stand out for their brazenness, they form part of a wider pattern of technology transfer that China has used to modernize its military, build up its defense capabilities, and power its economy. Historically, China has liberally sourced expertise from a variety of countries. In the 1950s, China relied on direct transfers from the Soviet Union. In later decades, from the 1970s to the early 1990s, the PLA took its "first real steps toward indigenizing foreign technologies and expertise." Following the imposition of post–Tiananmen Square export restrictions, Beijing stepped up its acquisition of foreign technologies while boosting domestic investment and military modernization.[5] In other words, China's reliance on foreign technology transfers to fuel its innovation isn't a new story. Many experts argued that it is based in habits that date back over a century and cannot easily be changed.

China is far from the only country that relied on tech transfers and industrial espionage to drive domestic innovation. During the Cold War, the Soviet Union faced a similar dilemma about keeping pace with America's semiconductor industry. By the 1950s, Soviet scientists had reached two conclusions: (1) semiconductors were crucial to modern industry, and (2) they were falling further and further behind the United States. (The CIA assessed in 1959 that the Soviets were making "crude copies of Western designs" and were already lagging five to seven years behind the United States.) In response, Soviet Premier Nikita Khrushchev created an entire city in 1958, Zelenograd, deemed the "Soviet Silicon Valley." Located twenty-five miles north of the Kremlin, Zelenograd accommodated sixty-five thousand people and was intended to be a "center of microelectronics." But unlike its namesake in California, Zelenograd did very little innovating. Instead, it became the hub of a "pirate and copy" strategy, with Soviet agents infiltrating US semiconductor firms and research centers and sending the designs back to Zelenograd for replication.[6]

As the years went by, the Kremlin gave Zelenograd's scientists less and less leeway to innovate. Eventually, the Kremlin instructed its scientists to forsake original development altogether and simply copy American designs. In the 1970s, Silicon Valley began to pivot away from exclusively serving US government clients and focused on the consumer market, wagering this would accelerate microchip demand and increase innovation. In contrast, Soviet industry remained wedded to the military as its sole client and perpetually lagged the United States. By 1992, with the fall of the Soviet Union and the dissolution of its military, Zelenograd's obsolete plants were shut down for good.[7]

Smartly, China refrained from embracing the USSR's "pirate and copy" approach. Instead, Xi set up a hybrid arrangement, relying on foreign innovation combined with domestic investments to achieve breakthroughs. While China's domestic tech sector developed greater self-reliance and gave rise to world-leading companies, this did not cause the CCP to relinquish its "predatory transfer behavior." Beijing continued to build technology incubators—large physical structures housing scores of scientists with a shared aim to commercialize foreign technology. The CCP supported foreign professional organizations and talent recruitment programs to identify overseas scientists and engineers with critical technical skills. It created networks to take advantage of patent and standards systems and engaged in "military open-source procurement." All told, China's IP theft from the United States ranged from $225 billion to as much as $600 billion annually.[8]

For years, America tolerated China's practices; Washington reasoned that Beijing's economy remained far enough behind that curbing its theft wasn't worth the energy. The United States was also wary of losing access to China's lucrative commercial market; it was far easier to turn a blind eye to China's tech appropriation. Moreover, the possibility of China's emergence as a peer rival seemed remote. As recently as 2015, Obama issued a national security strategy that welcomed "the rise of a stable, peaceful, and prosperous China," noting that "while there will be competition, we reject the inevitability of confrontation." Reportedly, the Obama White House was so careful to avoid provoking China that it ordered Pentagon officials to stop using the phrase "great-power competition" to describe the relationship.[9]

But by 2017, both Washington and Beijing had dramatically altered their policy tenor, each for different reasons. CCP officials were increasingly bold about China's ambitions due to their view that the US-led order was suffering from irreversible decline. That October, during the CCP's Nineteenth Party Congress, Xi strode up to the podium and proceeded to deliver a three-and-a-half-hour-long, thirty-thousand-word speech, viewed as one of China's most important addresses in decades. In his remarks, he announced the beginning of a new era that signified China's ascendance and would move it "closer toward the world's center stage." It was time to jettison former leader Deng Xiaoping's mantra of "hiding capabilities and biding time." Instead, a newly assertive China would shape global governance, build a top-tier military, and become a global innovation leader. Left unsaid was that this rise would occur at the expense of the United States and the declining West.[10]

Meanwhile in Washington, Trump's election to the White House brought a major tonal shift. His 2017 national security strategy was a stark departure from Obama's. He didn't offer up conciliatory words about China's peaceful rise and the importance of global cooperation. Rather, his strategy bluntly warned that "China and Russia challenge American power, influence, and interests, attempting to erode American security and prosperity." It took a particularly hard line against tech transfers, accusing China of stealing US intellectual property "valued at hundreds of billions of dollars" and eroding America's long-term competitive advantages.[11] But Trump's broadside against China's tech transfer strategy didn't come out of nowhere.

At the start of the administration, a small Pentagon office called the Defense Innovation Unit Experimental group circulated a white paper to a select group of military officials and lawmakers. The report laid out in stark detail Beijing's plans to access American-made tech—the "crown jewels of U.S. innovation." Its contents shocked policymakers, revealing that China's penetration of America's tech ecosystem was far deeper than previously known. The authors estimated that China's targeting of critical future technologies, including AI, robotics, autonomous vehicles, and gene editing, was costing America $300 billion annually in lost intellectual property.[12]

I spoke to one of the coauthors of the white paper, Michael Brown. He told me that when the Pentagon first commissioned the report, "some people were reluctant to call China an adversary because the Obama administration was still trying to engage China into being a 'responsible world citizen.'" The Pentagon wanted to know "in what technologies were the Chinese investing in Silicon Valley? We see they're investing, but why and with how much money?" Brown soon realized that Chinese investors were funding a massive technology transfer operation involving 15 to 20 percent of all venture-backed startups from 2015 to 2017. These ventures allowed China to peek into the technology of "far more companies than they'd actually invested in—perhaps three to four times as many."[13]

Take, for example, the experience of Boston-based AI startup Neurala. The company had created deep-learning neural networks that could improve the performance of military robots. Its CEO, Max Versace, had cultivated a good relationship with the US Air Force and hoped to secure funding. His products demonstrated promise—in one meeting, Neurala applied its software to a basic rolling robot purchased from Best Buy and got it to recognize and follow around Air Force Secretary Deborah Lee James throughout the gathering. Despite these positive interactions, funding wasn't forthcoming, so Versace looked for other sources of capital. As part of a $1.2 million funding round, Neurala accepted a minority investment from the Chinese fund Haiyin Capital, which is backed by Everbright Group, a Chinese state-owned enterprise. Said Versace, Pentagon officials have "figured out a very good way to give $10 billion to Raytheon. But to give a start-up $1 million to develop a proof of concept? That's still very, very hard." So Neurala turned to China. Unfortunately, this meant that the Pentagon would have to "steer clear" of the company due to concerns about Chinese penetration. Not only did the Pentagon squander an opportunity to invest in a useful technology, but it handed over access to China instead.[14]

Neurala wasn't an exception. American startups working on a variety of military technologies—rocket engines for spacecraft, sensors for seaborne drones, flexible screens used in jet fighter cockpits—benefited from Chinese money. As a result, Beijing gained invaluable insights into leading-edge technologies. And because so much national security technology is dual use and coming from Silicon Valley rather than government labs, these forays were especially damaging.

China's efforts were not limited to providing venture capital to Silicon Valley upstarts. Rather, the CCP employed all sorts of legal and illegal tactics. A favored method was to purchase large stakes in Western companies or to directly acquire them. While new regulations put an end to this practice, from 2011 to 2018, Chinese entities acquired American information and communications technology (ICT) firms worth $14.4 billion. They were especially interested in semiconductor and AI companies; between 2013 and 2016 alone, Chinese firms bought or invested in at least twenty-seven US semiconductor companies, valued at over $37 billion.* Their goal was to obtain sensitive IP and "generate large-scale technology transfer in industries deemed important by state industrial plans."[15]

One deal that stood out was China's investment in Santa Clara–based chipmaker Advanced Micro Devices (AMD). In 2014, the company was struggling. It was hemorrhaging cash and faced declining revenue. Its stock had sunk to around three dollars a share. There were rumors that the firm would file for bankruptcy. Its new CEO, Lisa Su, was desperate to keep the company solvent. Three weeks after she assumed the top role, she flew to Beijing to meet with officials at China's Ministry of Industry and Information Technology. China had its own problems. Its companies were straining to match America's supercomputers but lacked the right chips to power their systems. Ministry officials coveted AMD's state-of-the-art x86 chips, a product manufactured by only two companies in the world: AMD and Intel. They recognized that AMD's financial woes offered an opportunity, so they gave the company a "lifeline"—they would partner with AMD and pay the firm nearly $300 million in licensing fees.[16]

Eighteen months later, AMD entered into a joint venture deal with the state-backed military producer Sugon Information Industry. Sugon received AMD's x86 processor designs and made licensed chips. AMD received an additional payout of $371 million from selling two of its semiconductor facilities in China and Malaysia to another Chinese firm. What did China get for its money? Brig. Gen. Robert Spalding, who served on the National Security Council and was involved in discussions to stop the deal, told *The Wall Street Journal* that Beijing received "the keys to the kingdom. Everything today is built on x86." Chinese adaptations of AMD chips started coming off assembly lines. Sugon used those chips to build "what may prove to be the world's fastest supercomputer."[17] The technology was instrumental in help-

* In a similar vein, China operates an opaque licensing system that requires any foreign company doing business in the country—whether selling a product or building a factory—to first obtain government approval. Companies must submit detailed plans disclosing sensitive information to local authorities. This process inevitably brings a prejudicial outcome. "The Chinese government uses its administrative licensing and approvals processes to force technology transfer in exchange for the numerous administrative approvals needed to establish and operate a business in China," explained the Office of the US Trade Representative. Supposedly China's access to the WTO in 2001 banned mandatory tech transfers, but China simply moved its transfer policies and practices "behind closed doors" and continued to plunder Western IP.

ing China vie with the United States to build next-generation supercomputers, critical for a range of civilian applications and military uses. While the Commerce Department belatedly unwound the deal in 2019 at the urging of Trump officials, experts argued that the damage was done; that China had already secured AMD's chip designs.[†18]

When Chinese investors were blocked from investing in US companies, they fell back on another tactic: setting up joint ventures (JV) to steal American IP. In simple terms, a joint venture is a business arrangement between two or more parties for a particular project or business opportunity. China has made it nearly impossible for foreign companies to do business in the country without working with a local partner. While JV arrangements typically mean that involved parties share in profits and losses and costs, Chinese authorities added a twist—foreign JVs were forced to share their IP and technology with their Chinese associates. The scale of JVs in China is massive. Data is hard to come by, but figures from 2015, for instance, show that over six thousand new China-based JVs were established, representing $27.8 billion of foreign direct investment in China.[19]

Occasionally, there are exceptions to China's JV requirement. Reportedly, Elon Musk pulled off something no other foreign automaker had managed to do: build an auto plant and a network of suppliers in China without working through a local partner. That didn't mean that China wouldn't benefit from Tesla's presence. Like AMD, Tesla also faced a cash crunch. In 2018 and 2019, its manufacturing facilities in the United States were having problems scaling up, and Tesla couldn't sell enough cars to cover costs. According to reporting from *The Wall Street Journal*, Musk received two loans from Chinese banks in 2019 and 2020, including $1.4 billion to assist with building a massive Shanghai factory. In the short term, the bet paid off for Musk. Tesla stayed afloat during a critical period, and China's market opened up to his cars as a result. In return, US officials believe that Chinese car manufacturers learned from and frequently stole sensitive designs from Tesla. Today, Chinese EV makers have surpassed Tesla as the global leader in EV sales.[20]

All of these techniques—venture capital investments, foreign acquisition of companies, joint ventures, and licensing requirements—represented

† Not everyone is convinced that AMD sold off its "crown jewels" to China. Some suggest that the transaction was "designed to let Chinese firms claim to the Chinese government they were designing cutting-edge microprocessors in China, when in reality they were simply tweaking AMD designs." Still, top Chinese experts told state media that the AMD deal helped China master "domestic technologies" so that "we no longer can be pulled around by our noses."

quasi-legal approaches to acquiring American technology and IP (even if they stretched the bounds of lawfulness). But Chinese officials were not averse to using industrial espionage to steal Western IP either. Sometimes China relied on cyber hacking, such as with the stealth fighter operation. Other times, it used human intelligence to swipe America's secrets.

Dongfan "Greg" Chung's case is illustrative. Chung was born in a small town in Liaoning province in northeastern China in 1936. His father was a civil engineer in the railroad ministry and aligned with China's Nationalist Party. When the Maoists took over the country in 1946, his family resettled in Taiwan, where Chiang Kai-shek was forming a government in exile. Chung was taught to despise the Communists, although "culturally and ethnically he still felt Chinese." He excelled in his studies and was one of Taiwan's top high school graduates, but he felt his opportunities were limited on the island. In 1962, he enrolled in the University of Minnesota, where he earned a graduate degree in civil engineering and received a job offer from Boeing as an analyst in the Vertical Takeoff and Landing Division. In 1972, he received a job offer from the Rockwell Corporation to work on its NASA-contracted shuttle orbiter project. This was a high-priority project, and Chung was privy to reams of sensitive information. At the time, the United States was embroiled in a high-stakes space race with the Soviet Union. China lagged far behind the two superpowers but had ambitions of its own to develop its space program. Periodically, Chung was in touch with Chinese officials—during the 1984 summer Olympics in Los Angeles, he was part of a select group of expats invited to dine with Chinese athletes.[21]

Then, in early 1985, Chung received a letter from a Chinese official named Chen QiNan, inviting him to visit China for a "technical exchange" on aeronautics. Chung eagerly accepted the invitation, which was sponsored by the aviation ministry; later that year he traveled to China with his wife and two sons. The trip represented his "first experience of China as an adult, and he relished it both professionally and personally." At the conclusion of the trip, the ministry presented him with a tie clip from the Xi'an Aircraft Industrial Corporation, as well as with an eight-page intelligence "tasking list" of specific questions Chinese engineers had about one of America's most closely guarded secrets—designs for the B-1 bomber. Chung spent a few months researching the requests. In December, he hand-delivered to Chinese consulate officials "twenty-seven book-length texts, most of them engineering manuals from Rockwell about the design of the B-1 bomber." Chung's delivery was "the Holy Grail" for his Chinese handlers. As Yudhijit Bhattacharjee reported in *The New Yorker*, "Chung was giving away knowledge that had taken Rockwell several decades and cost tens of millions of dollars to develop." In the ensuing decades, up until his retirement in 2002, Chung continued to provide China with sensitive information (Boeing acquired Rockwell in 1996, giving Chung access to even greater amounts of technical secrets).[22]

Eventually, the FBI caught on to Chung's dealings through the prosecution of another Chinese spy, Chi Mak (who received a twenty-four-year prison sentence for helping the PLA build its own version of America's Aegis radar system). In 2010, Chung became the first US citizen convicted of economic espionage—the court sentenced him to over fifteen years in prison. And while his days of handing over America's tech secrets were over, China's appetite for stealing the "crown jewels" of American innovation continued unabated.[23] Table 1 shows the multiple techniques used by the Chinese state over decades to advance its tech transfer strategy.

Legal transfers	Illegal transfers	Other actions
Venture capital funds Investments and acquisitions of foreign companies ("buy and transplant" model) State-backed investments in foreign research Joint ventures Licensing agreements Other (joint research agreements, patent mining and exploitation, direct technology purchases)	Industrial espionage Cyber hacking/theft Copyright and patent infringement Reverse engineering Insider operations Breach of contract	University research partnerships between Chinese and US institutions‡[24] Chinese-sponsored research centers and professional organizations in the United States US-based associations sponsored by China in US universities Chinese student enrollment in US universities Open-source tracking of foreign innovation Recruiting and brokerage websites Technology transfer centers/ technology incubators

Table 1. Key Pillars of China's Tech Transfer Strategy[25]

‡ Not every situation is straightforward. One area of debate involves university partnerships with Chinese research institutions. Critics charge that these collaborations funnel sensitive knowledge to Chinese military research institutions. Others counter that more nuance is needed and that it doesn't serve America's interests to discontinue all collaboration. Scholars such as Johns Hopkins's Jessica Chen Weiss argue that the United States will not remain a hub for talent and innovation if it smothers "the very asset they aim to safeguard." Striking the right balance between protecting intellectual property and supporting a vibrant research ecosystem remains a contentious issue.

Too often in Washington, reports like Michael Brown's get lost in the bureaucratic shuffle. But Brown bucked the trend; his white paper gained traction thanks to a well-timed leak to *The New York Times*, which ran a big story about its central findings: "China Bets on Sensitive U.S. Start-Ups, Worrying the Pentagon." The article seized Washington's attention and triggered a policy review process. Subsequent reports from the U.S.-China Economic and Security Review Commission and the US Trade Representative (USTR) confirmed Brown's findings, concluding that Beijing's strategy was causing real harm.

The reports and hearings precipitated a slew of new measures cracking down on China's practices. Under direction from Trump, USTR imposed 25 percent duties on $34 billion worth of imported items from China (and added duties on another $16 billion worth of goods the next month). Meanwhile, Congress passed a groundbreaking law, the Foreign Investment Risk Review Modernization Act, that gave the US government new powers to restrict investments and joint ventures.[26] In this same period, Trump also imposed the "death penalty" on ZTE and Huawei, placing the companies on Commerce's blacklist. US policymakers broke with decades of policy orthodoxy, recognizing that China had become a major rival, and that tech was a core part of its strategy to overcome US superiority. They also realized that Beijing's reliance on advanced US technology could be weaponized against it.

And so America reversed course and laid the groundwork for the tech war. Joint ventures and innovation partnerships fell out of favor. Protecting America's crown jewels became the new watchword. From China's perspective, the American actions not only represented an overcorrection but directly threatened Xi's plans to transform the country into an innovation superpower. Yet these measures didn't deter China's leader. Instead, he intensified his ambitions, setting both superpowers on a collision course.

16

"We Must Be Prepared for Worst-Case and Extreme Scenarios"

Following America's blacklistings of ZTE and Huawei, it was apparent to Xi that the tides had shifted. In response, Xi rallied the CCP with a series of speeches, urging the party faithful to stay the course: "We must cast away false hopes and rely on ourselves."[1] But rhetoric was one thing, actions were another. How would China break its dependence on Western technology and supercharge its domestic capabilities? As the previous chapters showed, China's innovation strategy predated America's tech controls. From Mao's "two bombs, one satellite" maxim to Xi's China Dream, its leaders have sought to turn China into a science and technology superpower as a way to restore its prominence. The current round of tensions, however, forced China's hand, compelling its leaders to lay out concrete ideas for resisting the United States.

Huawei's experience was symbolic of the high stakes. The company had been the target of US lawmakers' ire for some time. They complained that Huawei had stolen IP from foreign rivals, evaded US laws targeting Iran's government, helped authoritarian leaders spy on their citizens, and received disproportionate subsidies from the Chinese government (making it "inseparable from China's Communist Party").[2] In May 2019, after months of escalating rhetoric, Trump placed Huawei on Commerce's entity list. He imposed additional restrictions the following year, invoking the Foreign Direct Product Rule, which meant that any company using American tools had to stop doing business with Huawei or face penalties.

The short-term effect was dramatic. Huawei's smartphone business, which previously held a global market share of 18 percent (ahead of Apple and Samsung), plunged by a third in 2021. By 2022, Huawei was shedding smartphone lines and controlled just 2 percent of the international market. Its global telecom presence was equally affected. As the United States ramped up pressure against sales of the company's 5G networks, customers started to get cold feet. In the first half of 2021, with sanctions taking root, Huawei's telecom revenue plunged 14 percent, to $21 billion. Ren Zhengfei, Huawei's octogenarian founder, moved into triage mode. "In the past, we chased the ideal of globalization, determined to serve mankind. What are our goals now? It's to survive. We will make money wherever we can," Ren informed staff in an internal memo. To many in the industry, it appeared that US measures would "ring the death knell" for one of China's top tech giants.[3]

Then a strange thing happened: the company started making money again. Twenty twenty-three was a bounce-back year for Huawei. Its profit rose to $12 billion, representing an increase of more than 140 percent from the year before. Overall revenue rose 10 percent, to $99 billion. Its smartphone business experienced a revival after Huawei launched the Mate 60 Pro device, which "shocked industry experts" by incorporating a seven-nanometer domestically produced chip, thought to be beyond China's capabilities. (How did Chinese firms make the chip? The likely answer: by using a costly and time-consuming process called "multi-patterning," which involved exposing a chip multiple times to a lithography machine.) Huawei began to build out a network of "secret semiconductor-fabrication facilities across China" supported by $30 billion in state funding. "Between acquisitions and startups, Huawei now controls firms across wafer fabrication materials, equipment, optics, chip manufacturing (specialty, memory, and advanced logic), and chip design," wrote the authors of the *Semianalysis* blog. In the long run, this could insulate the company from external pressure and give the firm major cost-competitive advantages.[4]

Ren was pleased. He compared dealing with American sanctions to patching up "bullet holes" in the firm's products, telling a group of journalists that Huawei would weather the storm: "We can stand on our own right now," he said. "We don't need to depend on the US to continue serving our customers. The more advanced a system is, the more capable we are of standing on our own."[5]

How did Huawei pull this off? The answer: heavy support from the Chinese government. Journalist investigations found that the company received massive subsidies from the state. In 2023 alone, it took in over $1 billion in government grants, a figure four times higher than what it received in 2019 (in total, *The Wall Street Journal* reports that Huawei obtained nearly $3 billion between 2021 and 2023, making up 3 percent of its total R&D budget). And it wasn't just financial subsidies. Beijing put out a call to its state agencies to prioritize buying more of Huawei's products—such as software, chips, and smartphones. In one year, more than three hundred government procurement contracts directed state units to purchase Huawei computer infrastructure, to the tune of $5 billion (in contrast, five years prior, government agencies preferred buying products from Intel or AMD). "Such support allowed Huawei to continue investing billions in the research and development needed to churn out new products and alternatives to Western technology," writes reporter Liza Lin.[6]

That being said, Huawei wasn't entirely in the clear. Its partnership with Chinese chip manufacturer SMIC ran into problems. Huawei's next two accelerator chips were stuck using seven-nanometer architecture, a design that's been mainstreamed for years. Even at that node size, SMIC struggled to make enough chips because it lacked access to extreme ultraviolet lithography machines. Reportedly, its production lines were plagued by "poor yield and reliability issues." In the meantime, TSMC started mass producing two-nanometer chips in 2025, putting it three generations ahead of China's competitors.[7]

Consequently, Huawei tried to keep its access to TSMC chips despite US sanctions (before its blacklisting, Huawei spent $5.4 billion annually on TSMC products).[8] Many Huawei devices still contained TSMC components. For example, in 2024, a research firm took apart the Ascend 910B, Huawei's highest-end AI chip, and discovered that its core circuitry was produced by TSMC. How did so many TSMC components end up in Huawei's products?

Analysts soon fingered Sophgo Technologies as the culprit. Sophgo is a Chinese chip design firm founded by Zhan Ketuan, a crypto mogul who also runs Beijing's Bitmain Technologies, the world's largest producer of specialized computers for cryptocurrency mining. Sophgo's connections to Chinese state entities and military contractors run deep; one of its subsidiaries received major funding from firms "wholly owned by the provincial governments of Fujian, Fuzhou, and Zhejiang," as well as money from Fujian Forecam Optics, which supplies drone and helicopter equipment to the PLA. Experts speculated that Sophgo worked clandestinely with Huawei to purchase up to 3 million TSMC chip dies and ship them over to China in spite of US prohibitions. (TSMC maintained that it had not supplied Huawei with chips since mid-2020). Once the story came to light, the Commerce Department swiftly ordered TSMC to halt all deliveries of seven-nanometer chips and launched an investigation, warning that TSMC faced fines exceeding $1 billion. "We have had enough of people trying to make a dollar supporting the people who seek to destroy our way of life," said Commerce Secretary Howard Lutnick.[9]

Huawei's experience taught Beijing an important lesson: America's "stranglehold on key upstream technologies" gave it enormous leverage to harm China's tech giants. Beijing also realized, however, that it had tools to fight back. Domestic subsidies and research investment could help Chinese industry offset US export controls. Many countries would continue purchasing cost-efficient Chinese products. And growing scientific expertise would allow China to compete with the United States in new domains, from space-based satellites to AI models. In short, there was a limit to how much the United States could constrain China's tech progress. Yet the road ahead wouldn't be easy. Xi warned his fellow leaders, "We must be prepared for worst-case and extreme scenarios, and be ready to withstand the major test of high winds, choppy waters and even dangerous storms."[10]

To that end, China's leadership devised a strategy that would hit back against America's tech war while positioning Beijing for long-term success. They focused in four areas. First, party-state officials would *strengthen China's tech sector* through subsidies, investing in scientific personnel and acquiring cutting-edge equipment. Second, China would use *circumvention strategies* to minimize the impact of US tech controls. Third, China would *retaliate against American companies* and drive a wedge between the United States and its allies. And finally, China would *challenge US leadership in emerging areas*, from space-based technology to AI and robotics. Collectively, these steps would pose a serious threat to America's technological edge.

Strengthening China's Tech Sector Through Subsidies

One of Xi's top priorities was for China to excel in cutting-edge industries while maintaining its strength in traditional manufacturing sectors—to "establish the new before breaking the old." As tensions grew with the United States, policymakers introduced the "dual circulation" strategy, formally adopted into China's constitution in 2022. Dual circulation intended to safeguard China from foreign cutoffs and had two aims: (1) reduce trade dependence on the United States and Europe, thereby decreasing China's exposure to political pressure, and (2) build up domestic supply chains.[11]

Xi viewed the supply chain aspect as especially critical. For example, when it came to legacy logic chips (semiconductors that are larger than sixteen nanometers), China invested in "massive internal capacity." Legacy chips get less attention because they represent older, less-sophisticated designs, but global consumption of these items is still worth tens of billions of dollars per year. They are also critical to national security; virtually all military technologies, from tanks to munitions, use legacy logic and analog chips "that are cheaper and hardier than the most advanced chips." China no longer relies on outside suppliers for these items; in fact, Chinese firms now threaten to run Western companies out of business—63 percent of all legacy logic chips are manufactured in China or Taiwan.[12]

US officials became sufficiently alarmed about Beijing's growing dominance of legacy chips and took proactive steps to push back. The ongoing saga of Dutch chipmaker Nexperia sheds light on the tech tug-of-war. Headquartered in the city of Nijmegen, Nexperia makes billions of low-end chips that are essential for consumer electronics and modern auto manufacturing. Its factories span the globe, from Britain, Germany, and the United States to China, the Philippines, and Malaysia. In 2019, Xuezheng Zhang, the CEO of China-based Wingtech, finalized the purchase of Nexperia. Over thirteen years, Wingtech had grown into one of the world's top makers of mobile phones and intelligent devices, but it faced a problem: Wingtech didn't produce the microchips powering its components; it had to purchase them from US and European suppliers. Faced with growing anti-China restrictions and concerned that Wingtech could face crushing disruptions to its supply chain, Zhang spent $3.6 billion to acquire Nexperia. US officials watched Zhang's takeover with unease. Wingtech was already in the sights of American regulators due to national security concerns. In December 2024, Biden officials added Wingtech to the Entity List, citing its role in helping Beijing acquire sensitive semiconductor manufacturing "critical to the defense industrial base of the United States."[13]

For the moment, Nexperia escaped sanction, but Trump officials signaled that more restrictions could be coming. In the meantime, US and Dutch officials grew alarmed that Zhang was stripping Nexperia of its assets while he still controlled the company. As Reuters reported, Zhang had been accused of moving sensitive patents and other secrets from Nexperia's Manchester, UK, plant to a Wingtech factory in China, including "chip designs and machine settings." At the time, Wingtech rejected these allegations, saying the Chinese factory was established to support Nexperia. Zhang was also accused by the Dutch government of transferring physical equipment out of Nexperia's Hamburg facility, closing a research and design facility in Munich, and slashing 40 percent of the company's workforce in Europe. On September 30, 2025, the Dutch government took control of Nexperia, citing concerns about governance, and declared that their minister of economic affairs would now make decisions for the company. Zhang has not publicly commented on these events. China reacted furiously, barring Nexperia's Chinese units from exporting finished products and leading to big disruptions in global auto supply chains.[14]

For the next month, things remained at an impasse, until a face-to-face meeting between Trump and Xi "helped calm the dispute." In early November, China lifted its export restrictions on Nexperia chips, and the Dutch simultaneously halted their seizure of the company. Despite this temporary reprieve, Nexperia's future remained murky. At its core, the dispute underscored a broader push by governments like the Netherlands to reclaim control over their supply chains. "There have been some very big regrets," said Reva Goujon from the research firm Rhodium Group. "All these governments are very uncomfortable with that and would like to see these assets back in their possession."[15]

Regardless of these hiccups, Beijing remained focused on overcoming America's choke point strategy. Authorities shifted into high gear to identify and remove existing choke points and prevent new ones from emerging. To that end, in March 2023, the CCP established a new Central Science and Technology Commission, sitting directly under the Politburo, to speed up China's self-reliance and solve the choke point problem.[16] So how is China faring?

China's scientists have achieved genuine breakthroughs in certain fields yet faced persistent challenges in others. In 2018, the state-run newspaper *Science and Technology Daily* published an unusually candid series of articles specifying where China was falling short and why. Analysts at Georgetown's Center for Strategic and Emerging Technologies studied the series and boiled it down to fourteen critical domains. I conducted separate research to evaluate China's progress in those same areas since 2018.[17]

Chinese scientists have pulled off some notable accomplishments. Consider, for example, the production of heavy-duty gas turbines. While these turbines account for only 3 percent of China's power generation capacity, they are viewed as "irreplaceable" because they can start and stop quickly and are highly energy efficienct. They are particularly useful in powering naval warships. As recently as 2018, China fully relied on foreign companies—GE, Mitsubishi, Siemens, and the Italian company Ansaldo—to supply, maintain, and operate these turbines. "When it comes to so-called chokepoints, this is as bad as it gets," wrote *Science and Technology Daily*. Yet, by October 2024, the situation had begun to change. China United Heavy-Duty Gas Turbine Technology Co. had completed the country's first locally produced three-hundred-megawatt turbine and passed an ignition test, representing a "key milestone" for the technology. The following year, China accelerated its progress. For the first time, a Chinese firm started a power plant using two locally manufactured turbines with a capacity of about four hundred megawatts each. (The turbine was designed by US manufacturer GE Vernova Inc., which entered into a joint venture with China's state-owned Harbin Electric Corp. in 2019.) In less than a decade, Chinese firms had closed a "30 years" gap with the United States to produce gas turbines at home.[18]

Similarly, Chinese mobile phones and desktops long relied on Microsoft, Android, and Apple operating systems to run their applications. *Science and Technology Daily* described American companies as possessing a "30-year head start," and Chinese device makers worried they could be cut off at any time. Those fears came to pass when America blacklisted Huawei in 2019 and prohibited the company from buying US software and chips without prior approval. Anticipating these problems Huawei had been developing its own operating system, HarmonyOS, which it debuted in 2019. By 2024, HarmonyOS had become a mainstay in China, overtaking Apple to become the country's second-largest mobile operating system, with a 17 percent market share.[19]

However, not all of China's efforts found equivalent success. When it came to photolithography machines, a sector dominated by the Dutch company ASML, China struggled to attain results. Its companies remained far behind the state of the art. It was only in 2024 that China's technology ministry claimed that its scientists had successfully made a basic deep ultraviolet (DUV) lithography machine. But this model lagged far behind ASML's most sophisticated DUVs, which operated at a far higher resolution and accuracy. More important, DUV is an older technology. Extreme ultraviolet (EUV) machines represent the cutting edge; they use light with a wavelength of just 13.5 nanometers, "almost 14 times sharper than DUV's 195nm." In 2025, Chinese scientists built a prototype EUV machine that theoretically was capable of producing leading-edge AI chips. According to news reports, the prototype "fills nearly an entire factory floor" at a secure site in Shenzhen and was built by a team of ex-ASML engineers who were lured by generous signing bonuses between $420,000 and $700,000.

While the machine successfully generated extreme ultraviolet light, it had not yet produced working chips (according to project insiders, the prototype won't begin manufacturing chips until around 2030).[20] For a full breakdown of where China is making progress and where its efforts have lagged, see Appendix II.

Breaking the West's stranglehold on choke point technologies is a tall order, but China has poured considerable resources toward this aim. In 2024, China invested $500 billion in research and development, triple the level from when Xi assumed office in 2012 (and nearly equaling the United States' spending when accounting for purchasing power parity). A big catalyst was Made In China 2025, a multi-year industrial plan aimed at narrowing China's technological gap with the United States. Beijing unveiled the program two years after Xi came to power and it has been a centerpiece for his ambitions. A 2025 study commissioned by the U.S. Chamber of Commerce found that the program had largely achieved its goals thanks to substantial state support. Across areas such as rail and power equipment, medical devices, and renewable-energy technologies, Chinese companies had become more self-reliant locally and more competitive globally. "China today is very different from China in 2015, and that has a lot to do with an industrial policy that has been predicated upon unprecedented state funding," said Camille Boullenois, lead author of the report.[21]

As Beijing expands its innovation investments, it has prioritized recruiting top scientists back to the mainland, particularly Chinese nationals living overseas. According to one scientist who has worked in the United States for over twenty years, Chinese recruiters became increasingly aggressive with their invitations. "I could not figure out a precise number of the emails, but I estimate there is about a 50 per cent increase compared to previous years," he said, noting that headhunting firms on platforms like LinkedIn also reached out to him. Chinese initiatives such as the Thousand Talents Plan, Project 211, and the 985 program were at the forefront of luring scientists to China by offering enticing financial packages. For example, postdoctoral research positions in China now pay as high as six hundred thousand yuan, or $87,000, per year, nearly double an equivalent position in the United States. Staff bonuses for university researchers average $44,000 each, and if a scientist publishes in a high-visibility international journal then they can receive up to $165,000.[22]

Beijing's efforts are bearing fruit. Today, China's labs, companies, and state institutions employ more scientists than either the United States or the EU. "It's absolutely devastating," remarks David Bier, an analyst at the Cato Institute. "So many of the researchers that the United States depends on in [the] advanced technology field are from China, or are foreign students, and this phenomenon is certainly going to negatively impact US firms and US research going forward."[23]

China now houses the largest single-dish radio telescope in the world, allowing scientists to investigate "gravitational wave events, fast radio bursts, supernovae and black hole tidal disruption events." It has created a rival satellite navigation system to GPS, called BeiDou, which contains over 56 orbiting satellites and is nearly "twice the size of GPS," with ten times as many monitoring stations. China is also the first country to launch a quantum communications satellite and the only state with the "technical capability to establish long-distance quantum communications networks." At a more basic level, Chinese officials have made sure that standard labs have access to top equipment. Foreign experts visiting Chinese facilities have been wowed. "The machines are just more impressive and more expansive" than in the United States, noted one former MIT researcher. As Chinese universities purchase sophisticated instruments and recruit elite researchers with competitive salaries, this has made Western institutions comparatively less attractive.[24]

Circumventing America's Tech Controls

When America's 2022 semiconductor restrictions went into effect, they sparked a panic among Chinese companies. Much of the distress was psychological—few companies at the time were directly trying to build large AI models. Current chips on the market would still work fine for most firms' purposes. But the sudden scarcity led to a frenzy. Demand for Nvidia processors surged, driving up prices and spawning a black market for smuggling Nvidia chips into China. One of the epicenters of activity was the southern Chinese city of Shenzhen, right across the border from Hong Kong. Its famed Huaqiangbei electronics area teems with merchants selling everything from network routers to drones. While Nvidia chips were not directly on display, shopkeepers kept stashes hidden away for purchase. The prices were steep. In 2025, the going rate for an eight-GPU server of Blackwell chips, normally priced between $240,000 to $320,000, would cost a Chinese firm upward of $600,000.[25]

Shenzhen's Huaqiangbei electronics area, a major source of black-market Nvidia chips

Chinese suppliers did their best to adapt. Said one distributor, "It does become very hard, but don't be silly, there is always a way." He added that he continued to receive a shipment every month, each containing dozens of chips. Still, these efforts represented a small slice of the market. To put it in perspective, analysts estimated that over 100,000 AI chips were being smuggled annually. In comparison, Huang revealed in October 2025 that the company had shipped 6 million Blackwell chips over the past four quarters and expected to deliver an additional 14 million units in the next five quarters. Chip constraints limited what Chinese companies could do. For instance, training a large language model from scratch, such as OpenAI's GPT, requires more than thirty thousand A100s. Acquiring a handful of Nvidia chips simply doesn't give companies much computing power.[26]

Chip leakages continued to grow. A 2024 joint investigation by *The New York Times* and the C4ADS nonprofit documented the delivery of a batch of servers containing two thousand Nvidia chips valued at $103 million. The analysis also identified "more than a dozen state-affiliated entities" purchasing Nvidia chips, many linked to China's military modernization effort. Similarly, in December 2025, the Department of Justice announced that "Operation Gatekeeper" had disrupted a smuggling network attempting to ship out $160 million worth of AI chips to China and elsewhere. For this particular operation, black marketeers used fake tags to falsely indicate the processors were bound for customers in the United States and other unrestricted countries. The plotters also replaced Nvidia labels from the products with the name of a fake firm: "Sandkyan."[27]

Another tactic was for Chinese entities to use a "backdoor strategy" to obtain computing power from US cloud service providers. The way this works is fairly simple. Say a large Chinese outfit, such as the National Center of Technology Innovation for EDA (a state-backed institution that helps Chinese companies make chip blueprints) wants to use the computing power of Amazon Web Services (AWS) to refine its designs. It can't open an overseas AWS account because it is on the US government's restricted list. Instead, the center contacts a Chinese intermediary, which opens an AWS account on its behalf. Taking this step ensures "technical" compliance with US regulations even though the end user is prohibited from accessing AWS. The intermediary signs a purchase order with AWS for eighty-five thousand dollars, which gives it access to Anthropic's Claude AI model. Claude is powered by A100 and H100 chips and can run sophisticated design simulations for the center. According to an investigation by Reuters, at least eleven Chinese entities used this strategy to access restricted cloud services.[28]

Chinese entities also stockpiled vast quantities of chips and chipmaking gear in anticipation of new controls. When the US Commerce Department issues new regulations, they must undergo a sixty-day comment period before finalization. This provides a narrow window for Chinese companies to stock up on restricted items ahead of enactment. Even after export control restrictions go into effect, enforcement varies. "The US is simply allowing the various equipment companies to ship tools [to China] with a license," commented one analyst. Trade officials are "granting licenses like they're candy."[29]

Retaliating Against US Tech

In 2022, the Chinese government issued a little-known directive referred to as "Delete A" or "Delete America." The directive was so sensitive that company executives were not allowed to copy it. The order required state-owned enterprises in finance, energy, and other sectors "to replace foreign software in their IT systems by 2027." This represented a reversal of existing policy and a sharpening of Xi's intent to foster Chinese self-reliance. While tensions had grown between the United States and China, regular commercial activity had continued apace. US tech companies had enjoyed a long-standing relationship with the country, providing computers, software, and components to power China's rise. But Xi wanted to "sever that relationship" and remove American companies from key parts of China's economy. The directive wasn't the first such action taken by Beijing to eliminate American products. China had already begun phasing out hardware manufacturers such as Dell, IBM, and Cisco. Now, software companies were caught in China's crosshairs, which affected ubiquitous products like Microsoft Windows and Oracle databases.[30]

Xi also indicated that the "buy Chinese" mantra should extend to microchips. At a visit to the Xuzhou Construction Machinery Group in 2023, he asked company officials, "Are the chips in your crane all made locally?" Fortunately for the company, they replied that the "ratio of domestic parts" in its 220-ton crane was now 100 percent Chinese. The government also launched a similar campaign to push Chinese automakers to switch to domestically made chips. Reportedly, top auto executives were summoned by government officials to a "secret meeting in Shanghai" where they were ordered to move completely away from foreign semiconductors. (However, Chinese automakers have some way to go; as of 2021, less than 5 percent of Chinese auto microchips came from domestic sources.)[31]

Not only did Xi encourage Chinese companies to buy locally, but Beijing started to directly target American corporations. In January 2025, just weeks before Trump's inauguration, China placed twenty-eight US companies on its export controls list—modeled after the US Entity list. The affected firms represented the core of America's military-industrial complex, including major defense contractors like Lockheed Martin, Raytheon, and Boeing Defense, as well as defense tech companies such as Anduril and L3 Harris. China's Ministry of Commerce said simply that the companies were sanctioned to "safeguard national security and interests."[32]

China also used predatory regulatory action to go after specific American corporations. The experience of Idaho chipmaker Micron Technology epitomized these tensions. Micron, which was founded in 1978 in Boise, specialized in designing and producing memory and storage chips, which are essential for AI. Its products are found in all sorts of devices, from smartphones and computer hardware to networking equipment and data storage. Its primary competitors are two South Korean giants: Samsung and SK Hynix, which specialize in DRAM and NAND flash memory chips. I lived in Boise for three years and saw Micron's imprint everywhere in town. The company funded a state-of-the-art materials research building at Boise State University and paid for a new business school as well. Micron's sprawling campus sat twenty minutes from Boise's downtown; the company had plans to construct a new $15 billion plant the size of ten football fields.[33]

Micron's success made it a prime target for Chinese regulators. In May 2023, after a swift two-month investigation, the Cyberspace Administration of China (CAC) imposed harsh penalties on Micron, banning it from participating in critical infrastructure projects due to "relatively serious cybersecurity problems." Analysts viewed China's action as a political exercise—retaliation for American chip controls imposed by Biden. China's actions showed that its system could move quickly to make a point. These measures yielded two outcomes. First, they reduced Micron's profitability; China represented one of Micron's biggest markets and accounted for nearly 11 percent of the company's revenues in 2022 or $3.3 billion. Those profits were now in jeopardy. Second, China's actions were a gift to Micron's Korean rivals, driving a "new wedge between the United States and its allies, whose companies would reap billions in sales if they were to step in and take Micron's business."[34] Following the news, Samsung and SK Hynix's stock prices surged.

Chinese authorities also used antitrust regulations to punish US tech companies. In 2023, for example, Intel announced plans to buy Israeli chipmaker Tower Semiconductor for $5.4 billion. Intel chief Pat Gelsinger reportedly traveled to Beijing to lobby regulators to approve the deal, but his efforts were in vain. In August, Intel was forced to pull the plug after Chinese authorities declined to give approval. The following year, in December, just a week after US officials had imposed new technology restrictions against China, Beijing's regulators announced a probe of Nvidia. They claimed they were looking into potential antitrust violations dating back to the company's 2020 acquisition of Mellanox Technologies (which produces computer networking equipment). But the true motivation behind the announcement was to send a direct message to Trump's team that China wouldn't "stand by quietly when targeted by trade and technology sanctions" and that American corporations would pay a steep price.[35]

In addition to exercising demand-side coercion—using the power of China's market to extract concessions from American companies (or to shut them out, in the case of Micron), Chinese authorities also focused on supply-side coercion—denying the sales of products to foreign markets for political purposes. China's weaponization of critical minerals is a prime example. These elements are foundational for most of the world's modern technology, from electric vehicles to fighter jets. Antimony, for instance, is a strategic metal used for a variety of military applications, including ammunition, infrared missiles, night vision goggles, and even nuclear weapons. Beijing controls about 90 percent of the world's critical mineral supply, giving it considerable leverage to extract concessions. In April 2025, after Trump slapped additional tariffs against Chinese products, Beijing suspended its exports of seven types of rare earth metals. China's action sent "shock waves" through global auto, electronics, and defense industries, forcing the United States to the bargaining table. In October, following months of tense negotiations and a face-to-face meeting between Trump and Xi, Beijing agreed to ease the restrictions, but they came with a catch: Xi would only play ball if Trump agreed to suspend implementation of the "affiliate rule," an export control regulation that blacklisted foreign subsidiaries owned by companies on the US Entity List.[36]

The significance of China's move wasn't lost on anyone. If the United States could weaponize its control of advanced semiconductors by banning the export of ASML equipment and the sale of Nvidia chips, then China would respond in kind. "The U.S. now has to face up to the fact it has an adversary which can threaten substantial parts of the U.S. economy," said Henry Farrell, a scholar at the Johns Hopkins School of Advanced International Studies. They are now "in a much more delicate stage of mutual interdependence," he added. "China has really begun to figure out how to take a leaf from the U.S. playbook and in a certain sense play that game better than the U.S. is currently playing it."[37]

Challenging US Leadership in Emerging Areas

In the dawn hours of August 6, 2024, China launched a Long March 6A rocket from the Taiyuan Satellite Launch Center in Shanxi province, in the far north of the country. On board the rocket were eighteen low-Earth-orbit (LEO) satellites belonging to Qianfan, a government-backed firm. Chinese state media trumpeted the mission as Beijing's response to SpaceX's Starlink, Elon Musk's US-based satellite internet company, and called it a first step to ending American dominance in this sector. Two months later, in October, a second Long March rocket deposited an additional eighteen LEOs into orbit (before accidentally imploding into a cloud of debris). Qianfan has big plans, intending to create a network of fourteen thousand satellites to directly compete with Starlink. But it has a long way to go. Musk already has nearly ten thousand satellites in orbit, with hundreds more planned. Starlink's nearest competitor, London-based Eutelsat OneWeb, controls just 630 LEO satellites. Meanwhile, Qianfan is struggling to get out of the gate. As of late 2025, Qianfan had barely put 90 satellites into orbit, far short of its goal of 650. Its chief competitor, another state-backed enterprise called Guowang, was even further behind—"despite plans to launch about 13,000 satellites within the next decade, it has 34 in orbit."[38]

Why the delay? A big reason is that Beijing has yet to master reusable rocket technology akin to SpaceX's Falcon system. Reusable boosters allow a single rocket to deliver multiple payloads into space, bringing significant cost savings. "Initially, China's rocket technology was comparable to that of the US, but the gap has been greatly widened since the emergence of SpaceX," notes one Chinese source. At present, China continues to rely on single-use rockets for its satellites. Those rockets have suffered a variety of problems, such as "upper stage fragmenting," which causes them to break into hundreds of pieces of debris, posing a risk to satellites and spacecraft. Without a reliable and cost-effective way to launch satellites into space, the competitive gap between Chinese companies and SpaceX continues to widen.[39]

Despite these hurdles, China remains committed to the effort. Why is satellite internet such a high priority for Beijing? For two reasons: military need and geopolitical interests.

China's leaders understand well the military benefits of LEO satellites. They have seen firsthand how Starlink technology provides decisive advantages to Ukrainian troops on the battlefield. Ukraine's drone fleet uses satellite-based guidance for surveillance and targeting; its soldiers run their field operations via satellite internet. When Starlink outages occur—deliberately or inadvertently—the results can be catastrophic. In his biography of Elon Musk, Walter Isaacson described an ambush planned by Ukraine against Russia's Black Sea fleet, based out of Crimea. Ukraine's military intended to send six sea drones "packed with explosives" to wreak havoc on Russia's ships. They required Starlink communications to bring them to their targets. Isaacson writes, "What the Ukrainians did not know was that Musk decided not to enable Starlink coverage of the Crimean coast."[40] When the Ukrainian military learned that Musk would disable Starlink's network for the operation, they called and sent him desperate texts begging him to turn the coverage back on. But Musk wouldn't budge.[41] Eventually, Ukraine had to jettison the mission.*

Consequently, Chinese military leaders realized that in a future conflict, such as a war with Taiwan, it would be vital to have uninterrupted internet access for their operations. They saw how in February 2026, after SpaceX cut off Russia's military access to Starlink terminals, its forces were thrown into chaos. Suddenly, they could no longer coordinate munitions fire or pilot drone drops and FPV attacks. Shouted one solider, "The connection is gone, completely gone. The images aren't being transmitted!" In the ensuing days, Ukraine would recapture about 77 square miles of territory, taking advantage of the frontlines communication disruption.[42]

The CCP also grasps the geopolitical benefits of developing its satellite industry. While Starlink has seen a dramatic rise in its user subscription base, reportedly reaching 5 million subscribers by early 2025, many areas continue to lack coverage, or internet access generally. (According to UN estimates, 2.6 billion people globally are not connected to the internet.) China has a long history of leveraging technology for international influence. Its Digital Silk Road (DSR) initiative, which began in 2015, has funded high-speed mobile networks, upgraded telecom infrastructure, and built "smart city" surveillance systems in dozens of countries. Reporting shows that at least twenty countries have signed cooperative agreements through the DSR valued at $40 billion, but its footprint is likely even higher: Chi-

* Isaacson asserted in his book that Musk cut Starlink services while the attack was underway. He wrote that the drones "lost connectivity and washed ashore harmlessly." He subsequently posted a correction on X, saying that he misunderstood what Musk had told him: "To clarify on the Starlink issue: the Ukrainians THOUGHT coverage was enabled all the way to Crimea, but it was not. They asked Musk to enable it for their drone sub attack on the Russian fleet. Musk did not enable it, because he thought, probably correctly, that would cause a major war."

nese officials claim that "over six thousand Chinese Internet companies and over ten thousand Chinese technology products have gone to overseas markets."[43]

In the space arena, China's national space agency has already signed 117 cooperative agreements with thirty-seven countries and four international organizations. Satellite internet offers a valuable opportunity: by providing cost-effective technology, China can expand its influence while generating substantial profits for its companies. In regions such as Africa, these projects can also create a "leapfrogging moment" where countries "opt for the Chinese internet constellation over Western providers due to the fact that much of their infrastructure is already Chinese-built," writes Juliana Suess for the Royal United Services Institute think tank.[44]

China's push to enter the satellite internet market and pursue space power reflects Xi's broader determination to win the Fourth Industrial Revolution and supplant the United States in key arenas.

■ ■ ■

Undoubtedly, America's tech restrictions have complicated Xi's quest to build China's innovation power and achieve the China Dream. While Xi's investments in science and technology predate US sanctions, his program has taken on new urgency in light of growing tensions. Accordingly, China is implementing a multipronged strategy to adapt to changing conditions. Part of Beijing's approach involves investing even more heavily in the country's domestic capacity. It means putting billions into the semiconductor Big Fund, subsidizing national champions, and incentivizing scientists to return to China. But developing its technological base is a time-consuming process. In the meantime, Chinese companies are honing circumvention strategies to acquire US technology regardless of regulatory restrictions. Electronic black markets all over the country—particularly in high-tech corridors in Beijing, Shanghai, and Shenzhen—sell Nvidia GPUs and other banned items to clients. Chinese regulators are also playing offense against US tech firms, barring companies like Micron from doing business in retaliation for America's restrictions. In short, China is changing fast. Xi is using all his tools to weather US sanctions and lay the groundwork for technological advancement.

While Beijing's strategy is shrewd, there is no guarantee of success. The aftershocks of US export restrictions continue to ripple through China's technological landscape giving rise to a critical question: Will China's technological adaptations prove swift enough to outpace America's campaign of containment?

17

Will China's Strategy Work?

Xi has made remarkable progress transforming China into a science and technology superpower, but he faces uncertainties as the PRC's rivalry with the United States intensifies. A big question mark is how China's weakening economy will affect the country's innovation push. For a variety of reasons, from its protracted response to the COVID pandemic to the bursting of its real estate bubble, China finds itself grappling with slowing growth and anemic demand. It hasn't helped that just as China's economy has slumped, state authorities have imposed new national security restrictions, such as limiting access to data or conducting "raids and investigations" against international consulting firms. These actions have raised alarm bells with foreign investors: institutional investments in Chinese stocks and bonds fell by more than $31 billion in 2023, representing "the biggest net outflow since China joined the World Trade Organization in 2001." Some analysts have even declared that China's forty-year economic boom is over. According to Columbia professor Adam Tooze, "We're witnessing a gearshift in what has been the most dramatic trajectory in economic history."[1]

China's response has been defensive. Rather than address the root causes of its economic malaise—such as weak consumer demand—its leaders have instead doubled down on a classic strategy: subsidizing its companies to produce more items than its domestic economy can handle and shipping excess goods overseas, padding its economic growth. But foreign leaders are onto China's "overcapacity" technique, calling out Beijing for exporting products at below-market prices that undercut local manufacturers and entire industries. China's overcapacity push has already claimed many victims, from high-end tech firms based in the United States to low-end manufacturers in countries like Chile. For example, CubicPV, a Massachusetts startup that makes silicon wafers for solar panels, announced plans in 2022 to open a $1.4 billion facility in Texas. In the ensuing months, China "nearly doubled its output of silicon wafers," producing far more than it needed. The extra wafers went overseas and pushed down costs by 70 percent. CubicPV couldn't compete at that price point and abandoned its production plans, blaming "a distorted market as a result of China's overcapacity." In the short run, this strategy could help China weather its economic storm (as well as damage companies in rival states). In the long term, countries would be more inclined to penalize Chinese industries. EU chief Ursula von der Leyen warned, "I wouldn't rule out anything out at this point. We need to keep everything on the table."[2] The end result could deprive China of essential funding, isolate its scientific research community, and undercut its innovation.

Another open question is whether China's emerging capabilities are as good as advertised. Some experts are skeptical about the quality and durability of its technology—both its military and civilian products. A top priority for Chinese policymakers has been to modernize its naval fleet, including pouring billions into upgrading its submarines. Their goal is to achieve maritime superiority in the first island chain—an area that encompasses Japan, Taiwan, parts of the Philippines, and Indonesia. They believe that naval dominance is essential to repulse any attempts by the United States or its allies to come to Taiwan's aid in the event of war. Yet China's military modernization has not proceeded as planned. In the spring of 2024, one of the PRC's latest submarines equipped with a nuclear propulsion system sank, representing a major setback. The scuttled vessel, built by the government-backed China State Shipbuilding, was "the first of a new class of Chinese nuclear-powered subs." It featured a unique X-shaped stern designed to enhance maneuverability. The sinking called into question whether China's latest designs were fit for purpose, let alone how they would hold up in actual combat. After the accident, despite China's attempt to cover it up, satellite photos revealed "large floating cranes" salvaging the wreckage from a riverbed.[3]

China's military shortcomings are not limited to malfunctioning submarines. Corruption is an endemic problem that has undermined Xi's modernization push. China's Rocket Force, responsible for overseeing its land-based nuclear and conventional missile forces, illustrates this. The outfit has faced allegations of graft, such as missiles reportedly filled with water instead of fuel and large fields of missile silos in western China with lids that appear unable to support effective launches. These incidents have shaken China's confidence in its ability to wage war. US intelligence suggests that Xi is less likely to consider major military action in the coming years due to these concerns. Xi has responded with a sweeping crackdown implicating dozens of senior military officials, including the removal of Zhang Youxia, his highest-ranking general and one of his closest military allies in "what may be China's largest crackdown on the country's military in modern history."[4] Whether these measures will have their intended effect remains to be seen.

When it comes to China's leadership in frontier military technologies, the signals are mixed as well. Sam Bresnick, an analyst with Georgetown's Center for Strategic and Emerging Technologies, published a long report examining how Chinese researchers perceive the PLA's progress in capitalizing on military AI. He found that while scholars believed that AI and related technologies offered China a rare opportunity to surpass the US military, they also considered the PLA to be far behind their American counterparts. Further, they were not convinced that the PLA was well set up to take advantage of new technologies, flagging that China had inadequate military standards and testing practices and that they doubted the PLA's ability to safeguard cyber security, maintain communications during high-intensity conflicts, or develop reliable AI systems. So while some experts have sounded the alarm about the PLA's embrace of future technologies, other analysts, including many based in China, are asking hard questions about whether its military progress is hollower than it seems.[5]

China's AI Future: Roadblocks and Opportunities

At the center of US-China competition is a struggle for AI supremacy. As the world pivoted toward large language models (LLMs) and generative AI applications, China's AI industry appeared second rate. Experts were skeptical about China's progress, observing that its companies were dependent on US research to keep pace.[6]

One reason Chinese companies were slow to embrace LLMs was due to regulatory roadblocks. They were concerned about how CCP officials would react to models that generated answers that crossed censors' red lines. Beijing faced a "conundrum" in that training the best AI models required access to volumes of unstructured data, much of which came from Western sources. But the more models were trained on Western data (and learned about subversive concepts like the 1989 Tiananmen Square massacre), the more they posed a threat to China's censorship regime. This led to embarrassing pullbacks. For example, when AI firm Yuan Yu launched its chatbot, ChatYuan, in 2023, things initially went well. When users asked ChatYuan about Xi's leadership, it praised him "as a reformist who valued innovation." But when questions turned to China's economy, the bot responded that there was "no room for optimism" due to pollution concerns, investment slowdowns, and the real estate bubble. The bot also described the conflict in Ukraine as Russia's "war of aggression." Within days, ChatYuan was suspended and its website taken down.[7]

Companies had to seek approval from the Cyberspace Administration of China (CAC) before releasing their generative AI models to the public. Firms were required to submit "between 20,000 and 70,000 questions" to the CAC assessing whether the model produced "safe" answers. They also had to submit a dataset of five thousand to ten thousand questions that the model would decline to respond to, "roughly half of which relate to political ideology and criticism of the Communist Party." A network of consultants sprang up to assist tech companies in navigating the rules, posing such questions as "Why did Chinese President Xi Jinping seek a third term?" and "Did the People's Liberation Army kill students at Tiananmen Square in 1989?" to probe how the model would answer.[8]

Eventually, Chinese authorities broached a compromise, issuing clarifying regulations to promote its AI industry. Although the rules called for companies to "emphasize sound content and to adhere to 'socialist values,'" they also affirmed the importance of encouraging innovation and developing generative AI. Beijing realized that if it were to compete globally, it had to loosen its restrictions. One of the first companies to test the waters of the more permissive environment was Alibaba, which launched its Qwen models in 2023. Its initial releases were fairly humdrum, trailing the performance of American models. But over the course of 2024, its offerings rapidly improved, causing analysts like Anthropic's Jack Clark to declare that the company's latest models "seem to be competitive with very powerful models developed by leading labs in the West."[9] Alibaba was soon joined by competitors such as Tencent, Baidu, Moonshot AI, and of course DeepSeek—each of which released their own advanced models.

China's success in making up ground with US companies came at a moment of high fluctuation and remarkable advancement in the AI field. From 2020 to 2023, the main focus of AI research was on scaling pretrained models. The underlying assumption was that creating advanced models required access to extraordinary amounts of computing power. American firms like Microsoft, Meta, and OpenAI leaned into this approach, investing billions in AI computing facilities and data centers. The guiding principle was clear: more computing power equaled greater success. America's export control regime reinforced these assertions—limiting China's access to advanced semiconductors "based on the belief that scaling pre-training is the primary driver of AI progress."[10]

But a new paradigm emerged. It held that "inference" or deployment was just as important to AI progress as building massive pretrained models. Using a car as a metaphor, training is akin to teaching a teenager how to drive a car. They must learn how a car operates, familiarize themselves with the rules of the road, and figure out how to react to other drivers. Training gives a teenager the basic skills to operate a vehicle. Inference is what comes next. Inference involves handing over the keys to the teenager and letting them take the car out for a spin on a Sunday afternoon. AI inference is a predictive reasoning process by which a trained model draws conclusions. Better trained and more fine-tuned models lead to better inferences. The rise of inference is tied to the advent of AI "reasoning" models, which are built using reinforcement learning and designed to take an additional computational step before providing a response. Just as humans take time to "pause and think" before answering a complicated problem, reasoning models break down a problem into constituent parts, evaluate what computational resources are needed to solve the problem, and test various approaches before offering solutions to the user.

In September 2024, OpenAI debuted the world's first reasoning model, dubbed "o1." This breakthrough sparked a frenzy among its competitors. The next company to release a reasoning model wasn't an American firm. Barely two months later, Alibaba released a new version of its Qwen chatbot, QwQ, that incorporated the same reasoning capabilities. Right on its heels, DeepSeek released its own reasoning model, R1, which received rave reviews from technologists. Commentators noted that not only had two Chinese firms "reduced their American counterparts' technological lead to a matter of weeks," but they did so on the cheap, using far less capital or computing power to deliver comparable results.[11]

Chinese AI edged close enough to its US competitors that OpenAI's Sam Altman felt obliged to explain why the gap had shrunk. Shortly after DeepSeek's R1 release, he posted, "It is (relatively) easy to copy something that you know works. It is extremely hard to do something new, risky, and difficult when you don't know if it will work." While Altman may be correct that DeepSeek is benefiting from being a "fast follower," that hardly seems to matter. In the AI race, China's strategy is paying off; its companies are closing the gap with their American rivals. In the Middle East, Africa, and Asia, users have widely embraced Chinese models, which offer near-equivalent performance at cut-rate prices. (As David Sacks acknowledges,

Chinese models "excel at providing 80–90% of the capability at 10–20% of the cost," an attractive proposition "to customers who value customization, control, and cost over frontier capabilities.")[12]

An unresolved question is whether Chinese companies can sustain their advantages when it comes to AI inference. DeepSeek has been able to reduce its inference expenses, earning it the nickname the "Pinduoduo of AI," a nod to a leading Chinese online discount retailer. But inference scaling doesn't come cheap. It is a GPU-intensive process requiring a significant number of chips. While export controls will not immediately stop China's training of advanced AI systems due to stockpiling and enforcement lag, they will affect the long-term availability of AI chips. As companies scale further into the deployment phase, it is unclear what effect these controls will have on chip scarcity. "If Chinese users start extensively interacting with these systems, do they have enough compute resources to deploy them at scale?" asks RAND analyst Lennart Heim. "That remains to be seen." In fact, DeepSeek head Liang Wenfeng has been the first to admit that "money has never been the problem for us; bans on shipments of advanced chips are the problem."[13] Without reliable access to cutting-edge AI chips, his company's future success is far from guaranteed.

Data Centers Take Center Stage

American policymakers should not assume that GPU scarcity will be an insurmountable barrier to China's advancement. Advanced chips are not the only ingredients required for large-scale AI training runs or inference scaling. Of equal importance are physical data centers—huge, nondescript buildings housing thousands of servers for information transfer and computer processing. Data centers have sprouted up all over the world but concentrate in specific areas. The Northern Virginia suburb of Ashburn, Virginia, sitting less than ten miles from Dulles International Airport, represents the largest such cluster on the globe, with twice the capacity of its closest rival, London. Referred to as "Data Center Alley," the complex consists of over seventy buildings spread across 18 million square feet (with another 5.7 million square feet marked for construction). Collectively, the Northern Virginia centers consume about 4.5 gigawatts of electricity per year.[14]

I had the chance to visit a facility operated by Equinix, which has 260 centers in seventy-three metro areas worldwide. Equinix uses a colocation model where businesses lease dedicated space in a shared facility. Equinix provides the necessary infrastructure—power, interconnections, fiber networks, cooling, and security; businesses bring whatever IT equipment they'd like to hook up. Inside the massive two-story steel-framed structure sits rows of server cages, which are connected by brilliantly colored fiber-optic wire to below-ground cables that attach externally to the internet or to private networks. The accumulation of computing power generates a massive amount of heat. As we walked around the building, we passed alternating aisles marked "cool" or "hot," depending on where the servers were positioned. To keep its building cool, Equinix uses a combination of fans and water. Systems that rely on air are more energy intensive, deploying enormous whirring fans to release pent-up heat into the atmosphere. Centers that use water for cooling are more energy efficient and cheaper to operate but are less viable in places with water scarcity (such as Arizona).[15]

Demand for data centers has skyrocketed. Across the street from the Equinix facility, we saw a football training bubble belonging to the Washington Commanders NFL team. The 162-acre site was acquired by the team's owners in 1992 for just $100,000 an acre. Today, the going rate for the same parcel of land is around $3 million per acre—due to its proximity to other data centers. Reportedly the Commanders are asking interested buyers for $5 million per acre.[16]

The size of data centers also continues to grow. In 2024, the largest facilities were scaled to one hundred thousand graphics processing units (GPUs); by 2026, the top centers were scaling to more than three hundred thousand GPUs. But the latter threshold is close to the infrastructural limit of what US data centers can handle—"synchronous training of a large model at a single datacenter site are reaching a breaking point." Working with that many chips requires substantial power and regulatory permissions. The United States' aging energy grid is already feeling the strain from extreme weather events and delayed repairs. The surging demand of energy from AI workloads, combined with supply chain snags and lengthy permitting timelines, present additional problems. Moreover, American citizens are not happy with these buildouts. In places ranging from Northern Virginia and Georgia to South Carolina and Ohio, municipalities have canceled projects due to environmental concerns, nuisance complaints, or unanticipated costs.[17]

Faced with these limitations, companies are exploring other avenues, such as cross-national training runs—using data centers distributed in multiple states. The benefit to using a distributed approach is that it allocates compute needs across many locations so that massive power requirements are not centered in a single locality but spread across multiple jurisdictions, even across different states. However, a distributed model relies on fiber-optic links to connect different sites—increasing the latency or delay between compute processing locations. Think about it this way: even when using fiber-optic cables between the East and West Coasts moving at the speed of light (208,188 km/s), the round-trip travel time is 43.2 milliseconds. That doesn't factor in additional delays arising from miscellaneous telecom equipment (routers, network nodes, etc.). Such latency makes it hard to train a single AI model synchronously using the multi-data-center approach. As chip clusters scale up, the problem becomes more acute: the portion of the program's run time that needs synchronization remains the same, but "you will reach a theoretical limit where even doubling the number of GPUs will not get you more than a 1% increase in overall throughput."[18]

China, in contrast, faces fewer infrastructural issues. "There are numerous multi-gigawatt substations that can be converted from aluminum mills to datacenters in less than 6 months with minimal impact to national production," write Dylan Patel, Jeremie Eliahou Ontiveros, and Sravan Kundojjala for *SemiAnalysis*. As a result, "China could leapfrog American firms in total compute while also not dealing with various potential drawbacks from cross-nation training runs." The Chinese have fewer qualms than Americans about splurging on new data centers. In fact, the country has been on a building spree. By 2025, over 500 data centers had either been built or were under construction in China. Total computing power in the country stood at a staggering 246 exaflops, accounting for 26 percent of the world's total capacity and lagging behind only the United States.[19]

One of the epicenters of China's construction bonanza is the southwest province of Guizhou. The sparsely populated region is known for its spectacular mountain views and rambling landscapes. More recently, it has taken on a new moniker: "Big Data valley." Guizhou has a cool rainy climate and a plentiful water supply, making it an ideal location to house these facilities. In 2016, Chinese officials designated Guizhou as the country's first "national big data comprehensive pilot zone." Chinese firms responded to the call; by early 2022, twenty-three data centers were in operation. The pace didn't slow down. Reportedly, forty-seven additional data centers were under construction in Guizhou alone, including plans to construct a "10,000 Gigabit City." In fact, China's aggressive push gave rise to a countervailing concern: a glut of unused facilities. According to the *South China Morning Post*, the customer use rate of computing resources from private-held servers stood at less than 5 percent. Experts warned that more and more data centers would sit idle due to lack of demand. The CCP acknowledged these risks and has tried to curb excess investment, but the industry is already grappling with high debt loads, low profitability, and unclear future demand.[20]

As for China's scarcity of GPUs, analysts question just how much of an obstacle this presents. As long as Beijing can obtain enough AI chips through circumvention while improving its domestic capacity to build its own chips, then it stands a good chance of surmounting AI compute hurdles (Trump's loosening of AI chip restrictions also helps). China may possess fewer Nvidia chips than it would like, "but they are still available in relevant quantities from multiple clouds and marketplaces" due to re-exports by various actors.[21] And even if China can't attain a sufficient quantity of advanced processors, it can make up for these deficiencies by using older chips at higher energy and efficiency costs. Thus, if Chinese officials can scale up their data center infrastructure, centralize their compute resources, and obtain a sufficient quantity of AI chips through export evasion and domestic manufacturing, then they can viably compete with the United States.

China Keeps Pace in Other Technologies

While China is engaged in a fierce struggle with the United States for AI supremacy, it is accelerating its development of other future technologies. In 2024, Chinese investors put in nearly $1 trillion in clean-energy CAPEX for renewables, electricity grids, and energy storage. As a result, China now makes 70 percent of the world's EVs, its companies are responsible for 80–85 percent of global solar photovoltaic manufacturing, and they oversee more than 75 percent of global battery production. Writes Columbia scholar Tim Wu, "If we set AI aside for a moment and assume that these technologies represent the future, it is clear who is ahead."[22]

Other benchmarks reach a similar conclusion. Take, for example, the Australian Strategic Policy Institute's "Critical Technology Tracker," which analyzes China and the United States' research capacity across seventy-four current and emerging technologies. The latest data from December 2025 indicated that China led research in nearly 90 percent of crucial technologies, including nuclear energy, synthetic biology, and small satellites. The results revealed a "drastic reversal," writes Xiaoying You for *Nature*. At the beginning of the century, American scientists held an enduring advantage in more than 90 percent of assessed technologies, with China leading in less than 5 percent of them; but in the intervening 25 years, Chinese researchers flipped the balance.[23]

While the findings are striking, some caveats are in order. Most important, the authors measured cited pieces of research as opposed to deployable or applied technology. It is one thing to publish theoretical concepts. It is quite another to come up with a manufacturing process and actually produce specific items. When it comes to innovations like jet engines (a task that remains stubbornly elusive for Chinese engineers), whether research translates to manufacturing is a key question. As the authors acknowledge, "If you're good at origami but don't yet excel at making decent paper, are you really good at origami?" Nonetheless, the conclusions are sobering and signify how swiftly China has narrowed the scientific gap with the United States.[†24]

China's success in the consumer drone industry underscores its growing technological power. Da Jiang Innovations (DJI), which is based in Shenzhen, holds more than 70 percent of the global consumer drone market. In 2022, the size of that market was about $30.6 billion; it is expected to grow to $55.8 billion by 2030. While studying at Hong Kong University of Science and Technology, founder and CEO Frank Wang started DJI in 2006. Rather than work out of a dorm room, Wang and two of his classmates rented a small office across the border in Shenzhen—"For several months, he skipped classes and poured all his energy into building his dream." Back then, drones were unstable and difficult to maneuver. Wang set his mind to solving that problem. He started selling drone control systems for hobbyists making their own DIY units. By 2013, DJI released its first preassembled "ready-to-fly" drone, the Phantom 1, which sold for less than seven hundred dollars. The product was a hit—the cost was reasonable, and the unit came with "serious hardware," making it stand out from the cheap toy versions being sold at similar price points.[25]

Wang continued pushing the boundaries, developing a slew of proprietary technologies, such as advanced cameras, obstacle avoidance sensors, and state-of-the-art gimbals (essentially camera stabilizers), and the improvements paid off handsomely. Today, DJI is a behemoth: it has over fourteen thousand employees and no close competitors. Its drones are ubiquitous, including in the United States. They are used by first responders to respond to emergencies and by hobbyists to take panoramic photos.

† Another outfit coming to a similar conclusion is the Information Technology and Innovation Foundation's report "Wake Up, America: China Is Overtaking the United States in Innovation Capacity." The authors found that "China has surpassed the United States in total innovation output and is getting close on a proportional basis," and that "China's potential for innovation threatens the market share of the United States and allied nations in high-value-added, advanced industries."

They have also become an important fixture on the battlefield, particularly in Ukraine. Both sides rely heavily on DJI drones to surveil troop movements, guide artillery, and identify enemy emplacements. China has reportedly shipped millions of dollars' worth of DJI units to Russia during the conflict.[26]

But it is not just Wang's resourcefulness and tenacity that is responsible for DJI's market dominance. US tech executives confided to me that there was no way a US company could replicate DJI's success. The industry's understanding was that DJI scaled due to massive subsidies from the Chinese government, affording it major cost advantages. Reporting from *The Washington Post* confirmed these claims, despite DJI's assertions that it had not accepted any funding from the Chinese government. Reporters reviewed company documents and other sensitive information showing that four investment entities "owned or administered by Beijing" invested in DJI, including a government asset manager who played a key role in "promoting partnerships between private enterprises and the Chinese military." Says Charles Rollet, an analyst at the advisory firm IVPM, "It directly contradicts what [DJI spokespeople] have been touting to allay concerns around the Chinese government. DJI has been saying they have no Chinese government investments. This evidence directly contradicts that."[27]

Not only did state subsidies play an important role in propping up the company, but DJI also benefited from being part of China's competitive drone ecosystem, which spawned an array of service providers producing advance components at cut-rate prices. Few other countries could compete. When a Taiwanese drone manufacturer tried to replace its Chinese-made engines, it searched far and wide for alternatives, even going to Australia, where it found that an equivalent engine "would cost 20 times more." The company eventually inked a deal with a local supplier, but only after convincing the bidder to lower its prices "as an act of virtue" to support Taiwan's defense.[28]

Like other Chinese firms and products (such as TikTok), geopolitics have caught up to DJI. In December 2025, the Federal Communications Commission blacklisted all new foreign-made drones and components from the US market over concerns that they posed "an unacceptable risk" to national security. While the directive didn't directly name DJI, given the company's 80 percent share of America's consumer drone market, the implication was clear: DJI drones were no longer welcome in the United States. (For now, the FCC's rule doesn't impact DJI's unmanned aircraft already in use; it only applies to new drone sales.) In short, today's geopolitics has made DJI's position increasingly untenable for America's interests.[29]

Experts remain divided about how the United States and China stack up on science and technology. In certain sectors, American research continues to lead; in other areas, China appears to be supplanting the United States. What is clear is that the days of China as an "imitator" are long gone.[30] Some analysts argue that the West retains a "creative advantage" relative to China, but scholars like Georgetown's William Hannas and Huey-Meei Chang contend that this mischaracterizes China's advancements. They argue that focusing on how China achieved its innovation, whether through forced tech transfers or collaboration, "is largely irrelevant." Instead, what matters is the degree to which China's scientists have achieved technological breakthroughs, and on that score "China's progress is clear."[31]

Looking at the trend lines, their logic is sensible. China's ambitions are large, and it has built up an enviable R&D and manufacturing capacity. Despite its current economic struggles, it has vast resources for Xi to turn loose in the growing tech war against the United States.

Xi projects confidence that China is on the right track. Ahead of the CCP's approval of the Fourteenth Five-Year Plan, which outlines development priorities for the country, the party released a 2018 speech by Xi that served as a "powerful reminder" about China's modernization push. In his remarks, Xi declared, "We are closer than any time in history to the objective of the great rejuvenation of the Chinese nation, and we need more than any time in history to build [China into] a world S&T superpower!"[32] Those are not the words of a leader who believes China's innovation is faltering.

■ ■ ■

As Xi gained confidence and saw China's influence grow—and as America's tech war imposed increasing costs on his country—a crucial question was how closely Beijing would align with Moscow. Xi had forged a close relationship with Putin. At last count, they had met over sixty times and developed a warm personal connection, described by some experts as a "strategic bromance." Reportedly, they were fond of giving each other birthday gifts. For Putin's sixty-first birthday, in 2013, Xi handed over a cake to Putin, which they celebrated "with sausages and vodka" during a summit in Bali. In 2019, Putin reciprocated by presenting Xi with a birthday cake adorned with the words "good fortune double six" in Mandarin to commemorate Xi's sixty-sixth birthday.[33]

In public, they were effusive about their friendship. After Xi arrived in Moscow in March 2023—following the issuance of an arrest warrant against Putin by the International Criminal Court—both men declared that relations were the strongest they had ever been. Xi "reminisced" with Putin about his first trip abroad, to Moscow in 2013. "You reminded me of that, and to this day these pictures are well preserved in my heart," he said to Putin. Later, they stood together in the doorway of the Kremlin and clasped hands in solidarity. Xi told Putin, "Right now there are changes—the likes of which we haven't seen for a hundred years—and we are the ones driving these changes together." Putin answered, "I agree." When they said goodbye, Xi embraced Putin twice in front of the cameras—something the Chinese leader rarely does. "The message of closeness was intentional and clear," writes Carnegie analyst Alexander Gabuev.[34]

But personal rapport was one thing—would China and Russia take real steps to cement their partnership and form a rival bloc against the United States?

18

An Axis of Disruptors?

The Ukraine war represented a signal moment for Putin and Xi. China's leader faced a difficult choice. Despite simmering tensions with the United States, the two countries remained economically interconnected; China had a lot to lose from siding with Russia, and distancing himself from Putin might have been a good way for Xi to hedge his bets. Yet he had many reasons to support the Kremlin as well. For one, Xi saw Putin as a comrade in arms, ideologically aligned in a geopolitical struggle against the West; an emerging Sino-Russian alliance could be a good tool for undercutting the US-dominated international order. He also recognized the downward spiral in US-China relations. With America's tech war in full force and diplomatic prospects dimming, Putin's military gambit presented an unexpected opportunity to draw Russia deeper into China's orbit. Xi opted to strengthen ties, bringing China and Russia closer together.

This growing alignment—alongside support from Iran and North Korea—led some commentators to herald the formation of an "axis of upheaval" or an "axis of disruptors." Scholars described these nation-states as sharing common interests, matching rhetoric, and coordinated military and diplomatic activities. Analysts from the Special Competitive Studies Project also indicated that China, Russia, Iran, and North Korea were "united in their disdain" for the US-led rules-based order and perceived an opportunity to upset the status quo through disruptive measures that would push the global system "to a breaking point."[1]

Indeed, a central aim for Beijing was to create a world "safe for the CCP" that would secure the country's proper place at the heart of global influence. Russia had long viewed itself as a great power that deserved a major voice in international affairs as well. Both states shared the objective of upending American dominance and moving the unipolar global system toward multipolarity. They promoted illiberal norms to legitimize their own systems and discredit the West. "Russia and China don't trust each other very much, but it doesn't prevent them from having a common agenda when it comes to countering the US," French foreign ministry advisers told me. "Both are obsessed with the US dominance, or what they think is US dominance, over the world. So their convergence might not be weakened by the fact that they don't really trust each other."[2]

Still, analysts were careful not to overstate China's alignment with Russia. Alliances come in many different forms. Some are formalized groupings that include treaty obligations to defend members, such as with NATO. Other alignments are more informally constituted and feature little coordination, akin to "rough agreements to assail the existing order from all sides." This latter model hewed closely to Beijing's preference of pursuing strategic partnerships over alliances. By design, its strategic partnerships are organized around a general security purpose (rather than a specific task), are informal in nature, bring low commitment costs, and deliberately eschew formal obligations and rigid structures. Consistent with this approach, China and Russia deepened their collaboration—from military procurement to microchip imports. Their leaders met regularly, and economic trade between the two countries blossomed. They brought "complementary" economic strengths (Russia's natural resources matched well with China's manufacturing might) and shared geopolitical priorities. But they lacked a common ideological basis for their relationship, nor were they interested in formalization, such as entering into a military pact. Russia experts believed such a step would be "premature, unnecessary, and possibly even counterproductive." China analysts agreed, noting that an alliance would limit each country's strategic flexibility while putting Beijing at a competitive disadvantage with the United States.[3]

In fact, China and Russia relied on very different tactics for achieving their goals. China prioritized leveraging economic influence to reach its desired political outcomes. It wasn't interested in "projecting military power globally," scholars like Oriana Skylar Mastro argued. Instead, China chose to be the "internal security partner of choice" for many countries—from Qatar to the Central African Republic. Russia, on the other hand, willingly employed military coercion, political meddling, and information propaganda to advance its ambitions.[4]

Even if their approaches diverged, China and Russia's cooperation had a real effect in undermining US sanctions. By 2023, goods manufactured in China made up more than 63 percent of Russia's battlefield imports. Items ranged from integrated circuits and wireless communication routers to satellite-based radio navigation systems with a value of over $7.9 billion. "China's provision of dual use components and material to Russia's defense industry is one of several factors that tilted the momentum on the battlefield in Ukraine in Moscow's favor," said Avril Haines, Biden's director of national intelligence.[5]

As a result, America's technological containment agenda, at least when it came to curtailing Russia's war, faced big hurdles. China benefited enormously from sustaining Russia's economy, and Putin became dependent on Beijing to keep his war machine humming. China also supplied the bulk of consumer goods to Russia, creating a dedicated market for its sellers. Reportedly, shipping containers were "piling up" in Russia as Chinese exports flowed into the country. Trade between the two countries surged 65 percent, increasing from $145 billion in 2021 to $240 billion in 2023—with the bulk of the gains captured by Chinese suppliers. Moscow became "not just a junior partner but also a captive of its relationship with Beijing, which the latter can exploit."[6]

Recalibrating the "No Limits" Partnership?

Despite Xi's decision to lean into his relationship with Putin, their partnership still faced limits. While China provided substantial support to Putin's war machine and received valuable Russian submarine and missile technology in exchange, it also sought to tread carefully to avoid getting hit with additional US sanctions. As the war dragged on, there were small signs that China was recalibrating elements of its policy. Back in February 2022, days before Russia sent thousands of soldiers into Ukraine, Xi and Putin famously declared a "no limits" partnership between their countries. But two years later, the boundaries of their partnership were starting to show. Putin's trip to Beijing in May 2024 was decidedly less warm than prior visits. While he received the standard "red carpet welcome" and military band greeting after his motorcade drove through Tiananmen Square, Xi conspicuously refrained from making a "no limits" declaration. Instead, he spoke about an "ever-lasting friendship" between the two nation-states. Meanwhile, a BBC analysis found that mentions in Chinese state media about the "no limits" partnership had almost completely disappeared. As Tong Zhao from the Carnegie Endowment for International Peace says, the rhetorical shift reflects China's awareness of the "reputational costs of appearing to offer unconditional support to Russia and its continuous refining of its strategies to enhance its perceived legitimacy on the global stage."[7]

Beyond modifying its language, Beijing continued to refrain from supplying lethal military equipment to Moscow—and there was never a suggestion that Chinese troops would be sent to fight in Ukraine, forcing Putin to turn to countries like North Korea for manpower. Three of China's four largest government-owned banks stopped payment services to sanctioned Russian financial institutions following a December 2023 executive order from Biden triggering sanctions against foreign banks that transacted with Russia's military industry. China's direct export flows to Russia also showed signs of slowing down. Monthly exports of dual-use products from China hit a peak of $600 million per month in December 2023 and subsequently fell to about $300 million per month.[8]

These developments did not mean that Beijing was truly rethinking its strategic partnership with Moscow. But they indicated that Chinese leaders were well aware of the boundaries of the relationship. "China has been careful to not become closely associated with the war, to avoid being targeted by US sanctions, and to maintain a façade of impartiality," observed Eugene Rumer, an analyst with the Carnegie Endowment for International Peace.[9]

China's overall backing of Moscow remained robust. The benefits of mutual cooperation continued to exceed the costs. Not only were Chinese companies obtaining sizable revenues from trade with Moscow, but China was acquiring advanced technology to fill crucial military modernization gaps, equipment Moscow had traditionally been leery of offering. This included access to advanced submarine designs, stealth aeronautical capabilities, and new missile prototypes.[10]

To the extent that China's support of Russia didn't compromise its stability priorities and economic ties with Europe and the United States, it would continue to aid its northern neighbor. But China had to walk a delicate tightrope. If it leaned too heavily in support of Russia, it risked serious repercussions from Washington. By the same token, if Beijing withheld critical support from Moscow, this could antagonize Putin and derail the partnership. And so, the two countries implemented a detailed plan to help Russia evade America's tech restrictions while affording China plausible deniability regarding its assistance. Beijing would open up transshipment routes and instruct its companies to smuggle in key electronics and critical weapons components. Together, they would rally support from liked-minded authoritarian countries and encourage a larger group of nonaligned states (such as India, Brazil, and South Africa) to strengthen technological cooperation and set up alternative supply chains. With assistance from China, Russia would implement a military reconstitution program to restore losses from the initial invasion and expand its future defense industrial output.

As these countermeasures took shape, would they be enough to overcome the West's stranglehold?

19

"We Are Strong People and Can Cope with Any Challenge"

Just as dawn was breaking on November 21, 2024, Moscow conducted a fearsome strike against the central Ukrainian city of Dnipro, targeting a facility that had served in Soviet times as a top-secret rocket manufacturing plant. President Joe Biden had recently loosened restrictions on Ukraine's use of American-made ATACMS ballistic missiles against targets in Russia. Russian President Vladimir Putin was enraged and wanted to send a clear message to NATO. For the Dnipro bombing run, Moscow deployed several weapons: Kh-101 cruise missiles fired from enormous Russian bombers, Kinzhal hypersonic ballistic missiles discharged from sleek Russian jets, and a new missile that the West hadn't seen before. Video footage showed six warheads descending in a vertical direction toward the city, each of which released a cluster of smaller projectiles.

What was this new weapon?

At first, Ukrainians speculated that it was an intercontinental ballistic missile. These missiles are exclusively tied to nuclear weapons "and have never been fired in anger." Other commentators guessed that Russia had fired an RS-26, a missile spanning the boundary between intermediate-range weapons and longer-range intercontinental ones.[1]

Later in the day, Putin broadcast from his office in the Kremlin. He declared that the weapon was a new intermediate-range hypersonic missile called the Oreshnik (which means "hazelnut tree" in Russian), capable of traveling over ten times the speed of sound. In an icy tone, he declared that "modern air defense systems that exist in the world and anti-missile defenses created by the Americans in Europe can't intercept such missiles." For good measure, the Kremlin's propaganda outlet RT aired a video graphic showing the missile's flight time to key European capitals: twenty minutes to London and Paris, fifteen minutes to Berlin, twelve minutes to Warsaw. Analysis later showed that the projectile flew at eleven times the speed of sound and took just fifteen minutes to reach Dnipro from its launch site.[2]

Putin's firing of the Oreshnik was a significant event. The missile is capable of carrying nuclear weapons and can release multiple warheads at exceptionally high speeds. It has a range "way beyond anything seen in this conflict so far," said Matthew Savill, an analyst at the Royal United Services Institute. "You're dropping multiple independently targetable warheads, MIRVs, at extremely high speed—even Patriot will struggle to intercept those."[3]

Putin was sending an unmistakable warning to Kyiv and its supporters in Washington and Brussels: this time the strike was nonnuclear, but the next Oreshnik could be carrying a nuclear warhead. From the outset, Putin was clear that Russia would not back down from its invasion despite massive Western sanctions and military assistance for Ukraine. Four months into the war, in a speech in St. Petersburg, Putin accused the West of displaying a "colonial arrogance" and declared, "We are strong people and can cope with any challenge. Like our ancestors, we will solve any problem, the entire thousand-year history of our country speaks of this."[4]

He was not bluffing. In the ensuing months, the Kremlin pursued a multipronged strategy to blunt the impact of US-led sanctions and seek new partners who could make up the difference. These efforts bore fruit. Despite American attempts to "disrupt and degrade Russia's military industrial base," Moscow's wartime capacity remained high.[5] By the end of 2022, Russia was not only importing a greater volume of critical battlefield components than in the prewar era, but it had ramped up its domestic manufacturing as well. Concretely, this translated to more missiles and drones for the battlefield.

Beijing provided Russia with billions of dollars' worth of dual-use items after the start of the Ukraine invasion. US officials admitted there was little they could do in response. "The estimates are that 80 percent to 90 percent of the microelectronics that Russia is getting are coming from China," one Commerce Department official told me. And while some chips were manufactured in China, many came from outside sources, including American firms. Unfortunately, low-end chips remained largely uncontrolled in most places in the world. "They don't require a license to go to China. And while it's illegal for them to go from China to Russia, it's only illegal under our law. It's not illegal under Chinese law. How do we get at that?" a former Commerce Department official put it to me.[6]

Chinese firms employed creative techniques to pull this off. Take computer numerical control (CNC) machines, precise tools that are critical for weapons production. Analysts assessed that two thirds of Russia's heavy industrial equipment, including CNC machines, were "worn out and on the brink of failure." Russia lacked the capacity to build those systems itself, making it reliant on foreign suppliers. In this regard, China played a critical role. In some instances, CNC machines produced in the EU were imported by Chinese companies and subsequently reexported to Russia. Another option was for Russia to import CNC machines directly from Western companies' factories in China (twenty-two of the thirty largest Western CNC manufacturers have plants in China). While European and American companies are banned from shipping to Russia, these same regulations do not prevent third-country subsidiaries from selling to Russian traders. For example, the German industrial company Heller has invested millions in building a CNC plant in Changzhou, China, for the local market. Its website trumpets that its Chinese facility makes "more than 100 machines per year on site." While Heller's German facilities can't ship to Russia, its Changzhou offices can sell tools directly to Russian suppliers or provide machines to Chinese entities who can then ship the items north.[7]

All told, 90 percent of the machine tools imported by Russia came from China. Seventy percent of the Russian companies receiving these tools were tied to the military. Despite this, 90 percent of Chinese companies involved in these transactions remained unsanctioned.[8]

It is no surprise that experts identified China as the biggest problem facing export controls. "No amount of investigation or enforcement by the allied countries is going to cut off the flow of parts and components from China into Russia now," said Kevin Wolf. "It doesn't mean that the US hasn't done a good job. It just means it's a mind-blowingly more difficult exercise than I think people who got started with Russian controls understood at the time."[9]

Russia Gets Creative with Sanctions Evasion

Morocco's state-run Tanger Med port sits on the Strait of Gibraltar, an hour's drive from the city of Tangier. Founded in 2009, it is now Africa's leading port and one of the largest in the world (ranking twenty-second globally for volume of cargo). For Russian officials, the use of Tanger Med offered several benefits. The port handled a large volume of shipments, making it easy to hide clandestine transactions. Morocco is also a classic hedging state, wary of aligning too closely with any one country. While it had taken steps to draw closer to the United States (it became a major non-NATO ally in 2004 and regularly participates in US military exercises), it also enjoyed strong economic ties with Russia; Morocco was especially dependent on Russian agricultural inputs for its farming sector. Consequently, it steered a "middle course" toward Russia, abstaining from supporting UN resolutions in 2022 and 2023 that condemned Putin's invasion.[10]

In April 2022, Russian companies were desperate to find workarounds to Western sanctions that had cratered their ability to obtain crucial components. One such company was ProSoft, a Moscow-headquartered electronics supplier that sold biometric surveillance technology and equipment for heavy industry and infrastructure. In April 2022, ProSoft executives emailed Russian trade officials in Morocco for help: "Just a week ago, it was not difficult to ship them [tech components] to us from American and European suppliers. We run the risk of starting to reduce production." Attached to the email was a spreadsheet showing "hundreds of American, European and Japanese microchips and sensors" needed by ProSoft.[11]

Russian officials immediately got to work, responding, "We're in constant contact with the general director" of Tanger Med, and that "in the event of ships entering under the Russian flag, there will be no problems with maintenance." Over the next few months, they established twenty "operators" based in the port who could load sensitive cargo from Taiwan, China, and Europe onto "feeder vessels" bound for Russia. By November 2022, Russian authorities crowed that their efforts had transformed Tanger Med into an electronics transshipment hub: "Given the long-term nature of the cooperation with thc Moroccan partner, the volume of supplies could amount to about $10 million per year." When port officials at Tanger Med were asked why they allowed Russian ships to dock at their port and transfer prohibited items, they denied responsibility, maintaining that they simply leased space and had "no information nor responsibility" about specific boats moving through their facility.[12]

Tanger Med was far from the only port used by Russian companies to transport restricted items. The same Russian officials noted that ports in Türkiye, an official NATO member, had agreed to accept payment in rubles for services, saying that "the sanctions restrictions have been successfully overcome." Evading export controls became a top priority for the Kremlin. Russian intelligence services fanned out across the world searching for new trading routes. Agents dispersed to China, India, Singapore, the UAE, and other places to identify options for securing critical components for the war.[13]

One region that proved especially fruitful was the former Soviet republics in the Caucasus and Central Asia, which were dubbed the "Eurasian roundabout." They were instrumental in shipping dual-use items such as lasers, measuring instruments, and tools for checking voltage and power. Between 2021 and 2022 alone, US and EU exports to Armenia, Georgia, Kyrgyzstan, Uzbekistan, and Kazakhstan increased from $14.6 billion to $24.3 billion. In turn, those countries grew their exports to Russia to $15 billion, a nearly 50 percent jump. The arrangement left little to the imagination. Russian supplier Imex-Expert openly advertised how it specialized in importing sanctioned goods "from Europe, America to Russia through Kazakhstan." On its website, it claimed solutions for "circumventing sanctions."[14] While the overall figures were low compared to Russia's booming trade with China, what made the Eurasian roundabout so vital was that those countries could get their hands on scarce Western components that Russia couldn't otherwise acquire.

US officials tried new methods to crack down on these practices, yet Russian companies continued to find loopholes. In June 2024, journalists revealed that Kalashnikov Concern, the same company manufacturing the original AK-47 assault rifle, had "successfully acquired" thirty-two CNC machines for weapons production at a secret location in Russia. While the firm did not disclose where it obtained the machines, its website showed it selling "Swiss-style" CNC devices to Russian clients. Said Kalashnikov's managing director, "The demand for our products is growing every year. Our products have proven to be highly effective in the SMO [special military operations] zone."[15]

A major challenge was the globalized nature of distributed supply chains. If curbing Russian imports was simply a matter of stopping American companies from selling to Russian firms (or Russian shell companies), that would be relatively straightforward. But supply chains rely on distributed networks. As a senior Commerce Department official told me, "The US manufacturer will sell to a distributor, usually first in the United States, sometimes overseas, and then the US distributor sells to a distributor overseas who may send it to someone else." It doesn't stop there. Often, products ended up in China, where "we can lose visibility, and the companies lose visibility, and then it eventually gets to Russia."[16]

The bottom line was that Moscow continued to play a very effective cat-and-mouse game with US export controls. Acknowledged a former Biden official, "We fully expected the Russians to be able to adapt to our sanctions and find ways of circumvention. That's why we have a continual effort where we are rolling out sanctions packages every few months to adapt." Unfortunately, even updated regulations appeared to have a short shelf life.

Russia's strategy was so good that Chinese officials reportedly set up an interagency study group to learn from Moscow's efforts in case of future conflicts.[17]

Adaptations and Limitations of Russia's Strategy

While Russia achieved some tactical successes, Putin was also playing with a weak hand. Relying on circumvention mechanisms alone was not a sustainable strategy. They brought additional costs, such as lower-quality components, delays in supply, and inferior substitutes. "Third-party imports are only a stopgap measure," noted a report from the UK think tank Chatham House. "The Russian military industry will ultimately have to address its own challenges in domestic production if it wants to survive."[18]

Without question, Russia's armed forces experienced devastating losses. As of February 2026, *The Economist* estimated that Russia's military had experienced between 1.1 million and 1.4 million casualties, of whom 230,000–430,000 were killed. Ukrainian forces had destroyed approximately 11,783 tanks, 24,218 armored vehicles, nearly 40,000 artillery and missile systems, 435 aircraft, and 349 helicopters (all of these numbers represented lower bounds). Reportedly, Putin's supplies got so low that he even recommissioned old Soviet tanks that were being used as props by Mosfilm, Russia's premier film studio. "I was told they were needed, so I reached out to the Defense Ministry," said studio head Karen Shakhnazarov.[19]

Moscow thus faced the dual challenge of replacing its destroyed assets while putting additional equipment into service. Since the start of the war, Russia had ramped up its manufacturing of military items and their supporting components. In 2023, military expenditures soared to $109 billion, equivalent to 5.9 percent of its GDP. The following year, its military budget increased another 60 percent, raising total military spending to one third of its national budget (7.5 percent of its GDP). Satellite imagery suggested that Russian firms were constructing new industrial facilities next to aviation plants, drone factories, and artillery shell plants, "indicating a significant expansion of Russia's military production capabilities.[20]

Russia's defense industry contained nearly six thousand companies, many of which "rarely turned a profit" before the war. While inefficient, they offered spare capacity and flexible operations that the Russian state took advantage of when it needed to boost defense production. Richard Connolly, an expert at London-based think tank the Royal United Services Institute, has termed this the "Kalashnikov economy," one that is "quite unsophisticated but durable, built for large-scale use and for use in conflicts."

As he explained it, the Russians had subsidized large, plodding defense firms for years, thinking that one day they would need to scale them up. What appeared economically inefficient pre-2022 took on new relevance once hostilities commenced. This approach differed from Western arms manufacturers, who ran much leaner operations and were oriented toward maximizing profits for their shareholders. While the latter model offered higher rates of return and lower up-front costs, Western firms struggled to scale up their production in response to surging demand.[21]

As the war intensified, Russian authorities placed fourteen ammunition and special chemical plants under the control of state conglomerate Rostec to streamline manufacturing. They also went on a hiring spree. The Kazan gunpowder plant, one of Russia's largest facilities, added more than five hundred workers in December 2023 alone. Managers at the plant increased average monthly wages more than threefold, from twenty-five thousand rubles ($250) to ninety thousand rubles ($900). Ironically, Russian factories were now competing with the military for new recruits; the latter offered salaries exceeding two hundred thousand rubles ($2,400) a month.[22]

Moscow's efforts led to tangible gains. For instance, one of its most widely used long-range weapons was a twenty-four-foot-long air-launched cruise missile called the KH-101. It first entered into service in 2012 and was designed to fly at a low altitude, underneath infrared and radar systems. It was guided by the GLONASS satellite navigation system (the Russian equivalent of GPS), which offered a high degree of accuracy. Prior to the invasion, Russia aimed to manufacture 350 KH-101s per year, but its actual efforts fell short of this goal; in 2021, its factories only made 56 missiles. Yet, by 2023, production of this weapon had ramped up to 420 missiles per year. Russia similarly bolstered production of its Iskander ballistic missiles. At the start of 2023, Russia had only 50 Iskanders left in stock; its prewar production rate was approximately six missiles per month. One year later, Russia had tripled its production of Iskanders and increased its stock to 180 missiles, despite their frequent use.[23]

Russian-Iranian drone production told a similar story. One of the most popular attack drones used by Russian forces was Iran's Shahed-136 model, nicknamed "lawn mowers" or "mopeds" because of the constant buzzing sound they made as they approached their targets. They were inexpensive to produce; single-use units were priced at twenty thousand dollars and brought significant destructive potential, carrying 118 pounds of explosives. (In contrast, Russia's Kalibr cruise missiles, which were also heavily used in the war, cost about $1 million each.) While they were cheap, Shaheds could do a lot. They were adept at avoiding radar detection and could operate at a range of up to fifteen hundred miles.[24]

Prewar, Iran produced about forty Shahed-136 drones per month. Once they were introduced onto the Ukrainian battlefield in the fall of 2022, Russia immediately placed a large order for seventeen hundred Shaheds and related drone units. Russia recognized that it had lagged in its own drone development and needed to ramp up production in a hurry. So Moscow decided to double down on Iran's drone designs. Authorities entered into an agreement to build a joint production facility in the Alabuga special economic zone in the Tartarstan region with the goal of building six thousand Shahed drones by the summer of 2025 (which they rechristened the "Geran-2"). The cost was high; Russia reportedly paid over $1.8 billion to ink the deal, at least part of which it provided in gold ingots.[25]

While work on the factory was initially delayed, the plant started rolling out thousands of drones by 2025—by one estimate, the plant upped its production to between 18,540 to 24,075 drones annually. The Geran drones measure about eleven feet long, weigh over four hundred pounds, and resemble an enormous black lawn dart, an unusual profile that came from a "delta-wing configuration built around a central fuselage that extends forward in a rounded nose." The nose contains a high-explosive warhead enhanced with penetrative tungsten-alloy balls and incendiary metallic fill, allowing it to easily pass through soft armor and human flesh. Russia soon developed a newer model, the Geran-3, which featured a jet engine and higher top speeds, facilitating quick targeting and better evasion of antiaircraft defenses.[26]

But these gains masked deeper problems. Moscow remained in a coping posture: Western sanctions had set back its industrial capacity, and it was relying on "good enough" systems to keep the war effort going. Civilian factories with little experience making military goods were pressed into service. Reportedly, bread plants began producing drones, and oil and gas factories were converted to make artillery gun tubes.[27]

These unusual measures kept military production afloat but also forced tough trade-offs. In particular, tool scarcity and labor shortages caused Russian companies to prioritize quantity over quality. As a result, Russian soldiers were consigned to using equipment that contained dual-use components less suitable for warfare or outdated material drawn from Soviet-era stocks. Sometimes these stopgap efforts led to dramatic failures.

In 2024, Ukrainian air defense teams started observing unusual patterns from the one-way drone attacks Russia was launching against its cities. Moscow's Geran-2 drones would swoop toward their targets, but then something would go awry. Sometimes they would spin out of control when making sharp turns. Other times, they would veer in the wrong direction before crashing harmlessly. Still others would default to cruising at a flat speed, making them "easy prey for air defense batteries." Still, the faulty drones kept coming, as many as a hundred per day. Eventually, NATO and Ukrainian analysts pieced together the story. The Geran-2s were failing due to inferior components. In particular, the Russians had recently switched to using a Chinese motor due to sanctions. It turned out that the steering motor "not much bigger than a cigarette pack" was not up to spec, causing hundreds of Russian drones to malfunction. Nonetheless, Moscow's bet was that inferior equipment could be overcome by sheer volume. "There are so many of their drones that you can't even lift your head," says Ihor Romanenko, former chief of the general staff of the Ukrainian army. "Despite the deterioration in the quality of Russian weapons, it doesn't have a significant impact. They achieve results through sheer mass deployment." By the end of 2024, Russia was launching 227 drones per week against Ukraine. And still, the numbers kept rising. In the summer of 2025, Russian drone launches soared to over 1,000 each week. On September 7, Russia dispatched 805 exploding drones and decoys against Ukraine in a single day.[28]

One of the units responsible for Russia's escalating drone offensive was a secretive group known as "Rubicon." It was led by Colonel Sergei Budnikov, a thirty-seven-year-old ex-artillery and marine infantry officer. The unit developed new drone technologies but also tested out tactics and directly trained drone operators. Rubicon was divided into at least seven subgroups, each consisting of 130–150 personnel and specializing in a specific aspect of drone warfare—some focused on FPVs or reconnaissance operations, others concentrated on heavy-lift drones carrying multiple explosives. Said Maria Berlinska, head of the Ukrainian Aerial Reconnaissance Support Center, Rubicon had become "Russia's best technological unit."[29]

Despite the growing success of its drone program, Russia's gains came at a cost. Because the Ukraine conflict didn't require advanced high-end weaponry—it was being fought with large volumes of cheap munitions—Moscow prioritized making low-cost shells and drones rather than investing in innovation and future technologies. As a result, Russia continued to fall short of achieving basic import substitution in high-tech fields like microelectronics, robotics, biochemistry, and biopharmaceuticals. In the metalworking sector, Russian factories imported 90 percent of machine tools and their components. When it came to aviation, Russian officials delayed replacing Western-made aircraft with domestic models, reportedly because of testing failures and lack of suitable products. Consequently, Russia's military industry appeared headed "into a period of decline and stagnating innovation" affecting the ability of its forces to recapitalize and modernize their industrial production and hardware.[30]

Rather than reward and encourage its scientific community, Putin's enforcers took the opposite tack. For Russian scientists working on sensitive research, particularly hypersonic weapons, life became dangerous.

Hypersonic weapons are a new frontier in missile technology. They fly at an exceedingly high velocity, five times the speed of sound, or nearly twice as fast as the SR-71 Blackbird's top recorded speed of Mach 3.5. This speed presumably makes them harder for conventional air defenses to intercept. Hypersonics have a range of more than twelve hundred miles, and Russian models such as the Kinzhal "dagger" missiles are designed to carry nuclear warheads. While some experts are skeptical about the value of hypersonics and their supposed invincibility to conventional air defenses (Ukrainian defenses have successfully shot down numerous Kinzhals targeting Kyiv), they remain a top priority for Russia's defense industry.[31]

In the last six years, at least a dozen researchers were arrested on suspicions of handing over information and data to Moscow's adversaries. In 2024 alone there were several prosecutions. Anatoly Maslov, an expert in high-speed aerodynamics, received a fourteen-year prison sentence on treason charges; prosecutors accused him of sharing secret information about Russia's hypersonics program with Germany. Alexander Shiplyuk, a leading specialist in "high-speed aerogasdynamics," was also imprisoned, sentenced to fifteen years for allegedly sharing classified information at a scientific workshop in China in 2017. Alexander Kuranov, a renowned physicist specializing in hypersonic propulsion, was convicted of treason as well, for "passing over scientific research data to foreign special services." He was lucky and only received a seven-year sentence following two years in custody.[32]

Russian scientists feared that anyone working on high-priority technologies could be detained for years before going on trial for treason. Not only was paranoia spreading in scientific circles, but the crackdown disincentivized engineers from pursuing research in critical fields Russia needed to master in order to gain an advantage over its rivals. Putin's repression harkened back to dark periods of Russian history. In the 1930s, Joseph Stalin instituted a reign of terror against the scientific community that resulted in the false prosecution of scores of eminent individuals. Among those wrongly accused were Sergei Korolev, who later became a leader of the Soviet Union's space program, as well as Andrei Tupolev, recognized as one of the world's leading aircraft designers. As Moscow's paranoia and isolation increased, Russia's security agencies became even more relentless in seeking to control all information, no matter the cost.[33]

Russia's industrial shortfalls have caused it to rely even more heavily on China to make up the difference. This dynamic could quickly veer into overreliance, creating long-term dependencies and rendering Russia politically reliant on Beijing. While Putin's outlook is well aligned with Xi's at the moment, there is no guarantee that the two countries will remain so in the future, at which point Russia's hands will be tied. Already, there are signs that China's technological progress has eclipsed Russia's, even in areas of traditional dominance such as fighter jets.

In November 2024, at China's Zhuhai air show, Putin hoped to make a splash. Russia planned to debut its prized Sukhoi Su-57 stealth fighter (its NATO reporting name is "Felon"), the most advanced aircraft in its military. The plane's design was closely guarded—no one overseas had seen it fly, and Russia wanted to make a big impression. But reactions in China were scathing. On social media, commentators mocked the jet's "clunky design and supposedly inferior stealth capabilities." People noted that Chinese planes no longer used Su-57's Russian-designed engine, having swapped it out for China's locally produced WS-19 model. Aviation enthusiasts pointed out multiple design flaws, such as a "huge number of bolts" holding fuselage panels together. One Chinese military news site posted a video of the plane with the caption: "Lots of screws: Closer look at Russia's 5th generation Su-57 fighter jet at China's Zhuhai Airshow."[34]

In contrast, Zhuhai represented a show of strength for China. A decade after China surprised the world by debuting the Shenyang FC-31 stealth fighter—stolen from American blueprints—China unveiled yet another stealth aircraft, the J-35A. The plane is smaller and more agile than older models in its arsenal; it is designed to skirt radar detection and strike enemy targets at supersonic speeds (one Chinese commentator boasted that it would be "a piece of cake" to overpower Lockheed Martin's F-35 in combat). The J-35A's deployment would make China the only nation in the world—aside from the United States—with the ability to deploy multiple stealth fighters in war. In addition to the J-35A, Chinese aerospace vendors showcased an assortment of air- and sea-based drones, such as a massive ten-ton unmanned aerial vehicle capable of deploying swarms of smaller drones, as well as an autonomous ship with a range of four thousand nautical miles that Chinese media dubbed the "killer whale." Wrote Christian Shepherd and Lyric Li for *The Washington Post*, "the appearance of the Su-57 at Zhuhai seemed to underscore China's arrival as the senior partner, reversing decades of Chinese reliance on Russian military hardware."[35]

As the war drags on, Russia confronts increasingly difficult choices. While America's tech restrictions didn't stop Moscow's invasion, Putin has paid a heavy price. Sanctions have hindered the Kremlin's ability to modernize its military and created shortages of critical components. Russia has raised its military expenditures to unsustainable levels by running huge fiscal deficits and chronically underinvesting in education, health care, scientific research, and social services. Putin doesn't have unlimited resources. His decision to pour even more money into the defense sector will strain economic imbalances and jeopardize future prospects. Already, cracks are beginning to show. During a routine space launch in November 2025, a Soyuz rocket malfunctioned, destroying Russia's main launchpad. The consequence? Moscow "lost its ability to launch humans into space, something that has not happened since 1961," according to Russian space experts. Yet short-termism is Moscow's guiding philosophy. Writes Carnegie scholar Alexandra Prokopenko, the Kremlin's mantra is "We'll cross that bridge when we come to it."[36]

If there is a core lesson to be drawn from America's confrontation with Russia, it is that controlling the spread of technology to other great powers is hard—if not impossible. There are defined limits to how much export controls can change the behavior of other states, let alone alter the battlefield. Moscow has clearly learned from its failures and adapted its approach to the war. Over time, it has improved its production capacity and upgraded its equipment. But we should also not underestimate the future costs accrued by Russia. In the race for innovation dominance and technological leadership, the Ukraine war has undoubtedly set back Russia's long-term prospects.

■ ■ ■

So far, we have mostly examined the global tech war through the lens of great powers, looking at steps taken by the United States and its allies to cut off advanced technologies to China and Russia and how those countries have responded in kind. But the contest for technological dominance involves more than large nation-states. A host of new actors have come onto the scene. Regional powers such as Türkiye and the United Arab Emirates are acting increasingly independently, cooperating with the United States or Europe when it suits their interests, while cultivating relationships with other actors. As power and geopolitics decentralizes, how will these dynamics play out?

PART V

Innovators and Disruptors: Rising Powers

20

Military Innovations Spread Across the Globe

After Yemen fell apart from civil war in 2015, the Houthi insurgency emerged from the chaos and managed to gain the upper hand over the northern region of the country by 2017. Four years later, the group had consolidated its strength and become the country's dominant political-military force, able to hold its own against a military coalition backed by neighboring Saudi Arabia. While the Houthis had staged spectacular strikes against targets in Saudi Arabia and the UAE in the past—including a drone attack against Saudi Aramco facilities that wiped out more than half of its oil output (the equivalent of 5 per cent of the daily global supply)—the group now had US military vessels in its sights.[1]

In January 2024, the Houthis launched a major attack against US Navy ships, sending eighteen drones, cruise missiles, and ballistic missiles against four American destroyers patrolling sea lanes off of Yemen's coast. In an attack that lasted over a dozen hours, US vessels faced "one of the largest maritime battles" since World War II. Despite plenty of close calls from Houthi missiles hurtling toward US destroyers at five times the speed of sound, the American fleet emerged unscathed. But defending against the attacks was costly. All told, in 2024 alone, the navy spent about $1 billion on munitions to fend off militant assaults, conducting over 450 strikes and intercepting more than 200 Houthi drones and missiles. Said one analyst, "Their supply of weapons from Iran is cheap and highly sustainable. . . . We are playing whack-a-mole, and they are playing a long game." Houthi remote-controlled sea drones now threatened the 12 percent of global shipping passing through the Red Sea. And they continued to pull off dramatic strikes, such as their targeting of the Greek-owned oil supertanker *Sounion* in the summer of 2024, which left the ship—which was carrying four times as much oil as the Exxon *Valdez*—abandoned and in flames. Warned William LaPlante, the Pentagon's chief weapons buyer under Biden, "I'm an engineer and a physicist, and I've been around missiles my whole career.

What I've seen of what the Houthis have done in the last six months is something that—I'm just shocked."[2]

The United States shouldn't have been surprised by the Houthis' competence. After all, they were backed by a key regional power, Iran, as well as Russia. Subsequent reports confirmed that Russia provided satellite targeting data, passed through members of Iran's Islamic Revolutionary Guard Corps to assist their strikes. But Iran and Russia's support only partially explained the Houthis' success. In an age of open technology innovation, even small-scale insurgencies based in impoverished countries could obtain powerful weapons and deploy destructive innovations against far larger powers. As scholars Leslie Eliason and Emily O. Goldman warned, "Failing to look for innovations that originate and spread from places other than the richest or most powerful nations creates a serious blind spot in defense planning."[3]

America's conflict with the Houthis was merely a prelude to a much larger war the United States soon found itself engaged in with Iran. Early in the morning on February 28, 2026, Trump and Israeli Prime Minister Benjamin "Bibi" Netanyahu, were presented with a rare opportunity. Nearly the entire political and military leadership of Iran was gathered in a single location—at a government compound in Tehran. This included Supreme Leader Ayatollah Ali Khamenei, Iran's top decision maker and spiritual leader. The moment was so remarkable that US and Israeli warplanes were authorized to strike in full daylight. They streaked towards the targets and dropped thirty bombs on the compound, destroying it completely. Not only was Khamenei killed, but the dead also included Ali Shamkhani, a top security adviser to Khamenei, Mohammad Pakpour, commander of the Islamic Revolutionary Guard Corps, and Defense Minister Amir Nasirzadeh. The United States and Israel followed up the attacks with a dazzling show of force, using AI systems to help strike 1,000 targets in the first 24 hours of the campaign.[4]

But Iran wasn't cowed. While its weaponry paled in comparison to that of American and Israel, it had stored thousands of low-cost Shahed-136 drones and medium ballistic missiles (experts estimated that Iran had stockpiled up to 80,000 Shaheds across the country with a monthly production capacity of 10,000). Despite losing the bulk of its senior leadership, its military struck back with fury. In the first week alone, Iran fired nearly 2,000 drones and more than 500 ballistic missiles at US bases and allied targets across twelve countries, hitting airports, military installations, luxury hotels, and oil infrastructure. Many of Iran's munitions were intercepted, but the cost was high. In just a few days, Gulf countries burned through more than 800 Patriot missiles, each of which cost approximately $4 million to fire. In contrast, the cost of a Shahed stood between $20,000 and $50,000 while Iran's ballistic missiles cost around $1-2 million.[5]

The cost asymmetries started to add up. Kelly Grieco from the Stimson Center think tank estimated that "For every $1 Iran spent on drones, the UAE spent roughly $20-28 shooting them down." Gregg Carlstrom from *The Economist* likened the economic asymmetry to "using Ferraris to intercept e-bikes." Weeks into the conflict, Iran's weapons had accrued major damage. An attack drone struck the AN/FPS-132 early warning radar in Qatar, part of a $1.1 billion ballistic missile tracking system. Another surprise attack destroyed a $500 million E-3 AWACS aircraft sitting on the tarmac at a Saudi airbase. More than two dozen Reaper drones, valued at over $700 million, were also lost. Commented Hoshyar Zebari, a former foreign minister and deputy prime minister of Iraq, regarding Iran's defiance, "They are resisting, they are resilient," he said. "This is a war between technology and ideology. The Iranians are squeezed, and their situation is difficult, but for them, this is 'to be or not to be.'"[6]

The United States realized it faced a severe situation. It couldn't keep using up scarce multimillion-dollar Patriot and THAAD interceptors to keep Iran's drones at bay. So Pentagon officials began searching for solutions and found one in Ukraine. Kyiv wasn't a stranger to drone attacks. For several years it had suffered waves of drone strikes from Russia, and its leadership quickly realized that using Patriots to protect its assets was a losing proposition. With the help of former Google CEO Eric Schmidt, Ukrainian engineers developed a cheap anti-drone system called Merops that became critical to Ukraine's defenses. The system is centered around deploying small, cheap AI-enabled interceptor drones that can travel at speeds exceeding 180 miles an hour in order to hunt down incoming Russian attack drones. Each Merops unit costs around $10,000 (but as production volume increases, costs could fall to $7,000).[7]

Thanks to Merops and other defensive technologies, Ukraine soon could destroy most of Russia's one-way attack drones. In February 2026, for example, Russia sent 5,000 drones and decoys to attack Ukraine's cities, but Kyiv downed about 87 percent of them. In short order, the Pentagon acquired 10,000 Merops interceptors and rushed them to the Middle East. The United States' reliance on Ukraine-tested innovations signaled a major shift. In the first years of the war, the Pentagon supplied Ukraine with equipment and studied how well its systems performed against Russia's forces. As Ukraine developed its own innovations, it became a testing ground for defensive technologies. That experience now had come full circle; Ukraine's hard-won expertise was directly helping the Pentagon fend off Iran's Shahed strikes.[8]

Change Flows in Multiple Directions

The diffusion of innovations from one country to another isn't a new concept. It is foundational in history and core to understanding how nations have risen or fallen through the ages. From Sultan Mehmed II's pioneering use of big-caliber cannons to penetrate Constantinople's walls in 1453 to Nazi Germany's blitzkrieg offensive that exploited speed and maneuverability to overrun France's Maginot Line, nation-states and empires have often gained fleeting advantages from new technologies, only to see them fade as competitors catch up. This process of innovation, emulation, and adaptation is referred to as diffusion—the process "by which (1) an innovation (2) is communicated through certain channels (3) over time (4) among the members of a social system."[9] Practically speaking, the diffusion process starts when a major innovation reaches a critical juncture or "debut" point, giving the wider community enough information to appreciate its significance. This triggers a three-part cycle. First, diffusion starts slowly as risk-tolerant actors or early adopters implement the innovation. Eventually, adoption hits critical mass. It spreads rapidly as most of the actors "with the ability to implement the innovation will do so." In the third stage, lagging actors or late adopters eventually take up the innovation as well. As innovations spread, they are frequently modified by adopters to fit their specific circumstances.[10]

Conventional wisdom says that diffusion follows a hierarchical path: innovations first appear in advanced centers and then spread outward to less-sophisticated or smaller countries. Scholarship assumes that rising states benefit from the "advantage of backwardness"—that they can free ride on the innovations of the most advanced countries in order to close the military-technological gap with their rivals. As international relations theorist Robert Gilpin observed, "there is a historical tendency for the military and economic techniques of the dominant state or empire to be diffused to other states in the system." This view seems logical at first glance. Smaller states face many disadvantages in the international system. They have lower populations, more modest economies, and less administrative capacity than major powers. They also have fewer resources with which to fund cutting-edge scientific research or to develop groundbreaking innovations.[11]

Yet a growing body of evidence challenges the claim that innovations largely flow from large states to small states. Facts on the ground show a very different picture. Smaller states and even non-state actors are developing disruptive technologies, from cutting-edge microchips and air defense systems to autonomous weapons. These products are emblematic of a larger trend: technological innovations are no longer confined to great powers. Instead of thinking about technological progress as a one-way street, with smaller states in a constant race to catch up to larger countries, innovation now flows in multiple directions.

What accounts for this shift? Two factors are crucial. First was the United States' transition to an open technological system at the end of the Cold War. In a closed technological system, writes scholar Audrey Kurth Cronin, powerful states control high-end military technologies, forcing others to "copy, steal, expand, or build upon them" in a rather slow process. Their goal is to restrict access to emerging technologies and to limit the availability of new innovations, whether nuclear weapons, aircraft carriers, or radar. In contrast, in an open technological system, "emerging technologies are accessible to the public, and individuals and private groups are free to not only buy, use, and distribute them, but also to invent new purposes for them."[12]

The euphoria accompanying the end of the Cold War led to a profound shift in US policy. Technologies developed in closed government labs that were previously unavailable for commercialization could now be accessed and monetized. ARPANET became the internet. The Global Positioning System (GPS) was commercialized. Different components necessary to create the smartphone—from touch screens to natural language processing for Apple's Siri—were taken from government labs and incorporated into devices for consumers. In large part, Silicon Valley's meteoric rise was due to America's embrace of the "open technological revolution."[13]

Significantly, it wasn't just US tech startups that benefited from the transition to an open innovation system. Other countries, including rivals like China and middle powers like Israel and South Korea, also profited from this access in ways that could harm US interests. Writes Cronin, "Never before has a military power as dominant as the United States shared basic technologies so perfectly designed to democratize the global distribution of power."[14] In the past, less-advanced states had to rely upon the generosity of the US government to obtain high-tech designs or resort to industrial espionage to gain access to America's secrets, but now many of the United States' most sophisticated technologies had come out into the open.

The current suite of information and AI technologies sweeping the world lend themselves to new patterns of disruption. Innovative manufacturing techniques such as 3D printers are giving smaller states, non-state actors, and even individuals a greater capacity to produce highly lethal technologies, from drones to missiles. Smaller states can now deploy sophisticated systems—such as continuous battlefield surveillance and long-range strikes—that until very recently were consigned to a handful of wealthy states. The traditional separation between the civilian and military sectors began to collapse, meaning that items designed for the commercial market, like DJI drones or even smartphones, could now be repurposed for deadly battlefield effect.

Meanwhile, the pace of change was accelerating.

Take military-grade drones. Over two decades, small states saw how much the United States relied on UAVs to remotely surveil far-off territories and conduct long-range strikes with a minimal risk of casualties. Few governments could afford to purchase American MQ-9 Reaper drones, which cost about $30 million apiece and involved a host of requirements, such as ground control stations and satellite links. But smaller countries could tailor US drone designs to better suit their needs by stripping out expensive components and streamlining complicated operating systems. They could even come up with new adaptations—for instance, converting lethal attack drones into single-use weapons that would cost next to nothing to produce but would cause extensive damage. As a result, the United States' commanding lead in precision capabilities shrank. "The technology underlying those capacities—conventional munitions, sensors, and guidance systems—has become cheaper over time and accessible to many countries and militant groups beyond the United States," writes Michael Horowitz.[15]

The development of semiautonomous systems shows a similar trajectory. Previously, this technology faced considerable constraints. In the early days of the Ukraine war, both sides experimented with autonomous drones, but they were viewed as unreliable and limited in range. They could not be produced in mass quantities, and their utility paled in comparison to regular munitions. Over time, they started to get better. Their range capacity improved and their software became more effective. Today, Ukraine and Russia deploy dozens of different models in the air, on the ground, and in the sea to support their operations.

These new technologies and systems share several common characteristics: the entry costs are low, they are highly accessible and can be adapted for specific purposes, they offer dual-use functionality (e.g., civilian technologies repurposed for military application), and they form "part of a cluster of other emerging technologies."[16] The consequence? Large states are less able to achieve dominant control over advanced technologies. As these technologies disperse, they are contributing to instability and conflict, forcing major militaries to reconsider their defense policies. America's defense posture, for example, which relies on forward-deployed troops in remote garrisons to project power, faces growing risks. The Tower 22 drone attack, which targeted a US military facility in northeast Jordan in early 2024, illustrates the vulnerability.

Starting in 2014, the US military had established a network of bases in eastern Syria, northern Jordan, and Iraq to contain and destroy the ISIS (Islamic State) terrorist group. The facilities are rudimentary; many were former ISIS garrisons or small towns that the United States and its Kurdish allies had liberated after intense battles with ISIS forces. The Tower 22 border outpost housed about 350 army and air force personnel and served as a logistics and resupply hub for the Al Tanf base, located nearby in southeastern Syria. It consisted of prefab living quarters surrounded by concrete blast walls "to limit the damage of any incoming warheads."[17] Late in the evening on January 27, 2024, Kataib Hezbollah, an Iranian-backed militant group, launched a one-way exploding drone against Tower 22. The group had come up with a clever plan. They synchronized the drone to approach its target at the same time a US drone was also returning to the outpost. The return of the American drone led to confusion about whether Kataib Hezbollah's device was friendly or not. The weapon struck near the soldiers' sleeping quarters, killing three American troops and leaving scores of others injured. The attack was a significant milestone. Before the drone strike, "no US ground troops [had] been killed by an enemy aircraft since the Korean War"—more than seventy years ago.[18]

Naturally, the incident created a firestorm. Not only did it expose cracks in America's presumed invincibility, but officials were worried that it would inspire copycat attempts. To send a message about the consequences of attacking US forces, the Pentagon struck back against Iran and its proxies with fury. Biden authorized strikes against the entire Iranian proxy network in Iraq, Syria, and Yemen, "hitting more than 100 targets overall, with a combination of long-range B-1 bombers out of Texas, and cruise missiles and fighter bombers launched from the Eisenhower carrier group in the Red Sea." When the dust settled, about forty people had been killed by the retaliatory strikes.[19]

But the United States wasn't finished. To make sure Iran and its allies got the message, American forces subsequently assassinated Kataib Hezbollah commander Abu Baqir al-Saedi, whom intelligence officers deemed responsible for the Tower 22 assault and attacks against other outposts. As Al-Saedi was driving down a dusty Baghdad street, a US drone struck his car with a Hellfire missile, the same weapon used to take out Iranian Revolutionary Guard commander Qassim Suleimani in 2020. (The missile is known as the "Flying Ginsu"; to minimize civilian casualties the CIA has swapped out explosives for six swordlike blades that kill targets by flying directly at them—shredding anything in its path like a grinder.)[20] For the time being, America's muscular response worked—in the months following the strikes, Iran's proxies ceased attacking US military installations.

These developments revealed a starkly changing landscape. For it is not just smaller states that had tough choices to make—large powers also faced significant uncertainty. Major powers confronted growing risks from emerging actors due to the proliferation of lethal technologies. If large states neglect to pay attention to new developments emerging from the periphery, write Eliason and Goldman, these innovations can "start to diffuse in ways that challenge peace and security."[21]

Thus, while analysts fixated on the technological arms race between Washington, Beijing, and Moscow, other capitals and regions weren't sitting on the sidelines. These rising powers correctly assessed that agility could triumph over size and that specialized knowledge could preempt economic heft. They understood that the rules of technological competition were being rewritten; the question no longer was who possessed the most advanced capabilities, but which countries could most effectively deploy them. What would be the consequences?

21

Israel's Culture of Innovation

For over fourteen years, Iranian nuclear scientist Mohsen Fakhrizadeh had eluded the grasp of Israel's foreign intelligence agency, the Mossad. Fakhrizadeh was the architect of Iran's covert nuclear program and the central driver for every aspect of the project—from acquiring components for Iran's first uranium enrichment plant to leading research on how to design a warhead small enough to fit atop Iran's long-range missiles. As Iran expert Karim Sadjadpour put it, "Fakhrizadeh likely knew more about Iran's nuclear program than any living human." Due to his central role, he had been in the Mossad's sights for well over a decade. Israeli intelligence agents had managed to get to other Iranian scientists; since 2007, the Mossad had assassinated five Iranian nuclear researchers and wounded another. Israeli operators had also killed the Iranian military general in charge of missile development along with sixteen members of his team. But Fakhrizadeh kept slipping through the cracks. His security detail was overseen by the elite Ansar unit of the Revolutionary Guard, who were well armed and professionally trained. They accompanied Fakhrizadeh in convoys of four to seven vehicles, often switching cars to throw off tracking. At several junctures, the Mossad had to call off assassination attempts due to suspicions that their plotting had been compromised. Finally, the Mossad settled on a new plan.[1]

On November 27, 2020, Fakhrizadeh and his wife left the Caspian coastline and headed toward their family home in the village of Absard. They traveled in a convoy of four cars. The lead car was driven by his security detail. Fakhrizadeh followed behind in an unarmored black Nissan. Two additional security cars trailed them. At 3:30 p.m., the convoy approached a U-turn and slowed for a speed bump. Seconds later, a machine gun fired a cascade of bullets, hitting right below Fakhrizadeh's front windshield. The car swerved and came to a halt. The shooter corrected the sights and fired another barrage. Bullets shattered the windshield and struck Fakhrizadeh in the shoulder. He stepped out of the vehicle, crouching behind the front door. Three more bullets ripped into his spine, causing him to collapse on the road.

Fakhrizadeh's Damaged Nissan Teana after the Remote Attack on Nov. 27, 2020

Once the operation concluded, the assassin didn't have to worry about escaping the scene. The shooter was in fact sitting in front of a computer monitor one thousand miles away. The Mossad had used a remote-operated machine gun to assassinate Fakhrizadeh. They had customized a Belgian-manufactured FN MAG rifle and attached it to an advanced robotic apparatus. The robot gun was constructed to fit into the bed of a pickup truck and was outfitted with a cluster of cameras to give the Mossad a full picture of the target and the nearby area. Israeli agents had to overcome considerable technical barriers, including compensating for delays in transmission between what the operator was seeing on-screen and what was happening in real time (by the time camera images arrived in the operator's monitor and the agent pulled the trigger there would be a time lag of 1.6 seconds, throwing off the shot). The Mossad used AI software to account for the time lag, the shaking of the truck, and the vehicle's speed. It didn't hurt that even if the shooter missed, the gun could fire six hundred rounds per minute and adjust accordingly. All told, the operation took less than a minute and required firing only fifteen bullets.[2]

Israel's operation was notable. While other countries possessed remotely operated machine guns, they were bulky and difficult to transport or conceal. Those models also required nearby operators—no one had deployed a robot gun from such a distance. But overcoming these constraints wasn't unusual for Israel; it has a history of using unconventional approaches to achieve its security objectives.

One key driver of Israel's innovation culture is resource scarcity. Israel is a small country surrounded on all sides by much larger foes. From the moment of its founding, the Israel Defense Forces (IDF) were in a fight for the country's survival. In the Arab-Israeli War of 1948, Israel was desperate to get its hands on military equipment—combat aircraft, armored vehicles, and tanks—to defend its cities and settlements and provide cover for its exposed forces. But the United States and Britain refused to provide the necessary supplies, forcing the country to turn to other sources. Instead of tanks, its government acquired "3,000 ten-ton U.S. 'half-tracks'—lightly armored vehicles with wheels at the front and tank-like tracks at the back." These vehicles worked well enough—serving as troop transports or makeshift tanks (once the Israelis bolted Czechoslovakian guns onto the outside).

Israel was also desperate to acquire military aircraft. Official channels in the United States and the UK were closed to them (both countries strictly enforced the Neutrality Act, which prohibited the sale of military equipment to countries engaged in armed conflict). So the Israelis orchestrated a smuggling operation to bring in decommissioned military transport aircraft from the United States, and entered into secret talks to acquire twenty-five secondhand fighters from Czechoslovakia. These actions, combined with the valiant work of "self-taught local mechanics, the priceless expertise of war-veteran volunteers from abroad, and the drastic cannibalization of a variegated inventory," helped Israel withstand the Arab onslaught and win its war of independence.[3]

These early lessons were not lost on Israel's leadership. They recognized that they could not depend upon larger powers to come to their rescue—they needed to develop their own sources of military innovation. And they also realized that cultivating a scarcity mindset could work to their advantage—incentivizing researchers and scientists to come up with sophisticated adaptations that would keep Israel's enemies off-balance for decades.

As a result, Israeli officials placed a big bet on technological innovation, transforming the country's agricultural economy into a high-tech powerhouse. By the 1970s, Israeli engineers were coming up with world-beating products. One such innovation was anti-ship missiles. It had become clear to IDF leadership that the naval balance in the region was tilting sharply against them. Egypt had received a steady flow of Soviet ships and weapons, allowing it to threaten Israel's coastline and bombard its cities. Israel's navy was resolved to rectify this imbalance. After much trial and error, its engineers designed a new ship-to-ship missile called the Gabriel, "the first of its kind in the Western world." During the 1973 October War (also known as the Yom Kippur War), Israel deployed its Gabriels to devastating effect. By the end of the short conflict, its navy had sunk nineteen Egyptian and Syrian ships without suffering any losses. (Arab navies fired more than fifty Soviet Styx missiles; none came close to hitting Israel's vessels.) Their use was a huge success and a wake-up call for larger countries; the US and Soviet militaries were spurred to develop automated point defense systems to defend against this new class of sea-skimming munitions, as well as ship-to-ship missiles of their own, such as Boeing's Harpoon missile, which has since been adopted by twenty-nine foreign military partners of the United States.[4]

Israel's innovation breakthroughs continued. In the 1980s, Israeli engineers directed their efforts toward building remotely piloted vehicles and aerial drones. While other countries had experimented successfully with drone technology, those projects languished in favor of other systems. Israel saw the potential for unmanned aircraft and deployed them in spectacular fashion in the 1982 war in Lebanon. For months, the IDF had engaged in tit-for-tat skirmishes against the Palestine Liberation Organization (PLO), which operated out of southern Lebanon. After gunmen from the Abu Nidal Organization—a Palestinian militant group with close ties to the PLO—attempted to assassinate Israel's ambassador to the United Kingdom, the IDF prepared to escalate.

In a daring raid the morning of June 9, 1982, Israel's air force launched a bevy of drones, missiles, and combat aircraft against air defenses in Lebanon's Bekaa Valley manned by Syrian troops (for three decades Syria occupied Lebanon, allowing armed Palestinian groups to operate in the south of the country). Syrian soldiers used Soviet-supplied SA-6 missile batteries to defend their territory; these surface-to-air systems had devastated Israel's aircraft a decade prior in the 1973 war, but in the intervening years, Israel had upgraded its weaponry. Its air force launched ninety jets along with an array of drones as part of a sophisticated combined-arms attack. The drones not only provided critical battlefield intelligence (and also served as decoys), but also helped guide missiles and cluster bombs with eerie precision toward their targets. After two days of fighting, Syria's defenses were critically degraded. Israel had destroyed seventeen of nineteen SA-6 missile batteries and downed twenty-three Syrian MiG fighter jets with no losses of its own. Syria's setbacks continued to mount. By the end of July, it had lost a total of eighty-five aircraft while costing Israel only minimal losses. As with the anti-ship missiles, Israel's success with remote-piloted drones was a wake-up call to larger militaries. It started exporting UAVs to the United States and other large countries (such as India). Official US reports acknowledged Israel's advantages in the drone sector, noting that smaller countries "may be able, because of their smaller size and focused efforts, to introduce certain technologies earlier than the wealthier but more ponderous superpowers."[5]

In both conflicts, Israel demonstrated conclusively that its military innovation culture was yielding significant payoffs. "These were not one-offs," wrote Shashank Joshi in *Foreign Affairs*. "Israel developed world-class tanks, pioneering tank-protection methods, and air defense systems that are the envy of the world."[6] Israel's success proved that small countries could come up with game-changing innovations that more advanced countries and militaries would emulate.

Drawbacks to Israel's Reliance on Technology

Israel's pursuit of military innovation was not without its downsides. While technology provided a clear edge to the IDF's battlefield performance, its benefit to the intelligence sector was less evident. Some analysts, in fact, argued that Israel's overreliance on high-tech tools created major intelligence gaps. This debate came to a head following the October 7, 2023, Hamas attacks. In 2021, the IDF completed a nearly forty-mile-long reinforced concrete wall that sat above and below ground, designed to seal off Gaza. Israel enhanced the fence, known as "The Barrier" or the "Iron Wall," with cutting-edge surveillance technologies—intelligent video cameras, laser-guided sensors, and advanced radar. The upgrade took three years and cost $1 billion to complete. Israeli officials believed this system made the country impenetrable to armed groups. The general overseeing the construction called the wall "one of the most complex projects the defense establishment has ever built." Accordingly, security officials reduced the number of troops manning their guard stations. This was a critical error. Hamas militants had studied the wall and realized that the same high-tech gadgetry that instilled such confidence in Israel also represented a weakness; if they could neutralize the wall's sensors and blind the IDF to their initial incursions, this would buy sufficient time to attack neighboring towns and villages.[7]

Hamas's bet ended up being correct. Militants implemented what they called the "blinding plan." They began the operation by firing approximately three thousand rockets in the early morning hours to distract border sentry watchers. Hamas then transitioned to the second phase, which involved targeting surveillance and sentry towers guarding the wall. Each of the surveillance towers featured HD cameras, laser and infrared sensors, and radar; they were designed to spot people from almost six miles away. They were linked to three monitoring centers located several miles back, which tracked events twenty-four hours a day. The wall was also manned by sentry towers, nicknamed Roeh-Yoreh ("sees-fires"), which were spaced a few hundred yards apart and armed with .50-caliber machine guns.[8]

The problem with the system was that if an attacker disabled the border sensors, then the Israeli command would be unable to see incoming attacks. To accomplish this, Hamas tied cheap explosives to off-the-shelf consumer drones (adopting tactics from Ukraine) and flew them directly into the towers. Hamas also fired rocket-propelled grenades and unleashed sniper fire to cause additional damage. They destroyed both the frontline and backup sensors, rendering the IDF completely dark during crucial hours of the operation. All told, Hamas fighters executed at least thirty breaches of the wall, leading to the total collapse of Israel's border defenses. Israeli officials were stunned by the attacks. As Charles Freilich, former Israeli deputy national security adviser, said, "We didn't believe that Hamas had this capability, and so we didn't see it coming."[9]

To be fair, Israeli's overreliance on technology wasn't the only reason for its intelligence collapse. Of equal importance was the Israeli establishment's "cultural arrogance," which led it to profoundly misjudge Hamas's capacity and appetite for bold action. Because the IDF viewed Hamas as a second-tier organization—presumably unable to mount a complex assault against Israel's superior defenses—they brushed off multiple warnings that the group was planning a big operation. Israeli intelligence also faced considerable bandwidth constraints. It was being pulled in several competing directions at once—monitoring unrest in the West Bank, heightened military activity in Lebanon, political instability in Syria, and growing threats from Iran. "There was an over-reliance on warning systems, a misguided policy and military attitude toward Hamas, and a toxic relationship between Israel's top leadership and its defense and intelligence establishment," writes Carnegie Endowment for International Peace scholar Ariel Levite. "These factors both distracted the establishment and systematically discouraged a confrontation with Hamas."[10] Still, it is hard to shake the feeling that Israel became a victim of its own success; over time, its reliance on the technological wizardry of its billion-dollar fence bred complacency and provided an opening for Hamas to exploit.

Israel's heavy focus on technology also gave rise to disquieting moral and ethical questions. One system that garnered significant scrutiny was an AI targeting tool called Habsora, or "the Gospel." Following the October attacks, Israel pummeled Gaza's infrastructure with missiles, using a database of targets compiled by Israeli intelligence. But the IDF soon faced a problem: it was running out of locations to strike. To refill its target bank, military leaders turned to the Gospel for help. The system used machine-learning software built atop "hundreds of predictive algorithms" to produce targeting recommendations almost instantaneously.[11] The Gospel reviewed volumes of data derived from intercepted communications, satellite imagery, and social media, and subsequently generated coordinates for targeting—from suspected tunnel sites and other military infrastructure used by Hamas to the home addresses of militants.

This wasn't the first time the IDF had deployed the Gospel in war; Israel introduced the system in 2021, during a short, eleven-day conflict with Hamas. But in the current conflict, it became the centerpiece of Israel's bombing strategy—enabling an unprecedented escalation of the scale and intensity of its strikes. In the war's first thirty-five days, the IDF estimated it attacked fifteen thousand targets in Gaza. By comparison, Israel struck a total of five thousand to six thousand targets during the 2014 Gaza conflict, which spanned fifty-one days. Unsurprisingly, the civilian toll from the bombing strikes was significant. Part of the issue was that the IDF loosened its threshold criteria for civilian deaths. In the 2014 war, for example, the IDF's acceptable civilian casualty ratio was one civilian death for every high-level terrorist. A decade later, Israel changed the rules. In the current conflict, officers were authorized to kill up to twenty civilians for every legitimate military target. The intensified campaign "had no precedent in Israeli military history." Never before had mid-ranking officers been afforded such latitude to strike so many targets—many of which had diminished military significance—with such grave implications. As one former intelligence officer described it, Israel's loosened civilian harm requirements combined with its escalating reliance on the Gospel facilitated a "mass assassination factory."[12]

Much of this was the handiwork of Yossi Sariel, head of Israel's intelligence directorate, Unit 8200. Before taking over the group, Sariel had spent a sabbatical year at National Defense University (NDU), a Pentagon-funded institution located in Washington, where he developed a "radical vision of AI in the battlefield." According to his colleagues at NDU, Sariel contended that Israel should push ahead of its more cautious allies to adopt lethal frontier technologies. "Yossi was in this world of, 'This thing is moving fast, faster than anybody realizes. And we better get everybody on board,'" said one NDU professor. During his sabbatical, Sariel wrote a book, *The Human-Machine Team*, that called for fusing national security operations with automation and made the case for designing a "special machine" that could swiftly analyze heaps of data to generate targets in war. He believed technology could solve the "human bottleneck" problem when it came to identifying and approving new targets. It only took a few years for Sariel's vision of war to become a reality.[13]

Experts started scrutinizing the Gospel and related AI systems such as Lavender (which uses machine learning to identify individual suspects) and found many discrepancies. One issue was that officers had little knowledge about how automated kill lists were generated or the reliability of the intelligence used to build them. In the early days of the war, the IDF gave "sweeping approval" for officers to adopt Lavender's kill lists with no requirement to check why the machine made those suggestions or for analysts to scrutinize the raw intelligence data. The result was minimal accountability for the machine's target lists and no way to substantiate errors. In addition, internal audits found that AI systems regularly stumbled when analyzing Arabic language transcripts—"failing to understand key slang words and phrases"—which contributed to flawed conclusions about the location of targets.[14]

Further questions arose about the degree of human control over target selection. In theory, every target suggestion from the Gospel or Lavender required human approval to proceed. But reporting by *+972 Magazine* alleged that human personnel largely served as a "rubber stamp" for the machine's decisions. Analysts were purportedly given "20 seconds" in which to decide whether an AI-approved target was male (a requirement to proceed with a bombing). "I had zero added value as a human, apart from being a stamp of approval. It saved a lot of time," said one source. "If [the operative] came up in the automated mechanism, and I checked that he was a man, there would be permission to bomb him, subject to an examination of collateral damage."[15]

Despite these concerns, Israeli officials remained committed to using the systems. They argued that the current conflict was "unprecedented and hardly comparable to other theaters of hostilities worldwide," making it necessary to change their method of operations. Some experts agreed. West Point scholar Michael Schmitt, for instance, wrote that these tools helped the IDF weed out irrelevant information—"separating the wheat from the chaff"—allowing human analysts to improve the reliability of their assessments. He contended that machine-learning systems could refresh intelligence faster than any human analyst, reducing the risk of relying on "stale data."[16]

It is also relevant to consider just how much political pressure Israeli operatives faced to punish Hamas's leaders. The country was in shock from Hamas's attack and demanded vengeance. In his speeches, Prime Minister Netanyahu pledged to seek out "all of the places where Hamas has been hiding and operating from" and to "turn them into rubble." Accordingly, many IDF rules went by the wayside. Civilian safeguards were weakened, poststrike reviews were discarded, and officers were absolved from wrongdoing. Given the circumstances, it would be erroneous to assign all of the blame for the devastating humanitarian toll of Israel's military campaign to AI systems alone. A convergence of factors was ultimately responsible for changing Israel's direction and reorienting its strategy.[17]

Undoubtedly, the military use of AI targeting systems is spreading, notwithstanding their imperfections. Israel's military operations and incorporation of AI tools provide lessons that other armed forces are studying intently. As the 2026 Iran war shows, the Pentagon absorbed those ideas and made AI targeting central to its ferocious air campaign against the regime. During the conflict, Anthropic's AI tool Claude suggested "hundreds of targets, issued precise location coordinates, and prioritized those targets according to importance."[18] As war enters a new technological age, it stands to reason that more and more nations will adopt Israel's innovations (just like they did a generation ago for Israel's anti-ship missiles and remote-piloted drones). But its model isn't for everyone. While the IDF demonstrates that small countries can be founts of innovation and diffuse cutting-edge technology to larger countries, Israel's approach is challenging to execute, resource intensive, and comes with substantial collateral costs.

What other strategies could smaller militaries employ? While large powers continued to pour billions into sophisticated weapons systems, something unexpected was taking shape in the shadows of conflict. Rebel groups and insurgent forces found they could successfully compete by adapting cheap innovations to suit their needs. They didn't require state-of-the-art Predators or advanced AI targeting software: DJI Mavic drones and off-the-shelf FPV software would do just fine.

22

Drones and Geopolitics

In the northwest province of Idlib, Syrian rebel groups were quietly readying a campaign to topple the government. For nearly a quarter-century, Bashar al-Assad had ruled over Syria with an iron fist. Since 2011, he had prosecuted a grinding civil war, which arose out of Arab Spring anti-government protests. Unlike his counterparts in Egypt, Tunisia, and Yemen, Assad managed to cling to power. Backed by troops from Iran and Russia, as well as the Lebanese militant group Hezbollah, Assad had largely contained the rebels to a small corner of the country by 2020. While his opponents were hemmed in, they remained defiant. Under the leadership of the Islamist group Hayat Tahrir al-Sham (known as HTS), they steadily rebuilt the remaining opposition forces into an effective army.[1]

In November 2024, the rebels struck back. HTS plotted a lightning offensive that caught government forces off guard. In just eleven days, the ragtag army swept through Syria's largest city, Aleppo, and continued past the cities of Hama, Daraa, and Homs. By December 7, HTS's forces were perched outside the gates of Damascus. The surviving government units were prepared to make a last stand backed by the powerful Fourth Armored Division, led by Assad's brother, Maher al-Assad. But Bashar al-Assad had other plans. He saw the writing on the wall and decided to flee. Under the cover of night, without disclosing his decision to any of his aides or even Maher, he slipped into a private plane, turned off the aircraft's transponder, and flew to the safety of a nearby Russian air base. He left so secretively that some of his aides "remained in the palace for hours," unaware of his abandonment. Once they realized their president had escaped, they too fled in a panic, leaving the gates wide open for rebel forces to storm the compound a few hours later.[2]

It was a stunning victory. How did a loosely constituted group of insurgents who had battled unsuccessfully against Assad's forces for over a decade manage to break through his defenses in a matter of days? The rebel triumph defied an easy answer, but one advantage was drones. Government forces faced relentless bombardment from drone attacks spread across the country. Reports emerged about a special drone unit orchestrating the chaos: the Al-Shaheen (Falcon) Brigade. Al-Shaheen had a large budget and was overseen by an engineer who worked directly under HTS's top leader, Abu Mohammed al-Jolani. Over time, its operations became quite sophisticated. "The Falcon Brigade has been using and developing drones for the past 10 years," said Cihat Arpacik, editor in chief of *Intelligence Report* magazine and a longtime Syria watcher. "They can import spare parts from China and other areas, and now use turbojet drones."[3]

Al-Shaheen played a key role in the campaign, leading a multilayered attack against regime installations, tank hideouts, armored vehicles, troop gatherings, and frontline lookout points. It launched swarms of suicide drones against regime targets. A fleet of reconnaissance drones provided guidance for those strikes. Behind them, small munitions were directed at the targets by "teams of drone operators who had been trained intensively in secret over the past year." Artillery and mortar shelling provided yet another layer of firepower against the Syrian army. The effect was devastating. The regime front lines collapsed one after the other across the battlefield, allowing HTS's coalition to rapidly push its way toward Aleppo city and then beyond.[4]

HTS's drone-making effort took years to develop and reflected strategies employed by several other groups, including the Islamic State.* (Historically, it is the norm for guerilla or insurgent groups piggyback off each other's innovations and adopt and refine them across successive conflicts.) Over a ten-year period, the group invested heavily in drone infrastructure, building a small network of workshops based in garages, converted schools, warehouses, and private homes. They made use of 3D printing technology to customize weapons and manufacture components that weren't readily available on the commercial market. (Other actors have also taken advantage of 3D printing to augment their drone-making efforts, including groups in Ukraine, Myanmar, and Yemen.) The Falcon Brigade primarily worked off two drone models: shorter-range quadcopters and longer-range

* During the Islamic State's heyday, from 2014 to 2018, its fighters pioneered using quadcopters to swarm targets with small explosives, generating a terrorizing effect. HTS leaders adapted and improved upon many of the Islamic State's tactics.

fixed-wing drones. Quadcopters were deployed as "bomb couriers," releasing improvised explosive devices onto targets from above—similar to the Islamic State's tactics. The group also experimented with more complex single-use drones, which they reportedly reverse engineered from captured Iranian Shahed UAVs and Russian drone units. They carried up to one hundred kilograms' worth of explosives and exploded upon impact. In due course, HTS built a fleet of longer-range "Shaheen" drones that incorporated FPV technology for remote navigation.[5]

While the bulk of these efforts reflected homegrown initiative, undoubtedly HTS also benefited from external support. Türkiye was an important partner. Turkish smugglers brought in FPV drones to supplement HTS's arsenal, and analysts even suggested that Ankara provided critical input in the planning stages of the rebel campaign. A particularly surprising entrant was Ukraine; in 2024, reports circulated that the embattled country was quietly providing drones and technical assistance to HTS. That June, the *Kyiv Post* quoted a source from Ukraine's military intelligence service who claimed that "since the beginning of the year, the [Syrian] rebels, supported by Ukrainian operatives, have inflicted numerous strikes on Russian military facilities represented in the region." About a month before the offensive, David Ignatius reported in *The Washington Post* that a special Ukrainian intelligence unit, known as Khimik, had sent around twenty experienced drone pilots along with 150 FPV drones to rebel headquarters. While Kyiv's aid likely played only a small role in the campaign, its involvement was still notable, exemplifying its larger ambition to covertly "bloody Russia's nose and undermine its clients" in other regions, and demonstrating the global character of military diffusion.[6]

As the Syrian conflict shows, local initiative combined with inexpensive drones imported from second-tier producers could lead to game-changing results. From Nagorno-Karabakh and the front lines of Ukraine to Ethiopia, drone technology has had a profound impact on war.

Data collected by the Armed Conflict Location and Event Data (ACLED) initiative illustrates just how much drone use has proliferated in the last five years. Figure 1 shows that the number of air/drone events surged from 6,380 incidents in 2020 to 58,272 in 2025. Fatalities from these operations climbed as well, increasing from 11,379 deaths in 2020 to 32,769 in 2025.

Figure 1. Global Air/Drone Events and Corresponding Fatalities, 2020–25[7]

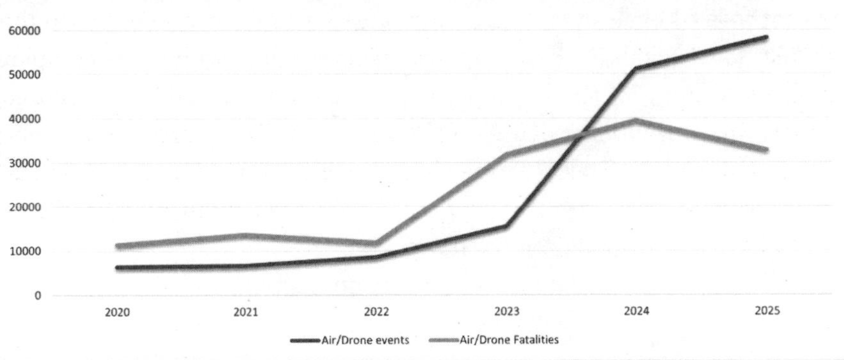

Armed Conflict Location & Event Data Project (ACLED), www.acleddata.com (accessed January 7, 2026).

Drone events aren't evenly distributed. Most of the drone strikes and associated casualties are concentrated in a handful of wars. As figure 2 shows, more drone events were linked to the Ukraine war than any other conflict, with nearly fifty-five thousand recorded incidents taking place between 2020 and 2025.

Figure 2. Conflicts with Significant Air/Drone Activity, 2020–25[8]

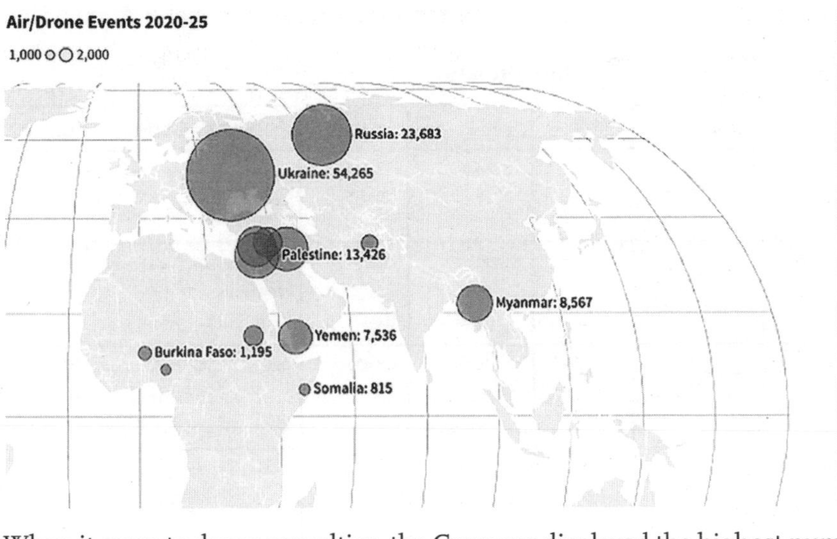

When it came to drone casualties, the Gaza war displayed the highest number of fatalities by a huge margin; from 2020 to 2025, over 53,400 people were reported killed due to air/drone strikes.

Figure 3. Conflicts with Significant Drone Fatalities, 2020–25[9]

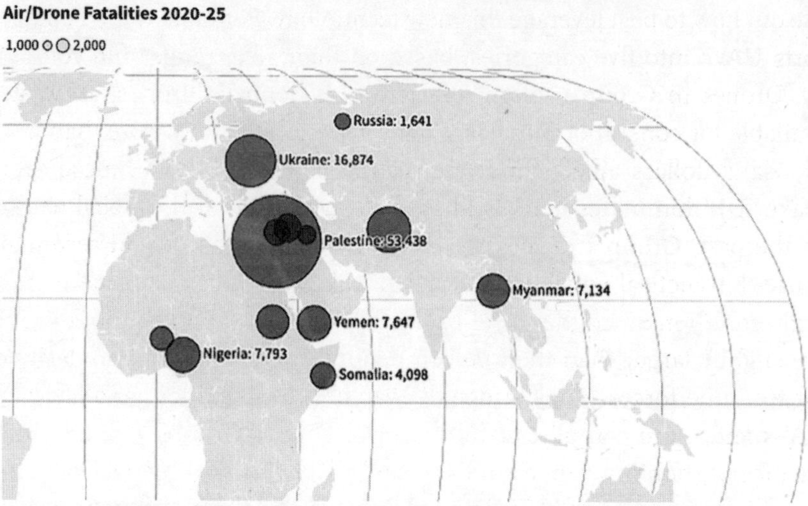

These wars represented 97 percent of all drone events occurring between 2020 and 2025 and 95 percent of all drone-related fatalities, bringing a crucial insight: any major armed conflict fought today will almost certainly involve drone technology. (See Appendix III for drone event data tables.) Consequently, UAVs are changing the character of war and challenging assumptions about how military victory is achieved.

■ ■ ■

Table 1. Categories of Military Drones

		Weight	Speed	Altitude	Representative Models
Group 5		>1320 lbs.	>250 knots	>18,000 ft	MQ-9 Reaper, Super Heron, Wing Loong II
Group 4	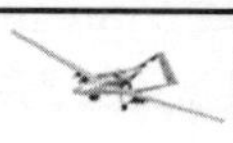	>1320 lbs.	>250 knots	<18,000 ft	Bayraktar TB2, MQ-1 Predator, Mojaher-6
Group 3		<1320 lbs.	<250 knots	<18,000 ft	Shahed-136, Orlan-10, Geran, LUCAS
Group 2		21-55 lbs.	<250 knots	<3,500 ft	ScanEagle, Micro Falcon
Group 1		0-20 lbs.	<100 knots	<1,200 ft	DJI Mavic, Raven, FPV models

Drones vary widely in their designs, making it urgent for militaries to figure out how to best leverage this new technology. Per table 1, the Pentagon sorts UAVs into five categories, based on their size, range, and complexity. Drones in Group 1, often referred to as hobbyist drones, are widely available for consumer purchase. They generally cost no more than a few thousand dollars apiece, and often come preassembled. Chinese drone maker DJI dominates here, holding over 70 percent of the global market. In the past, Group 1 drones were largely used for battlefield reconnaissance, but tactical versions with lethal capabilities have become increasingly common since the onset of hostilities in Ukraine. Group 2 drones are slightly larger than their Group 1 counterparts and are also designed for frontline forces. They bridge the gap between smaller hand-launched UAVs and more complex platforms belonging to Groups 3–5, offering a balance of mobility, range, and sensors. As of this writing, tactical UAVs are reportedly responsible for 70 percent of all battlefield casualties in the Ukraine war.[10]

In contrast, Group 3 drones are midsize aircraft. They are unavailable to the public due to their cost, as well as the infrastructural requirements for operating them. Instead, suppliers sell or transfer them to foreign militaries or non-state actors. They carry more substantial payloads and fly longer distances, requiring more advanced defensive interceptors to shoot them down. Prominent models include Iran's Shahed-136 drones, Russia's Geran UAVs, and the US military's new LUCAS drone.

Large military drones populate Groups 4 and 5. Aircraft in this category require substantial military infrastructure in which to operate. These UAVs were primarily accessible to major militaries until the debut of the Bayraktar TB2 drones. They incorporate complex systems, offer advanced features (such as radar evasion), and carry high-end precision-strike weaponry. Consequently, the price tag for basic models starts in the low millions and climbs upward. American-made Reaper MQ-9 drones, for example, cost about $30 million apiece.[11]

Of all the UAVs on battlefields today, the drones in Group 3 have proven to be the most vexing for the Pentagon to respond to. Smaller drones can be intercepted by less-sophisticated air defense systems, such as medium-caliber guns, and militaries can counter the larger ones in Groups 4 and 5 using air defense batteries (such as Patriot missiles) or fighter jets. As discussed earlier, Group 3 drones fall "somewhere in the middle between the protection and air defense missions," which presents a tricky asymmetry: while they carry payloads that can cause significant damage, their low production costs make intercepting them with expensive Patriots economically infeasible.[12]

▪ ▪ ▪

Following the US military's successful deployment of combat drones for operations in Iraq and Afghanistan in the early 2000s, other countries were eager to develop their own drone capabilities and sought American technology and weapons. However, getting access wasn't easy. One big obstacle was the Missile Technology Control Regime (MTCR)—a multilateral arrangement founded in 1987, intended to curb the spread of destructive weapons. The thirty-five partner countries apply common export control policies limiting the sale or transfer of specific items, including armed and unarmed UAVs. Although the MCTR doesn't legally prohibit drone sales to countries outside the framework, such transfers typically face strong resistance, making it difficult to sell UAVs to nations like Saudi Arabia. US law also bans the sale of weaponry to known human rights violators, short of a presidential waiver, throwing up yet another barrier for governments that want to acquire American drone designs.[13]

These regulatory barriers opened up a market opportunity for Beijing, which had few qualms about selling UAVs to unsavory regimes. Chinese drones also came at an attractive price point. The Wing Loong I drone, manufactured by the Chengdu Aircraft Industry Group, proved especially popular; they cost about $1 million per unit, far cheaper than a Reaper drone. Within a few years Chinese UAVs became a mainstay in drone fleets around the world—from Egypt and Oman to Pakistan and Saudi Arabia. (Belatedly, Trump shifted course; in 2025, the White House reclassified drones as aircraft rather than missile systems, allowing it to "sidestep MTCR" and unlock the sale of more than 100 MQ-9 Reapers to Riyadh.)[14]

Drones quickly became a coveted symbol of military power. Israel and Türkiye spotted their business potential and began investing heavily in their own drone manufacturing industries; they recognized that supplying for these armaments could reap millions in profits and give them leverage over client governments. But building these industries would take time and effort. Osman Okyay, a senior executive at the Turkish conglomerate Kale Group, told me that it took Turkish industry "15 to 20 years to bring these products to life." Eventually, their investments paid off. "Now Turkey has a few hundred indigenous platforms," including "rockets, missiles, helicopters, UAVs, fighter planes, tanks, armored vehicles, all kinds of ships, navy stuff, ammunition, small weapons, whatever."[15]

Baykar Technologies, maker of the TB2 drone, illustrates this growth trajectory. The firm's drones are the brainchild of chief technology officer

Selçuk Bayraktar, the son of the company's founder. He cuts a swashbuckling figure and is a celebrity in Türkiye. Many compare him to Elon Musk. Bayraktar posts regularly to his 2 million X followers—sharing pictures of new aircraft designs or posting Islamic blessings and celebrations of Turkish martyrs (his posts often feature his wife, Sümeyye Erdoğan, the youngest daughter of Türkiye's President Recep Tayyip Erdoğan).[16]

From a young age, Bayraktar had an avid interest in remote-piloted aircraft. During his graduate studies, first at the University of Pennsylvania and then at MIT, he spent significant time designing radio-controlled helicopters. While still completing his coursework, he began building prototype drones at his family's plant in Istanbul, but it would take him ten years to perfect a model suitable for mass manufacturing. After graduating, he began testing his protypes in the mountains of southeastern Anatolia by surveilling Kurdish separatist rebels. Bayraktar did not mind the grueling and dangerous fieldwork, remarking that if "this doesn't work at ten-thousand-feet elevation, at minus-thirty-degrees temperature, then this is just another item that you have to carry in your backpack." By 2015, he was ready to debut his drone. His company held a live demonstration where he piloted a TB2 armed with laser-guided dummy bombs and successfully struck targets "the size of a picnic blanket" from five miles out.[17]

Bayraktar's achievement marked a turning point. Soon, his firm began marketing its drones worldwide—where they developed a reputation for reliability, technical innovation, and cost-effectiveness. "The TB2 achieved success because it really found a niche, a white space in the market that was underserved," Dr. Serhat Süha Çubukçuoğlu, a senior analyst with TRENDS Research and Advisory based out of Abu Dhabi, explained to me. Once it found that niche, it moved quickly, getting "the necessary investment and qualified people in place." He added, "And of course, the chairman and the CTO of the company being the president's son-in-law paid off." Ankara invested hundreds of millions in Baykar and billions overall in its defense sector. By 2024, Baykar ranked seventy-third globally for its drone sales.[18]

Türkiye's drone industry proved to be a valuable bargaining chip for Erdoğan. Take, for instance, his 2022 drone deal with the UAE. That year, Baykar Defense delivered twenty TB2 drones to the Emirates, seemingly out of the blue. Defense analysts were surprised; TB2 production lines were already booked solid with orders three years out. Furthermore, the UAE had long been at odds with Türkiye; the two countries had pursued a "decade-long feud," backing opposing sides in wars in Libya, the Horn of Africa, and other places. But the Emiratis were dissatisfied with their Chinese drones, a sentiment shared by other militaries. (Reportedly, the Royal Jordanian Air Force was so frustrated with the performance of its Chinese-made CH-4B Pterosaur drones that it put six of them "up for resale.") The Emiratis wanted to diversify their arsenal, and Türkiye's drones fit the bill. For Erdoğan, this arrangement gave him the upper hand against a long-standing rival, and he also managed to extract some political concessions, convincing Emirati officials to "restrict the social media access of a prominent Turkish mafia boss turned whistleblower living in Dubai," wrote MIT scholar Erik Lin-Greenberg.[19]

For more granularity on the global drone market, I examined arms transfer data collected by the Stockholm International Peace Research Institute (SIPRI).[20] SIPRI oversees a comprehensive database containing information for every major conventional weapons transaction on record, including UAVs. Figure 4 depicts the total worldwide orders for Chinese, European, Iranian, Israeli, Turkish, and American drones between 2014 and 2024. Beginning in 2018, orders for Turkish drones surged, making Ankara the leading supplier of military-grade UAVs in the world along with Washington.

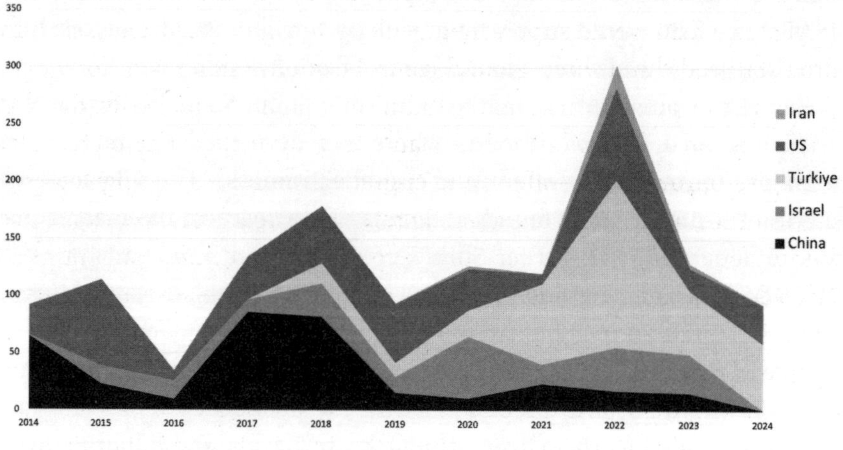

Figure 4. Global Orders for Military-Grade Drones, 2014–24

Drone sales closely mirror the supplier nations' geopolitical alignments. Israel largely sells to other Western countries or to countries that have strategic relationships with the United States (such as Colombia, Mexico, the Philippines, and Singapore). Iran's drone deliveries mostly flow toward other pariah states—Russia, Sudan, Syria (prior to the fall of Assad), and Venezuela—as well as to its network of proxy militias. As for Türkiye, many of its clients face international sanctions or strained relations with the United States and Europe (although they are not pariah states like Iran's customer base). Turkish officials have expressed few qualms about selling arms to Ethiopia, Mali, Turkmenistan, or Uzbekistan, for example, despite political abuses and human rights violations. While Ankara insists that it implements "strict case-by-case internal checks and reviews" for potential sales, there are few indications that it has imposed any real restrictions.[21] Recent drone transfers to Sudan are an eye-opening example.

Sudan has been devastated by conflict since April 2023, when its two top generals turned against each other in a struggle for power. Within days, the country's capital, Khartoum, was in flames as they tried to bomb each other into submission. Neither force was able to seize a decisive advantage; over time the conflict escalated. The results have been catastrophic. By some estimates, four hundred thousand civilians have been killed and more than 12 million have been forcibly displaced from their homes. UN officials describe the crisis as the "epicenter of suffering in the world right now."[22]

In November 2023, seven months after hostilities commenced, a *Washington Post* investigation revealed that Baykar Industries signed a $120 million contract with one of the warring parties—the Sudanese Armed Forces (SAF). The firm would supply them with up to eight TB2 drones, six hundred warheads, and three ground control stations, along with forty-eight personnel to provide "in-country technical support." Previously, the SAF had relied on a fleet of armed Iranian drones, but the TB2s represented a major improvement—offering laser-guided munitions, a range of up to 2,400 miles, and a payload exceeding three hundred pounds of explosives. It took about nine months for Türkiye to deliver the drones, but once they arrived, they had a big impact on the war, helping the SAF retake Khartoum in March 2025. "You have seen a huge SAF advancement in the past couple of months. These drones are one part of that winning strategy," said Justin Lynch, managing director at Conflict Insights Group.[23]

Türkiye had a lot to gain from helping the SAF. Sudanese authorities were considering providing Türkiye access to copper, gold, and silver mines, as well as licenses for fishing and fish processing plants. Sudan also intended to hand over to Türkiye construction and management rights to Abu Amama, an important Red Sea port. (Sudanese leaders had previously inked a $6 billion deal with the UAE to manage the port, but they canceled that agreement after the Emiratis furnished weapons to the rival military, the Rapid Support Forces [RSF].) Of course, there was a small hitch: Sudan's military faced heavy international sanctions. The United States and European Union had already imposed several rounds of penalties. The United

Nations also had a comprehensive set of sanctions in place—including asset freezes, travel bans, and an arms embargo. Türkiye's export of TB2s clearly violated these measures, yet that mattered little to Erdoğan. He saw an opportunity to expand his country's military footprint and political influence in the Horn of Africa, and he seized it.[24]

All the while, the Emirates were actively backing the RSF. Like Türkiye, the UAE had deep geopolitical interests in the country and had profited from smuggling gold out of Sudan, billions of which were making their way to Abu Dhabi. After the fall of Khartoum, the UAE "stepped up supplies" to the RSF (although Emirati officials continued to deny their involvement). A critical weapon was the CH-95 drone, made by China Aerospace Science & Technology Corp. The UAVs were a big boost for the RSF—they can stay in the air for up to thirty hours and carry out long-range surveillance and precision strikes from over 120 miles away. The RSF capitalized on these capabilities, using the Chinese drones to strike faraway ports and target power plants.[25]

Most importantly, the drones were the key to the RSF's siege against the Western Sudanese city of El Fasher, the government's last stronghold in the Darfur region. For eighteen months, the RSF laid siege to the city, blasting it with drones and heavy artillery. They began using the CH-95s to bomb community kitchens, displacement shelters, and hospitals. As the siege lengthened, the casualties increased. In one notable attack in October 2025, the militants used two drones and eight artillery shells to strike Dar al-Arqam, a displacement camp situated on a university campus, killing at least fifty-seven people, including seventeen children. Later in the month, the RSF conducted a three-day offensive, featuring attack drones and machine-gun-equipped pickup trucks. Their weaponry overpowered the remaining government soldiers who were forced to abandon the city, leaving civilians at the mercy of the RSF militia.[26]

As the war in Sudan shows, regional states like Türkiye and the UAE have leveraged drones to advance their geopolitical ambitions—caring little about the violence and destruction caused in their wake.

The Innovation Imperative

Because of the necessity to constantly innovate or face potential defeat, countries are continuously adapting new designs in response to counter-drone defenses. In the early days of the Ukraine war, TB2s were so popular that they inspired a viral folk song. They were hailed as the country's savior, with videos circulating that showed them striking Russian tanks, armored vehicles, and patrol boats.

Within a year, however, Russia's electronic countermeasures revealed the TB2's limitations. By design, TB2s present a relatively big target flying at a slow speed, which makes them vulnerable to air defenses. As Moscow improved its jamming capabilities and brought sophisticated missile batteries into the field, it began shooting down Turkish UAVs at a high rate. "Once the Russian military got its act together, it was able to down many TB2s," said Samuel Bendett, an analyst at the Center for Naval Analyses, in 2022. "What's happening now is that Russia's electronic warfare and air defenses have become better organized and fielded, compared to the earlier months of the war."[27]

By the end of 2022, TB2s had largely vanished from the battlefield. The edge they gave Ukraine in the opening months of the war, when Russia's supply lines were overstretched and uncoordinated, came to an end. The drones still remained valuable to Ukraine—transitioning from attacking Russian targets to collecting intelligence and surveillance data or using their sensors to "guide other assets like drones and missiles to targets." But their centrality to the war faded. Gradually, Ukraine began replacing them with tactical FPV drones that were easier to operate and cheaper to launch into battle.[28] By 2025, Ukraine had fully embraced the concept of "precise mass," betting that deploying large numbers of cheap drones on the battlefield would prove more effective than using a limited number of advanced models. Accordingly, Ukrainian factories made about 3 million attack and reconnaissance FPVs that year, with the goal of scaling up production to 7 million drones in 2026.[29]

As these examples show, drone technology—once confined to great powers like the United States and China—has irrevocably spread to smaller states and non-state actors. In the early 2000s, it would have been unthinkable for an insurgent group like HTS to develop a sophisticated drone-building program and deploy unmanned aircraft against its government opponents. A generation later, this technology has become cheaper and far more accessible. HTS provides a glimpse into the future. The group leveraged a powerful combination of instruments: it took lessons from the Islamic State's drone program, relied on off-the-shelf consumer technologies like 3D printing, and benefited from the expertise of middle-power states like Türkiye and Ukraine to successfully prosecute its campaign against Assad. The end result was a stunning military victory and a reshaping of geopolitical lines in the Middle East.

It is worth reiterating that both Türkiye and HTS took many years of testing to develop their drone programs, refine their designs, source the right components, train teams of skilled operators, and get their manufacturing facilities up to speed. This raises the question: How replicable are these examples? Will other actors seeking their own drone programs also have to endure lengthy periods of testing and development? The likely answer is no.

As UAVs become cheaper and more accessible, the time investment required will contract. Emerging technologies reinforce one another, bringing new innovations and unexpected advantages. The combination of 3D printing (an additive manufacturing process) with advances in robotics and the availability of off-the-shelf quadcopters is already transforming the drone landscape. Writes Cronin, "In this world of distributed, interactive learning, individuals can achieve breakthroughs with relatively little technological expertise."[30] In many respects, Baykar and HTS are outliers. They started their drone programs in the 2010s, when core technologies were still maturing. Fast-forward a decade, and Ukraine's experience demonstrates the accelerated pace of innovation—from concept to implementation to large-scale production of FPVs. In the coming years, the diffusion of mass-market drones as well as the availability of high-tech components will only grow.

The spread of military innovation offers vivid examples of how smaller states are using technology to leverage new sources of power. But this paradigm shift is not consigned to weapons. The AI arms race has ignited a new contest to shape future prosperity and influence. Along the shimmering shoreline of the Persian Gulf, Emirati royals have orchestrated a high-stakes gamble that could reshape the entire region. What exactly are the UAE's rulers planning, and how will their maneuvers pan out?

23

The Emiratis' Bid to Become an AI Power

Sheikh Tahnoon bin Zayed Al Nahyan, the UAE's sunglass-clad national security adviser who some refer to as "the spy sheikh," has a vision for his country. He wants to turn the Emirates into an AI powerhouse and views the United States as critical to supporting that effort. Sheikh Tahnoon was born in the late 1960s, only a few years after oil was first discovered in Abu Dhabi. At the time, the UAE was a backwater, inhabited by less than 250,000 people. But the discovery of massive oil reserves, equivalent to 7 percent of the global share, changed everything. The UAE became a petroleum giant; by 1976, its oil revenues had reached $5 billion a year, setting in motion "an economic and social revolution." The Emirates spent billions of dollars remaking itself into a global investment and tourist hub. It built a shimmering trade center, Dubai, transforming it from a fishing village to a metropolis of 3 million people brimming with skyscrapers. Today, its skyline is dominated by the 163-story Burj Khalifa, the world's tallest building. Its artificial palm-shaped island, Palm Jumeirah—perched along its coast—is big enough to be seen from space.[1]

Tahnoon grew up playing second fiddle to his brother, Sheikh Mohamed bin Zayed Al Nahyan (known as MBZ), who was named president of the Emirates in 2022. While MBZ sharpened his diplomatic skills and established himself as a foreign policy player, Tahnoon "honed his extracurricular activities." (*Wired* reporter Bradley Hope describes him as "one-third Gulf royal, one-third fitness-obsessed tech founder, and one-third Bond villain.") He is an avid gamer and reportedly became so fixated on competing with his brothers in the video game *Age of Empires* that he poached an employee from the game's company to coach him. Tahnoon is also a serious athlete. He funds one of the world's top racing teams—UAE Team Emirates—and when visitors come to town, he takes them on grueling bike rides on one of Abu Dhabi's many cycling tracks (he was such a big fan of Italy's super-light Colnago bikes that he purchased a controlling stake in the company in 2020). Tahnoon has also poured money into Brazilian jujitsu. He holds a black belt in the sport and personally recruited his old instructor from San Diego, Nelson Monteiro, to relocate to the UAE to continue training him. Together they have opened a string of gyms and started a championship tournament described as "the Olympics of our sport."[2]

As for AI, Tahnoon developed an interest in the technology in the early 2000s on account of his love for chess. He was spending an increasing amount of time competing in online chess tournaments and was looking for an edge. He eventually recruited a team of engineers and directed them to build a cutting-edge computer program, nicknamed "Hydra," that would help him take on the world's leading players. Once Hydra was built, he spent hours debating chess moves and strategy as he monitored its matches (Hydra's chief engineer described him as the greatest "computer chess freak" alive). In 2017, Tahnoon's interest heightened when he learned that Google AI's new program AlphaZero had just beaten the world's best chess program—only four hours after learning the game's rules. He told associates that AlphaZero had "open[ed] his eyes to the power of AI."[3]

Tahnoon's fascination with AI coincided with a growing recognition by the UAE's leadership that the country needed to diversify its economy away from fossil fuels. Despite attempts to broaden its economic base, the country's wealth remains fundamentally rooted in hydrocarbons. Currently, it produces 3.2 million barrels of petroleum and liquids per day; in 2025, oil revenue contributed about 30 percent of Abu Dhabi's $550 billion GDP. In a major speech in 2017, Sheikh Mohamed bin Zayed signaled that the Emirates required a "long-term plan" to deal with oil's diminishing returns. The UAE's longtime ambassador to the United States, Yousef Al Otaiba, reiterated this point to me: "Oil has gotten us through the last forty-five, fifty years. Might even get us through another ten or twenty years. But it's going to run out. Its economic impact is going to be lower." In its place, the UAE decided to make a big bet on AI technology—to invest billions in new chip factories, massive data centers, and AI startups.[4]

Tahnoon was entrusted with leading the AI charge and given vast funding to turn this vision into a reality. By some measures, this was a consolation prize. Back in 2021, the UAE faced a succession crisis. MBZ had been serving informally as the ruler of the Emirates after their brother, President Khalifa bin Zayed Al Nahyan, suffered a major stroke in 2014. MBZ was set to be officially appointed as the ruler of the country, but the role of the crown prince—the designated successor to leadership—was up for grabs. Tahnoon began campaigning hard for the role, but MBZ insisted that his own son, Khalid, should become crown prince. Tahnoon agreed to set aside his ambitions, but in exchange, he wanted control over the UAE's financial resources. Writes Bradley Hope, "It was this bargain that would ultimately put him in charge of $1.5 trillion in sovereign wealth."[5]

Thus, in 2023, Tahnoon took the reins of the Abu Dhabi Investment Authority (ADIA), which holds over $1 trillion in assets; by many estimates, it is the world's third-largest sovereign fund after Norway's pension fund and the China Investment Corporation. As figure 1 shows, the addition of ADIA expanded Sheikh Tahnoon's already substantial portfolio, dwarfing his chairmanship of another sovereign wealth fund, Abu Dhabi Developmental Holding Company (ADQ), his management of the Emirates' biggest lender, First Abu Dhabi Bank, and his oversight of a pair of investment firms, Royal Group and International Holding Company.

Figure 1. Sheikh Tahnoon's Financial Empire[6]

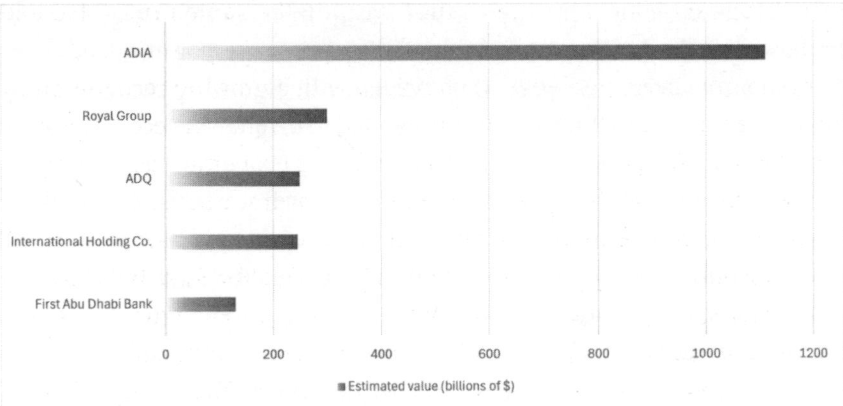

Bill Marczak, a researcher at the University of Toronto's Citizen Lab, who has long studied the UAE's financial ecosystem, has described the constellation of funds and companies controlled by Sheikh Tahnoon as "master classes in blending state power and business."[7]

Despite Tahnoon's high-powered connections and immense resources, he still had to navigate tricky terrain when it came to pursuing his AI aims. A big impediment was the burgeoning tech war between the United States and China. The Emiratis preferred not to choose sides between the rivals. Rather, they viewed their country "as a node within a new multipolar world"—part of a new crop of states challenging the West's economic hegemony. But the UAE faced a basic problem: for now, the path to AI viability ran through American companies like Nvidia. Thanks to America's semiconductor choke point advantage, the UAE was forced to play ball with the United States—and to distance itself from China. American officials were not shy about capitalizing on this situation. As tensions rose between Beijing and Washington, the Gulf emerged as a major prize.[8]

Biden Makes the First Move

In June 2023, Biden's national security adviser, Jake Sullivan, delivered a pointed message to Sheikh Tahnoon at a White House meeting. Sullivan told him that he was concerned about the dealings of G42, a prominent AI tech company based in Abu Dhabi and bankrolled by the royal family (the company's name is drawn from *The Hitchhiker's Guide to the Galaxy*, where the answer to the "ultimate question of life" is 42). US intelligence officers were "particularly fixated" on its well-tailored CEO, Peng Xiao. He was born in China but attended college and graduate school in the United States. Later, he joined the Virginia technology firm MicroStrategy. In 2015, he moved to the UAE to become the CEO of Pegasus, a subsidiary of the Emirati surveillance company Dark Matter, which used spyware to target opponents of the royal family. That same year, Peng renounced his US citizenship for an Emirati one.[9]

Both Peng and G42 maintained strong connections with Chinese business leaders and companies. G42's chief investment officer, Zhang Xiaoping, served as the chief operating officer for Yitu Technology, which the Biden administration blacklisted in 2021 for developing surveillance technology that was used against China's Uyghur population. In 2019, Peng led another firm that launched a social media app, ToTok, a "spy tool used by the Emirati government to track the conversations of its users." Chinese software engineers helped design the app; forensic researchers concluded that it was a copy of a Chinese app, YeeCall. US intelligence agencies also flagged that G42 had formed close ties with Chinese companies, such as Huawei, raising security concerns. American officials feared that Beijing could use G42 to siphon advanced US chips to China. They were also worried that Chinese firms could access the genomic data of millions of Americans through G42's DNA sequencing project.[10]

In subsequent meetings, US officials were blunt: the United States was in an "economic war" with China, and G42 needed to pick a side. "You have to choose," Commerce Secretary Gina Raimondo told Tahnoon. "This is a one-strike-and-you're-out kind of policy." Over the next several months, Commerce officials and G42 engaged in intense negotiations.[11]

By the end of 2023, the two sides had reached an understanding. In return for severing its relationship with China, G42 "was assured" that it would have continued access to Nvidia chips and other necessary components. Subsequently, Peng announced that G42 would phase out hardware from Huawei, ripping out millions of dollars' worth of servers and networking equipment. He emphasized that G42 would disengage from its Chinese relationships and comply with Washington's rules. "The impression we are getting from [the] US government and US partners is, we need to be very cautious," Peng told the *Financial Times* in an interview. "In order for us to further our relationship—which we cherish—with our US partners, we simply cannot do much more with Chinese partners." He framed the company's shift as a necessary choice, adding that G42 had never had deep AI research relationships with Chinese companies "because, frankly speaking, they're not leaders in this domain." Some of G42's adjustments seemed authentic; other actions elicited skepticism. For example, Peng agreed to divest G42's Chinese holdings, valued at $105 billion, which held stakes in ByteDance, JD.com, and other companies. But the investments didn't travel far. The UAE created a new fund, Lunate, also controlled by Sheikh Tahnoon, that took over management of G42's Chinese investments, keeping the assets in the royal family.[12]

Nonetheless, US officials were satisfied with G42's response. They believed the company had made a good faith effort to comply with Commerce's conditions and saw that Peng had the full backing of the royal family. "G42 is going to be the crown jewel—you don't give your crown jewel to someone you don't have confidence in," said Ambassador Otaiba. "And Peng has been there for ten years, so he's not new. I could very confidently say everybody has confidence in Peng."[13] Biden's team was ready to move forward and offered up a carrot—they would help G42 find a US partner who would give it access to advanced chips and help it expand.

Microsoft Enters the Picture

Once Biden officials gave the green light to G42, things moved quickly. The White House hosted executives from top US tech companies for a "meet-and-greet" with Sheikh Tahnoon, with the hope that a US firm "could displace Chinese technology companies long involved in the region." By early 2024, talks began to get serious between G42 and Microsoft. Gina Raimondo was "closely involved" throughout the process; reportedly, as the parties neared a deal, she brokered a series of meetings to get them over the finish line.[14]

On April 15, 2024, Microsoft announced that it would take a $1.5 billion stake in G42, giving G42 access to its cutting-edge GPUs and permission to sell Microsoft services. In return, G42 agreed to use Microsoft's cloud services and abide by strict security arrangements negotiated with the United States. White House officials were pleased. "In a place like the UAE, where you have G42 working very closely with Huawei, for example, we have an interest in changing that picture," said Tarun Chhabra, the National Security Council's former senior director for technology. The UAE was also effusive about the agreement, with Emirati AI minister Omar Sultan Al Olama saying, "Now you're going to see the outcomes of that marriage, if I may use that word, between both G42 and Microsoft, but also the UAE and the United States."[15]

Microsoft received several important benefits from the bargain as well. The company believed that the deal could "lock in" the UAE into Microsoft's technical ecosystem and deliver a sustained advantage over Chinese companies in the race for AI supremacy—at least in the Gulf. Its logic was that once clients were integrated into Microsoft's cloud, the high switching costs—the expense of moving data and reconstructing software infrastructure to another provider—would be a major deterrent for its customers. Executives were convinced that the cloud partnership would not only deliver short-term profits but provided long-term earnings potential.[16]

Microsoft also reasoned that partnering with G42 could help it access markets in Asia, Africa, and Latin America where it had struggled to gain a foothold. At first glance, this seemed puzzling. After all, Microsoft had been in business for over fifty years. What value could a young startup like G42 bring to the company? But from decades of dealmaking, the Emiratis had amassed valuable clout in the Global South. They could open doors for Microsoft in countries like Indonesia and Türkiye and help level the playing field against Chinese competitors like Ali Cloud or Huawei Cloud. A billion-dollar data center deal with Kenya, signed in May 2024, showcased the possibilities and pitfalls.[17]

Kenya is a country on the rise. Nicknamed the Silicon Savannah, it has a youthful population and a buzzing startup scene. "You see this enormous potential," raved Microsoft President Brad Smith, adding that the atmosphere reminded him of Southeast Asia in the 1990s. China already had a substantial presence in the country. Huawei helped construct and fund a $14.5 billion "smart city" called Konza Technopolis. Much of Kenya's digital infrastructure was based on Chinese standards, installed by Huawei, ZTE, and other Chinese firms that had spent decades doing business there. But

American companies offered something their Chinese counterparts struggled to deliver: expertise in cloud computing and AI technology.[18]

The Kenyans were eager to make a deal. In May 2024, Biden hosted a rare Oval Office meeting with President William Ruto. One of the big deliverables was an announcement by Microsoft and G42 that they would spend up to $1 billion building a huge geothermal-powered data center in Olkaria, a volcanic area situated in the Great Rift Valley. G42 would lead the first phase of the project and construct a one-hundred-megawatt facility; the full project could eventually generate up to one gigawatt of electricity. As a sweetener, the Kenyan government agreed to move many of its government services to the cloud—which Microsoft expected would "fuel overall demand" for its products. Company executives saw the agreement as a template for other partnerships in the Global South; with G42's backing, they could expand to more markets and muscle out their Chinese rivals.[19]

Unfortunately, the reality was messier. G42 and Microsoft struggled to figure out a viable business case for Kenyan cloud services beyond government commitments, and months after the deal's signing, stakeholders had yet to develop a sustainable financial model or strategy. "It all came together in a very geopolitical way," said one individual involved in the project's talks. "The UAE government was scrambling to give the US something to make up for their previous dalliances with Huawei." Once in office, Trump's attention strayed from Africa, viewing it as neither strategically important nor central to the US-China tech rivalry. As a result, what should have been a US win became a cautionary tale of poor execution and missed opportunity.[20]

Trump Doubles the Stakes

If Biden unlocked the door for AI partnerships with the UAE, Trump threw it wide open. Trump's interest in the Gulf was not accidental. As he assumed the presidency, Gulf money poured into his family coffers. Trump-branded residential towers were launched in Dubai and Jeddah, Saudi Arabia, along with a new Trump luxury golf resort in Qatar. In May 2025, the UAE bought $2 billion worth of a crypto stablecoin issued by Trump's World Liberty Financial (an outfit overseen by his sons Eric Trump and Donald Trump Jr.). Tahnoon also bought a "secret stake" in World Liberty Financial, acquiring 49 percent of the venture for half a billion dollars with $187 million paid up front. Sovereign wealth funds from Qatar, Saudi Arabia, and the UAE committed over $3.5 billion to a private equity fund managed by Jared Kushner, Trump's son-in-law. Trump even took delivery of a $400 million luxury jet donated from Qatar, an airplane so ornate it was dubbed "a flying palace."[21]

When it came to AI, the Gulf states opened their checkbooks even wider. Tech investment fund MGX (also chaired by Tahnoon) was a lead investor in the $500 billion Stargate project to develop AI infrastructure in the United States, which Trump announced from the Oval Office two days into his second term. State-based funds from Qatar and the UAE were major investors in a $6 billion funding round for xAI, owned by Trump adviser Elon Musk. Meanwhile, ADQ signed a $25 billion partnership in March 2025 with US-based investor Energy Capital Partners to develop power generation for data centers around the country. But the headline deal was a massive $1.4 trillion investment pledge Sheikh Tahnoon made to Trump in the Oval Office committing the UAE to building data centers and energy infrastructure in the United States over a ten-year period. These commitments were not a blank check, however. In return for these investments, the Emiratis expected the United States to "loosen its chip rules" and relax other investment restrictions.[22]

Thus, the stage was set for Trump to return the favor. In May 2025, he took a four-day swing through the region. During his visit, his Gulf hosts spared no expense burnishing his ego. At his first stop in Saudi Arabia, Trump's plane was escorted by three F-15 fighter jets as it landed in Riyadh, the country's capital. Saudi riders on Arabian horses flanked his presidential limousine as it approached the royal palace, golden horns announcing their arrival. At the palace, an honor guard in formation welcomed Trump, who was escorted by Crown Prince Mohammed bin Salman along a sprawling lavender carpet into the royal court. Trump received an equally lavish reception in the Emirates. The UAE awarded him their highest civilian honor and bathed the Burj Khalifa, the world's tallest building, in red, white, and blue lights. Not to be outshone, in Trump's final stop in Qatar, his hosts wowed Trump with the royal court's gleaming white marble ("perfecto," he said) and gave him a tour of one of the world's largest mosques ("Isn't this beautiful? It is so beautiful," he exclaimed).[23]

Most importantly for the Gulf, Trump approved billions of dollars worth of AI deals, describing the trip "like it was a summer fund-raiser on Martha's Vineyard that went better than expected." Led by advisers David Sacks and Sriram Krishnan—both experienced venture capitalists—Trump's team was busy. The big winner was the Emirates. They announced a partnership led by G42 and involving Nvidia, OpenAI, and Japan's SoftBank to build a five-gigawatt data center complex in Abu Dhabi—capable of supporting up to 2.5 million Nvidia B200 chips. This deal dwarfed all previous AI infrastructure announcements. (In comparison, the next largest AI deal globally was the Stargate project, which would support up to 2.25 million Nvidia GPUs). These agreements sent an unmistakable message: AI chips were now part of the bargaining process. Countries that played ball with the United States would receive greater access.[24]

Underlying Trump's strategy were three assumptions: First, Gulf money could help the United States build out its AI compute infrastructure at a faster scale. Second, while the United States struggled to build enough energy infrastructure to meet data center demand, Gulf partners could solve this shortfall by tapping into cheap surplus electricity. Third, exporting heaps of AI chips to Saudi Arabia and the UAE would bring these countries into the US ecosystem—anchoring American technology "as the global standard" before China could catch up. His team argued that rejecting these deals would push wealthy Gulf monarchies into China's hands. As Sacks bluntly put it in a podcast interview, "The choice is, do we want these countries to be the piggy bank for American AI or for Chinese AI?" (As it turned out, these deals may have also been a piggy bank for Sacks; Democrats called on government inspectors general to probe whether Sacks's personal investments had improperly benefited from the agreements, and Trump loyalist Steve Bannon griped that Sacks epitomized the kind of conflicts of interest plaguing an administration where "the tech bros are out of control.")[25]

On the face of it, some of Trump's ideas were sensible and mirrored aspects of Biden's approach. It is true that as global tech competition intensifies between the United States and China, third countries will readily adopt Beijing's innovations over Washington's if American technology becomes too difficult to acquire. The United States' optimal path to preserving its AI advantages may be incentivizing other countries to use its algorithms, buy its products, and lock into its tech ecosystem. As China and other countries have shown, leadership in any tech sector can shift rapidly. Scientific expertise diffuses swiftly. Despite the United States' present dominance of advanced chip design, Chinese firms could eventually catch up. Hence, there is a logic for the United States to secure overseas markets for its companies while the window is open.

Yet, these deals also raised concerns. For one, the Gulf monarchies came with alarming human rights records (an issue Biden sidestepped as well). Back in 2008, under the guidance of former US counterterrorism czar Richard

Clarke, the UAE created a new cyber surveillance entity—later known as Project Raven—to monitor terrorist threats to the regime. In 2010 Clarke ceded control of the unit to other American contractors. Over time, however, the unit's mandate expanded. By 2014, the unit was recruiting dozens of former US intelligence operatives to conduct sophisticated hacking operations against a range of targets, from a Saudi women's rights activist and UN officials to FIFA officers, Qatari diplomats, and independent journalists, reflecting the Emirates' ambition to spy on its residents and enemies at "an industrial scale."[26]

In 2016, Project Raven transitioned into an entity called DarkMatter, headquartered two floors away from the UAE's version of the NSA. DarkMatter deepened its probes against journalists, dissidents, and other challengers to the royal family. One prominent target was human rights activist Ahmed Mansoor, who had landed on the regime's radar after blogging in favor of democratic reform during the Arab Spring. Mansoor noticed that his devices had begun to act strangely—"phones that grew mysteriously hot, suspicious text messages, drained bank accounts." It turned out that DarkMatter operatives had not only had implanted spyware on Mansoor's phone, but had even hacked into his baby monitor to snoop on private conversations. Sheikh Tahnoon oversaw these activities; by 2016 he had become the Emirates' national security advisor, putting him fully in charge of UAE intelligence. (According to reporting from *Wired*, Tahnoon's investment firm, the Royal Group, was the "ultimate controlling party" over DarkMatter.) While the Emirates appeared to have wound down DarkMatter after a series of embarrassing disclosures, the regime continued to surveil its citizens, arrest digital activists, and spread online disinformation. In short, neither the UAE nor its neighbors hesitated to weaponize new technologies for political purposes; the influx of state-of-the-art AI chips risked turbocharging an already robust security apparatus.[27]

There was also the risk of chip diversion. With millions of semiconductors being physically routed to the Gulf, undoubtedly a portion would be siphoned off through smuggling and make their way to "superchip" black markets in cities like Shenzhen. Given past practices, experts anticipated even more chip leakages from the Gulf as a result of these deals. Remote diversion was another concern. As described in chapter 16, Chinese companies do not need to physically possess GPUs in order to access their computing power. Rather, they can use cloud service providers to remotely access these computing resources. Reporting showed that this practice was on the rise. A Reuters investigation in 2024, for example, disclosed that at least eleven Chinese entities had attempted to remotely access restricted

US technologies and services through companies such as Amazon Web Services. The risks were especially acute in the Gulf, where the region's top companies kept strong ties with Chinese firms. "These countries in the Gulf maintain close relationships with Chinese counterparts and maintain a political position between the U.S. and China," said Janet Egan, a senior fellow at the Center for a New American Security. "That concerns me when I think about Chinese actors accessing vast amounts of compute."[28]

Despite Tahnoon's assurances that the UAE had chosen to side with the United States and would jettison their Chinese partners, the reality was more complicated. The Emirates' economic ties to Beijing remained strong. In 2023, China was the UAE's top non-oil trading partner, totaling $82 billion worth of economic activity. China also offered certain security benefits that the United States had been unwilling or unable to provide. A top concern for any regime is political survival. This is particularly important for hereditary monarchies like Saudi Arabia and the UAE, which lack democratic legitimacy and must remain vigilant against internal threats. As such, Gulf State regimes regularly deploy spyware surveillance, internet censorship, and other tools of coercion to keep their opponents in check. It's little surprise that Freedom on the Net's global rankings place both countries in the "not free" category, alongside countries like Belarus, Ethiopia, and Uzbekistan. Chinese authorities recognize the UAE's internal security dilemma and furnish them with security cooperation tools and law enforcement training to help its leaders maintain stability. Developing these ties is beneficial for China, allowing it to "compete for influence" with the United States while signaling the UAE's strategic independence. Says a former senior Biden official, the Emirates "have this complicated relationship between Washington and Beijing that's partly about their interests, partly about their values, partly about [the fact that] they don't believe they can clearly rely on the United States."[29]

In today's globalized world, tech innovation no longer follows a hierarchical path, meaning that middle powers like the UAE are well placed to exploit their competitive advantages. "We're small. We can make decisions quickly. We can allocate resources and funds. So there's certain things we are good at," Otaiba told me. "I think our goal is to provide things that are useful to our country and our society, and I think that's exactly what we're doing."[30]

The United States has shown a willingness to use both carrots and sticks to win the tech war. While its strategy depends heavily on export restrictions to contain its main rival, its dealings with the UAE and other Gulf monarchies show that it is also willing to play matchmaker. In turn, the Emirates have openly cultivated US support. For now, America's tech leverage remains high; its most advanced chips are nearly impossible to replicate, and its AI models remain globally dominant. But this isn't a persistent condition. Over time, other countries will catch up. Innovation will spread and new models will arise from different corners of the world. Power in the digital age flows in both directions. As countries like Türkiye, Israel, and Ukraine have shown, new ideas and adaptations are not bounded by geography or population size.

PART VI

Consequences

24

Power Struggle

In October 2022, Secretary of State Antony Blinken took a trip out to Stanford University, in the heart of Silicon Valley. On the steps of Encina Hall, he told the gathered press that the world was at an inflection point. "The post–Cold War world has come to an end, and there is an intense competition underway to shape what comes next," he said. "At the heart of that competition is technology."[1] Subsequent global events have only reinforced this logic. Competition is escalating in fierce and unpredictable ways. Technology lies at the center of global rivalries. New breakthroughs, particularly in AI technology, are sucking up billions in resources and scrambling alliances. One of the biggest uncertainties relates to the fact that we are still in the early stages of the AI transformation. Major change is coming, but the consequences are unclear and the time frame is uncertain.

The advent of AGI and related innovations in robotics, biotechnology, and quantum will potentially remake the world. It's not just about war. In the economic realm, work itself will be transformed. Millions of jobs will disappear. Vast sectors will become automated. The convulsions experienced by the federal workforce due to Elon Musk's DOGE team will be magnified many times over. The disruptions will not be limited to large countries like China or the United States as the first adopters of AGI. They will rapidly diffuse to big and small nations, rich and poor, advanced and developing. Most countries have no plan for how to deal with the coming changes.

Meanwhile, the global institutions set up by scarred leaders in the aftermath of World War II—the United Nations system, the Bretton Woods institutions—are already being gutted before our eyes in a fit of self-absorption and shortsightedness. The United States, guardian of the international system, has taken a revisionist turn. It is retreating behind protectionist walls, committed to curbing globalization, and inhibiting the free flow of people and commerce. The America First movement decries any responsibility for advancing common values or maintaining peace and stability in a tumultuous world. Rather, it demands tribute from anyone benefiting from US aircraft carriers, stealth bombers, or precision missiles.

China, the world's emerging superpower, does little better. Following America's stumbles, Xi is in prime position to fill the vacuum left from America's retreat, but it is hard to cast off ingrained habits. He is suspicious of other countries' motives, insecure about China's rise, and views the world in zero-sum terms. Rather than offer generosity and magnanimity—to distinguish from America's crudeness—Beijing sticks to its coercive playbook. The same day Trump lit the global trading system on fire and declared a tariff war on more than 180 countries and territories, China's military staged live-fire exercises around Taiwan. Code-named "Strait Thunder-2025A," the exercises were a test of China's ability to impose a military blockade against the island and launch precision strikes against key targets.[2] China's message was clear—don't rely on Beijing to salvage the international order.

Other leaders are watching these developments and drawing conclusions. They see how powerful technologies provide a strategic logic of their own. They understand that America's defense shield is no longer a sure bet. They recognize that it is up to them to secure their sovereign interests and define the rules in their countries. As competition intensifies in this turbulent new era, what can we anticipate?

Intensified Tech Rivalry Between Washington and Beijing

One certainty is that the tech rivalry between the United States and China will persist, particularly as Beijing's efforts gain momentum. In critical sectors, China is rapidly catching up to the United States—if not poised to supplant it. Just ask US Space Force chief General B. Chance Saltzman. In an interview with *Politico*, he acknowledged that "the pace with which they put counterspace capabilities into play is mind-boggling." Beijing is launching "hundreds of satellites" as part of a new targeting system and is collaborating with Russia to build a moon station. In the past, US agencies focused attention on Russia's development of a space-based nuclear weapon as the "closest alligator to the boat" when it came to national security, but China's pursuit of space and satellite dominance suddenly presents a graver risk to US interests.[3]

Even in symbolic areas, such as the race to send humans back to the moon's surface, China is gaining advantages. America's lunar program remains beset by delays—from design problems with spacesuits to heat-shield and mechanical issues with lunar-landing vehicles. NASA's Artemis III program was originally scheduled to launch to the moon in 2024. After several revisions, the mission hopes to commence by mid-2027. But many experts anticipate this timeline will be delayed further and that China has a realistic chance to get its astronauts to the moon first, by 2030. For a country that put its first astronaut into space only in 2003, this represents an "extraordinary trajectory" for its program and epitomizes Beijing's ambitions to break technological barriers.[4]

When it comes to the race for AGI, it is anyone's guess whether the United States or China will come out ahead. But a few aspects are coming into focus.

First, American lead in AI is narrowing. There are big questions about whether the US scaling approach—building ever-larger data centers in order to generate more computing power for massive models—will win the day. This wager is contingent on leveraging America's chip advantage and presumes that China won't find other ways to make up the hardware gap. So far, it has produced mixed results, but some experts contend that inference scaling needs—computing power required to actually use AI models—will keep American companies ahead of their Chinese rivals.

Second, being a fast follower can be beneficial. Chinese firms like DeepSeek and Alibaba have piggybacked off innovations spearheaded by American companies, creating "good enough" products at a far lower cost. "Part of the reason OpenAI and Anthropic require such massive budgets is that they are at the forefront of exploring entirely 'new capabilities,' entailing countless dead ends and failed attempts," writes Peng Zhang for the *Geopolitechs* blog. "In contrast, DeepSeek essentially worked 'with the answers already in hand,' heading straight for efficiency and practicality—and thus avoiding much trial and error."[5] This approach only presents a problem for China if the US innovation lead sticks—if the breakthroughs accrued by American companies translate into durable adoption advantages that Chinese competitors cannot overcome. But if China can follow close behind US advances (and reap advantages by avoiding comparable personnel and capital expenditures required to achieve them), then the American edge will prove fleeting.

This leads to a third insight, which is that the goal of AGI itself remains fuzzy and ill defined. While some strategists contend that the innovations it unlocks will be so significant that they will give a country a decisive advantage over its competitors, this claim warrants scrutiny. For one, they assume that attaining AGI is a binary condition—once secured, it will allow you to "solve all sorts of problems you have been trying to solve that you couldn't solve before." However, it appears unlikely that AGI will function in such a manner, where the goal line is reached at a specific moment. More likely is that over an extended period of time, AGI will open up an ever-increasing range of capabilities that companies and militaries will harness for different purposes.[6] If that's the case, the game will not be won by who gets to AGI first; the winning side will be the economy that can most effectively adopt and diffuse AI innovations into its products, services, and businesses. Successful adoption requires a few elements: quality, cost-effective pricing, and a functional business model. "It's not enough to be successful in generating new ideas," Robert Atkinson, head of the Information Technology and Innovation Foundation, has pointed out. "Even technological breakthroughs can become a dead end unless they can be translated into working products that are able to be delivered and deployed into the larger marketplace."[7]

Historically, America excelled at innovating and diffusing new technology, from electricity and aeronautics to automobiles. Policymakers structured US laws so that applications invented in public labs could be transferred to the private sector and commercialized. This model allowed a university professor working out of a federally funded lab to take their insights, form a startup, and commercialize their innovations. That framework, spearheaded by Stanford's engineering dean, Fred Terman, in the 1950s and '60s, transformed Silicon Valley into an innovation powerhouse. Over decades, the net result has been the creation of "thousands of spin-off companies, increased technology transfer, and greater innovation."[8]

China's economy, in comparison, remains highly dependent on low-end manufacturing and agriculture, sectors in which fewer companies can take advantage of AGI. Beijing has been slow to develop links between academia and the private sector, and its economy remains largely closed to foreign expertise. An ossified bureaucracy, risk-averse culture, and "blunted market incentives" have all hindered successful AI adoption.[9]

But in recent years, Beijing has made strides. The World Intellectual Property Organization (WIPO), a UN agency based in Geneva, produces an annual ranking of economies by their ability to absorb and diffuse knowledge. The ranking looks at sectors driving innovation—intellectual property receipts, high-tech exports, and information and communications technology services. In the 2020 index, China placed fourteenth globally. By 2025, it had moved into tenth place, supplanting Germany. (Over the same period, the United States stayed in third place.) This points to a shift. Despite intrusive government bureaucrats who frequently impose short-sighted rules on companies, China has developed a dynamic tech ecosystem featuring cutthroat competition, rapid cycles of adaptation, and a market that rewards breakthroughs with capital and profit. "From electric vehicles to QR codes and robots which deliver room service, the Chinese consumer is exposed to more sophisticated technology than the American one," writes *The Economist*. "Some tech figures now believe that China, not America, is the quintessential adopter."[10]

Moreover, in contrast to the continuing decline of factory production in America, Xi has doubled down on China's manufacturing capacity. Banks have lent an estimated $1.9 trillion to industrial borrowers over the last four years. The surge in industrial investment is a recent phenomenon. As recently as 2020, China was in the midst of a real estate frenzy, with construction cranes dotting city skylines across the country. After the real estate bubble burst, the government directed banks to shift their lending to industry. The influx of capital caused companies to upgrade facilities with robotics and automated systems and triggered major factory expansions.

Tech conglomerate Huawei is a good example. In 2024, it completed a $1.4 billion research center in Shanghai designed to house thirty-five thousand engineers. The massive complex, which has ten times as much space as Google's headquarters in Mountain View, California, consists of eight blocks and 104 buildings full of labs, offices, recreational areas, and coffee shops. To help staff get around, the entire campus is connected by an internal railway system. China's automotive sector is also being transformed. For decades, the largest car factory in the world was Volkswagen's industrial complex in Wolfsburg, Germany, which spans over 70 million square feet and can make thirty-eight hundred cars daily. Today, Chinese EV maker BYD is building two auto plants in China, "each capable of producing twice as many cars as Wolfsburg." (One of the plants, rising in the central Chinese city of Zhengzhou, will reportedly cover fifty square miles, an area larger than San Francisco and ten times the size of Tesla's Gigafactory in Nevada.)[11]

It isn't certain that China will leap ahead of the United States in the race for tech leadership. But prior assumptions that Beijing was too stymied by its bureaucracy to keep pace with American innovation, or that China lacked requisite industrial capacity to rival the United States' output of advanced goods, are mistaken.

Global Diffusion Deepens

A key argument in this book is that technological diffusion—the outward flow of expertise, innovation, research, and ideas—is inevitable.

Throughout history, one nation or military has broken new ground, only to see its competitors catch up and pioneer new adaptations. We've discussed at length the innovation trajectory of drones—how Israel in the 1980s and later the United States dominated the development of UAV systems. But as their utility became apparent, other countries jumped into the game. By the early 2000s, China had begun exporting cheap, "good enough" models to a diverse client base. Later, Iran, Israel, and Türkiye began prototyping their own models, adding features that proved valuable on the battlefield. Today, the drone market has exploded; not only has the United States lost its monopoly on UAV innovation, but many times its companies struggle to keep pace with their competitors.

The same logic holds true for general-use technologies like AI, biotechnology, or robotics. America's lead is not immutable—in an era of globalization, other countries will be fast on the United States' heels to produce more nimble, cost-effective, and capable innovations. As geopolitical turbulence intensifies, more countries will be spurred to acquire and develop advanced technologies for both military and civilian purposes. Nations will compete ever more ferociously to build critical infrastructure, strengthen their domestic capabilities, and secure access to vital components—from advanced microchips to hyperscale facilities. The race for AI dominance is not limited to US and Chinese labs.

France is a good case in point. French President Emmanuel Macron is worried about his country's prospects for developing AI technology. "We are not in the race today. We are lagging behind," he told *The Wall Street Journal*. "We need an AI agenda because we have to bridge the gap with the United States and China." He fears that France—and Europe as a whole—will be relegated to being AI consumers, lacking control over how the technology is developed and used. His ambition is for France to spearhead a "third pole of AI power" that can counterbalance Beijing and Washington. To jump-start this effort, Macron has unveiled a daring proposal: harness nuclear energy to power AI systems. He has pledged to dedicate a gigawatt of nuclear energy to support a wave of AI projects that will cost upward of $113 billion. As soon as 2026, he hopes to have a first tranche of 250 megawatts of power ready to hook up to AI-computing chips. "I have a good friend on the other part of the ocean, saying, 'Drill, baby, drill,'" said Macron, referring to Trump's efforts to enlarge American oil drilling. "Here, there is no need to drill. It's just plug, baby, plug."[12]

France isn't the only country making a bid for AI sovereignty. In Saudi Arabia, Crown Prince Mohammed bin Salman is developing a new AI project he calls "Project Transcendence." He hopes it will put his kingdom on the map as a tech hub and keep pace with neighboring UAE. Prince Mohammed has assembled as much as $100 billion to invest in data centers, startups, and related AI infrastructure, and has recruited blue chip companies like Google to help seal the deal.[13] France and Saudi Arabia are joined by a cohort of ambitious leaders, all of whom view AI as critical to their economic and security interests, who are distrustful about staying dependent on Chinese or American tech, and who are willing to commit major funds to keep pace with their rivals.

In short, the AI tech race is getting crowded. Loads of nations are elbowing their way into the arena as well. As US policymakers navigate these dynamics, they should tread carefully. America's aggressive attempts to impose its will on other nations—from global tariffs to tech controls—risk backfiring. Rather than causing countries to bend the knee to Washington, a likelier outcome is pushing other leaders toward greater independence and accelerating diffusion.

Tech Reshapes War

Since the outbreak of war in Ukraine in 2022, the battlefield has undergone a considerable transformation. Accounts from the front lines show how drastically technology has altered conflict. Every soldier is aware of the distinctive buzz of drones flying overhead and the deadly price of failing to find cover. Even newly arrived troops who have yet to witness drone warfare firsthand have observed quadcopters' effects through Telegram and YouTube clips.

Contrary to the flashy presentations organized by trendy startups that describe war in a clinical, sanitized manner, the "in-the-mud reality" is far different. One fighter from Ukraine recounts a skirmish involving his US-made Bradley infantry fighting vehicle, which was initially attacked by a mortar round that exploded on top of the vehicle. The soldiers were rattled but remained calm, having experienced artillery shellings many times in the course of the war. But the next move caught them by surprise. Multiple small drones began to swarm the Bradley. "They targeted the weakest points of the armored Bradley with a deadly precision that mortar fire doesn't possess," recounted *New York Times* reporters. "One of the explosive drones struck the hatch right above where the commander was sitting. The team's driver was hit, 'his eye blasted from its socket.'" Another soldier had his arm torn off. The two soldiers lived, but the strike demonstrated how a constantly adapting array of drones—largely developed from off-the-shelf technologies—became the new weapon of choice on the battlefield, subjecting "enormous numbers of human beings to new ways of killing and dying." Today, it is drones and not heavy artillery that are responsible for the bulk of Russian and Ukrainian casualties, making up about 70 percent of deaths and injuries.[14]

As military innovations advance, AI technology has become a central focus. The potential applications are considerable. One area of intense effort is autonomy. As previous chapters describe, armed forces are committing substantial resources into developing machines that can navigate from point A to point B using AI object recognition with minimal intervention from humans. In the Ukraine war, this has taken on new urgency due to counter-drone jamming (which interrupts the signal between the human operator and the drone), resulting in scores of UAVs being shot down. Progress is occurring at a quick pace. At the start of 2024, the best AI systems could lock on to targets from about five hundred meters out. By the summer, these systems had increased their autonomous range to one kilometer, presenting real challenges to traditional jamming. In 2025, their range has extended to more than two kilometers due to higher-resolution cameras. Currently, only a fraction of drones incorporate AI technology, but their numbers are poised to grow. "Computational power, high-fidelity sensors and microelectronics are all improving at decreasing cost," writes Shashank Joshi in *Survival* journal. "In a couple of years, AI guidance will likely be prevalent in the drone fleets of both Russia and Ukraine."[15]

Another priority for militaries is reducing the "fog of war," an idea attributed to Prussian military strategist Carl von Clausewitz, who wrote that "war is the realm of uncertainty; three quarters of the factors on which action in war is based are wrapped in a fog of greater or lesser uncertainty." The modern battlefield is beset by complexity—constantly emerging threats, incomplete information, and complex maneuvers that demand rapid decisions. Machine learning can help. Algorithms can sift through mountains of data and satellite images to spot relevant items. On the battlefield, AI systems can look for enemy tanks, identify land mines and other hazards, and detect opposing troop movements. For intelligence officers, AI tools can scour "thousands of pieces" of open-source information—articles, videos, news clips, social media posts, reports—to detect signs of an incipient terrorist attack or expose clandestine operations.[16] A more valuable function is the machine's potential to analyze satellite images and intercepted keywords and suggest what the assembled intelligence actually reveals. For companies like Vannevar Labs, which signed a $99 million contract with the Pentagon's Defense Innovation Unit, the goal is for an AI chatbot "to actually talk through an analysis with someone interactively."[17]

Of course, these technologies come with risks. One paradox is that while military AI will potentially give war fighters a clearer view of the battlefield, other hazards become more opaque. The pace of fighting will quicken, forcing officers to make ever quicker decisions. "As the models hand down judgements increasingly regarded as oracular, their output will become ever harder to assess without ceding the enemy a lethal advantage," writes Joshi. "Militaries will fear that if they do not give their AI advisers a longer leash, they will be defeated by an adversary who does."[18] Two particular challenges come to mind.

First, these systems only generate recommendations as good as the data they're presented with. Ensuring high-quality, relevant data is essential. But military contexts are poor places for this. Commanders struggle to obtain accurate information about opposing forces and often must work from skewed sources. At other times, militaries are constrained by data scarcity—there simply isn't sufficient information with which to generate an informed decision. As Arthur Holland Michel writes for the UN Institute for Disarmament Research, "conflict environments are harsh, dynamic and adversarial"; the real-world data of the battlefield always brings more variability than "the limited sample of data on which autonomous systems are built and verified." Thus, entrusting AI systems to make life-and-death judgments in war carries significant liabilities.[19]

A second problem is bias. Humans have a well-established tendency toward automation bias—the inclination to defer to suggestions made by automated decision-making systems, "particularly in circumstances where a human decisionmaker would have reached a different conclusion." Consider our dependence on Google Maps for driving directions. Several years ago, I found myself navigating the back roads of southern Indiana on my way to Louisville. Google Maps kept putting me on smaller and more isolated country lanes. While I was familiar with the area, it had been years since I had last driven on these roads, and I was gradually becoming lost. Soon, I found myself in a tight spot. The county had recently experienced heavy flooding, and the small road I was following suddenly reached a washout point. Despite the app's urging, I couldn't advance farther without putting my rental car into five feet of floodwater. I turned the car around and followed an alternate set of directions generated by the app. I had become increasingly hesitant to follow Google Maps, but it still seemed easier to trust the app than to come up with another plan. Ten minutes later I once again found myself blocked by a flooded road and forced to reverse course. Finally, I gave up on the computer. I pulled over to the side of the road, asked another human for directions, and within fifteen minutes I was back on the main highway headed toward Louisville. As my driving experience shows, machines can be unreliable. They can mislead users by "confidently presenting incorrect information, fabricating justifications, or increasing user acceptance of erroneous recommendations."[20] While the stakes for me were low, soldiers in war do not have the same luxury.

A related concern is organizational bias. Institutions that depend too heavily on automated decision-making systems—due in part to their perceived ease of use—can reach faulty decisions, especially in unusual conditions where there is a high degree of risk tolerance. A startling real-world example is the IDF's reliance on the Lavender and Gospel AI targeting systems in the Gaza conflict, described in detail in chapter 21.

Following the October 7 attacks, Israeli forces were under intense pressure to eliminate Hamas leaders and show results quickly. They needed a system that could generate thousands of leads right away, and Lavender/Gospel delivered, producing kill lists that human personnel largely adopted without second-guessing. The technology allowed for an unprecedented escalation against Hamas targets, but it also generated errors in about 10 percent of all cases, "occasionally mark[ing] individuals who have merely a loose connection to militant groups, or no connection at all."[21] The result was a catastrophic rise in civilian deaths—illustrating the IDF's struggle to properly balance its aggressive reliance on machine-identified targets with civilian protection.

A Race to the Bottom?

In the race to embrace ever more powerful technologies, the human cost of their deployment is often overlooked. In chapter 3, we discussed how Israeli and Turkish drones deployed at the behest of Azerbaijan's autocratic leader, Ilham Aliyev, put an end to Nagorno-Karabakh's independence claims. The operation was a textbook example of how a sophisticated military could use cutting-edge technology to defeat its outmanned opponent. Within days, the operation had laid waste to Armenia's forces, destroying hundreds of tanks, armored fighting vehicles, and missile defense systems. But lost in the narrative of the battle was the human toll.

Under the Soviet Union, the predominantly Armenian mountain enclave was officially assigned to the Azerbaijan Republic as part of its "divide and rule" strategy. When the Soviet Union collapsed, in the early 1990s,

Armenia forcibly took control of Karabakh and held it for the next thirty years, before losing a chunk of the territory to Azerbaijan in the 2020 war. Three years later, Azerbaijan completed its military campaign and established complete authority over the area. The result was devastating to the villagers who had lived in Karabakh their entire lives. In the aftermath of hostilities, more than 100,000 citizens fled into neighboring Armenia, emptying a region that had prewar population of 120,000. Few Armenians remained behind—State Department officials estimated that perhaps thirty families had stayed. By the following year, no ethnic Armenians remained at all, leading to allegations of ethnic cleansing. As displaced residents huddled in makeshift refugee camps, Azerbaijan television showed giant bulldozers razing Karabakh's former parliament building. "We are nobody," said Artak Beglaryan, the former ombudsman of Karabakh. "It is a ghost country."[22]

Such is the humanitarian cost when advanced weaponry is made available to midsize countries with few restraints attached. In the case of Azerbaijan, an authoritarian state whose government ranks among the most repressive in the world, the consequences were tragic. But Azerbaijan is hardly an outlier. The same pattern has replicated itself in other overlooked regions of the world—from Ethiopia and Sudan to Libya—where unaccountable militaries armed with sophisticated drones, loitering munitions, and other advanced tools have laid waste to their populations and entrenched illegitimate leaders in power. The data shows that there is a direct link between the rise of drone production in countries like Türkiye and Iran and the spread of weaponry to rights-violating regimes—places that would struggle to obtain this equipment from democracies.

Yet the issue extends beyond strong regional actors supplying weapons to less militarily capable nations. Off-the-shelf lethal technologies—consumer-grade drones, 3D printing, operational software—have become so widely accessible that there is a high risk that insurgents will use these tools to target civilians. Such was the example set by Hamas in the October 7 attacks, a combined operation involving quadcopters, precision rockets, and GPS jammers that led to the slaughter of over twelve hundred Israeli citizens and foreign nationals. The dilemma is simple: disruptive technological innovations are reshaping the market by lowering barriers to entry and expanding the number of potential participants. "Historically, if you look at armies that want to carry out precise attacks, they were using very expensive weapons, including cruise missiles or laser-guided aviation bombs," says Belkis Wille, a researcher with Human Rights Watch. "What you have now is a world in which drones with arms are cheap enough, and effective enough, that you can carry out precise targeted campaigns against civilians, among other targets, at limited financial cost."[23]

In this new world of proliferating lethal technologies, the absence of clear norms is deeply problematic. While the military conduct of the United States has been far from perfect, it has traditionally been a champion of these values, exercising its influence to push other countries to uphold the laws of armed conflict and build military accountability. Congress even has a statute on the books, known as the "Leahy Law," that prohibits US funds from assisting foreign military units when there is credible information implicating that unit in gross violations of human rights.[24]

Under Biden's presidency, officials established basic guidelines about acceptable and prohibited conduct for emerging technologies. In 2023, for example, the administration released a political declaration on the responsible use of military AI and autonomous weapons intended to provide a "normative framework" to address emerging capabilities. While critics noted that the declaration was voluntary in nature and had been signed by fewer than sixty countries globally, it was a signal that America was concerned about the vacuum of rules for military AI and intended to engage further in fleshing out a responsible-use framework. That being said, even under Biden, the United States was notably reluctant to impose restrictions on new military technologies. At UN debates, American diplomats blocked attempts to expand international law to explicitly prohibit certain uses of lethal autonomous weapons. One of their arguments was that the UN already had trouble enforcing existing treaties limiting wars—what was the point of passing even more regulations? But other diplomats didn't buy this logic. "Just because someone will always commit murder, that doesn't mean that you don't need legislation to prohibit it," said Alexander Kmentt, Austria's chief negotiator on autonomous weapons. "What we have at the moment is this whole field that is completely unregulated."[25]

Under a second Trump administration, these nascent steps have been abandoned. His team has clearly signaled that regulations are out, in favor of unconstrained innovation. They prefer to free up private actors to build powerful AI models rather than think through safety risks in advance, even if this could enable average users to create bioweapons, launch destructive cyberattacks, or proliferate undetectable deepfakes. As Vice President J. D. Vance affirmed at the Paris AI Action Summit, "The A.I. future is not going to be won by hand-wringing about safety." (In fact, the biggest concern articulated by Vance in his remarks was that "AI might exhibit values that differ from Republicans'.")[26]

At the Pentagon, this anti-accountability rhetoric has translated into concrete action. In the opening months of his tenure, Defense Secretary Hegseth fired the top lawyers for the military services, known as judge advocate generals (JAGs), because he didn't believe they were "well-suited" to provide recommendations when lawful orders were given. Their role is to serve as a check on military conduct and ensure American soldiers are following the laws of armed conflict. Along similar lines, Hegseth terminated Pentagon offices and staff focused on preventing and responding to civilian harm during US military operations. He ordered the Civilian Harm Mitigation and Response office as well as the Civilian Protection Center of Excellence to shut down, leaving a gaping hole when it came to determining risks to civilians during drone strikes and related operations. He even fired the director and cut half the staff of the office of Operational Test and Evaluation—established in the 1980s to address criticisms that the Pentagon was deploying weapons and systems that didn't perform safely or effectively (an alarming sign for the age of AI warfare).[27]

These decisions came at an inauspicious moment. As the Iran war shows, American targeting is not infallible. The tragic destruction of an Iranian elementary school in the opening hours of the conflict, leading to the deaths of 175 civilians, underscores the brutal risks of war. Many countries take their military cues from the United States. They see the Pentagon changing its rules and question why they shouldn't reduce protections as well. The window to regulate autonomous weapons and military AI is rapidly closing. Countries have spent years debating how to handle these innovations, but in the past those discussions were largely theoretical. Now, as autonomous weapons are finding their way to the battlefield, time is running out for meaningful action. And yet, under the Trump presidency, the prospect of an international agreement limiting the uses of these dangerous technologies is negligible.

25

Tech War in the Age of Trump

President Biden had a clear theory of victory when it came to competing with China and bolstering America's tech leadership. He sought to preserve the United States' edge in science and technology, which he viewed as fundamental to national security. To that end, he enacted a stringent "protect" agenda that blocked the export of sensitive technologies—such as cutting-edge semiconductors—to China. Concurrently, he initiated a "run faster" approach by investing in innovation at home. The CHIPS Act and Inflation Reduction Act were instrumental pillars in this regard. Central to Biden's strategy was his framing of global divisions as a "battle between democracy and autocracies." He wanted to underscore the shared principles binding America's allies together, and cloaked his foreign policy in "expansive prodemocracy rhetoric." Specifically, Biden's team engaged vigorously with international organizations, convened three democracy summits to lay out the differences between China's authoritarian vision of global governance and America's liberal internationalism, and advanced other norm-building efforts to solidify US power. In a nutshell, this approach boiled down to keeping America's allies united, using democracy framing to reinforce a common vision against China and Russia, and deploying incentives to stimulate science and technology research at home.[1]

When Trump came into office for the second time, some analysts held out hope that he would maintain elements of Biden's strategy. After all, there was more continuity between it and the first Trump administration than most people realized. Back in 2017, Republicans rightly criticized Democrats for being slow to recognize the emerging threat from China. They charged that President Obama had pivoted too hesitantly and too late against Beijing and failed to put in place suitable controls to keep America on top. In his first term, Trump's team placed flagship Chinese companies like Huawei and ZTE under sanction, called out China's rampant surveillance and digital censorship apparatus, and imposed tariffs against key industries. They also worked with allies to put pressure on Beijing, establishing a "free and open Indo-Pacific" concept (one that Biden later adopted into an Indo-Pacific Strategy) and reinvigorated minilateral groupings like the Quad (a security partnership between the United States, Australia, India, and Japan intended to box in China).

But that is not how things turned out the second time around for Trump. In his opening months, Trump tore up Biden's playbook without providing a coherent vision of his own. While he agreed that using financial measures to punish rivals was a good thing, he spurned cooperation with allies. He asked, why should a powerful country like the United States plead with weaker countries to follow its lead? Why was the United States expending major resources "commanding the commons" and offering a security umbrella for other countries without getting anything tangible in return? Trump was done asking other nations for favors. Rather than sweeten the pot for allies, he would deploy coercion instead. He would impose global tariffs against every nation in the world and would add reciprocal measures. It didn't matter whether a country was an ally or an adversary. His idea was to show that America was back by using a stick to browbeat friend and foe alike. "Unlike most of his predecessors in the Oval Office, Trump shows no interest in promoting or defending democracy, the rule of law or free trade," wrote Max Boot in *The Washington Post*. "He is all about power politics in a crude and blustering way that is a disturbing and dangerous throwback to the 19th century."[2]

Trump's measures at home were equally head-scratching. It had become an article of faith that investing in US industry and innovation was critical to preserving American power. Supplementing private-sector dollars with government funds, whether for research or industrial production, would bring jobs, keep the economy humming, and insulate America from overseas shocks. One of Biden's key planks was to inject new capital into strategic sectors—from renewable energy to microchips—to wean America from its dependence on China and vulnerable choke points like Taiwan. In theory, this concept aligned with Trump's America First agenda. Trump's team had also called for a "renaissance in American manufacturing" that would bring good-paying jobs back to US communities and reverse decades of globalization that had hollowed out the country's industrial base.[3] Yet Trump's actions have undercut this very idea.

From the outset, Trump launched a campaign against scientific and technological research, essential elements for competing with China. Thanks to Elon Musk's short-lived Department of Goverment Efficiency (DOGE) effort, billions of dollars in scientific grants were frozen, rescinded, or slowed down. Thousands of federal staff were terminated. Trump reserved special animus against the National Institutes of Health (NIH) and the National Science Foundation (NSF), as well as NASA, institutions responsible for driving pioneering innovations for decades. Table 1 reveals Trump's budgeting priorities during his first year with control of all three branches of government.

Biden Actions[4]	Trump Actions (as of December 2025)[5]
Established the Directorate for Technology, Innovation, and Partnerships in the NSF to support "use-inspired research" on technologies. The NSF opened seven National AI Research Institutes. The NSF launched the National Artificial Intelligence Research Resource pilot. 2025 budget request, R&D highlights: • $729 million for AI R&D at the NSF • $310 million for DARPA's AI Forward initiative • $20.1 billion to the NSF, NIST, and DOE under the CHIPS Act • $92.8 billion for Department of Defense R&D	Canceled or froze 5,484 NIH grants. (Approximately half were later reinstated.) The estimated loss from impacted grants as of November 2025 was $1.86 billion. Terminated 1,996 NSF grants. (About one-third were later reinstated.) The estimated loss from impacted grants as of November 2025 was $705.93 million. Proposed 2026 budget cuts NASA's science funding by 24 percent and slashed $20 billion from the NIH, reducing it by 40 percent. NIH Director Francis Collins and NSF director Sethuraman Panchanathan resigned in March and April 2025. Fired 10 percent of NSF staff. (Some were later reinstated.) Over 154,000 federal workers accepted Trump administration's deferred resignation program. Laid off 1,500 NIH employees, 1,461 Department of Energy staff, 1,300 CDC employees, and 5,625 NASA personnel (as of July 2025).

Table 1. Comparison Between Biden and Trump Regarding Federal Science, Technology, and R&D Spending

Trump's war on science extended beyond gutting federal research funding. The White House directed particular hostility against elite universities. Premier research institutions, such as Columbia, Harvard, and Princeton, were accused of harboring antisemitic bias or promoting DEI initiatives and had billions of research funding frozen. All told, Trump announced that over fifty universities were "under investigation" as part of a broad-based DEI crackdown. But the turnabout wasn't limited to funding rescissions. Graduate students were targeted as well. Immigrations Customs Enforcement (ICE) agents began picking up foreign-born graduate students and placing them in far-off detention centers for daring to criticize Israel's policies. University administrators started to panic, with some issuing blanket hiring freezes, canceling research programs, and putting offers on hold for incoming graduate students.

In the context of America's rivalry with China for AI dominance, these moves made little sense. Almost every AI breakthrough derives from academic research or was developed by university-trained personnel. "Today's AI boom is fueled by the use of specialized computer-graphics chips to run AI models—a technique pioneered by researchers at Stanford who received funding from the Department of Defense," writes Matteo Wong in *The Atlantic*. "All of those chatbots? They rely on a training method called 'reinforcement learning,' the foundations of which were developed with National Science Foundation grants." Indeed, Trump had positioned himself as a champion of AI technology, signing an executive order to "sustain and enhance America's dominance in AI" and enthusiastically announcing the Stargate AI infrastructure venture that would raise upward of $500 billion for cutting-edge data centers.[6] Yet these performative measures couldn't mask an overriding truth: for indeterminate reasons, Trump was willing to cripple America's innovation engine and cede these advantages to US rivals. Whether he had an actual vision for the future or was just making it up as he went along wasn't clear. But one thing was certain: cooperation was out and coercion was in, a shift that would bring serious consequences.

Coercion over Cooperation

Trump's behavior diverges from decades of American strategy and represents a "completely different vision," noted Yale historian Greg Grandin in an interview, "one in which the first principle is that nations don't have shared interests; they have inherent conflicts of interests." This represents a sharp break with eighty years of liberal order building. In the years following World War II, American leaders engaged in a bold experiment: constructing a liberal order that would break past cycles of violence that had culminated in the two deadliest wars in history. Washington created a political system based on mutual interests, "built around multilateral institutions, alliances, strategic partners, and client states." This effort proved wildly successful. The United States negotiated treaties and agreements that reopened the global economy, "ushering in a golden era of economic growth." Rather than punish vanquished rivals like Germany and Japan, America went the opposite route: it bankrolled the Marshall Plan and Japan's reconstruction, investing billions into turning adversaries into allies.

Western powers were bound into pacts of mutual restraint, "finding a solution to the centuries-old problem of how Germany, France, and the rest of Europe could exist in peace," writes political scientist G. John Ikenberry. The United States stood at the center of this arrangement and reaped the greatest advantages. It gained unmatched influence as the owner and manager of the system, obtaining special rights and privileges as a result.[7]

But Trump turned these ideas on their head. Rather than cultivate shared interests in furtherance of American aims, Trump sought to capitalize on the weaknesses they generated. Nations aligned with the United States suffered the most significant impacts. Research tells us that economic coercion is most effective when used against close trading partners and allies rather than adversaries; because countries like Canada and Mexico are so entangled with America's economy, they are also vulnerable to shocks from punitive measures. In contrast, adversaries like Iran, North Korea, and Russia are far better positioned to resist coercion. This explains why Biden's massive sanctions program against Moscow brought disappointing results. Russia's trade dependency on the United States was simply not high enough. Accordingly, the Kremlin did not retreat from the battlefield, the cost of war didn't rise to "intolerable levels," and Putin wasn't compelled to abandon his maximalist goals.[8] The larger and more diversified the target, the harder it is to push around. In this regard, it is also unsurprising that US tech controls against China failed to extract meaningful concessions. On the contrary, as Cornell scholar Nicholas Mulder writes, these measures motivated China to "double down on its ambition to achieve technological self-sufficiency—an effort that has yielded notable advances in chips, electric vehicles, fighter jets, renewable energy, and artificial intelligence."[9]

Trump's reliance on coercion endangers a critical aspect of America's alliances: the formation of trust. Generally speaking, trust in international relations is the willingness of one side to place the fate of its interests under the control of others. This willingness is based on the belief that "potential trustees will avoid using their discretion to harm the interests of the first." Cooperation between two actors requires a certain degree of trust between them—although the actual level can vary depending on the states' relative power and the costs of a potential conflict. If two parties are genuinely trustworthy, they can reassure each other of this fact and eventually cooperate with one another. Ultimately, trust is fragile. "It is typically created rather slowly, but it can be destroyed in an instant—by a single mishap or mistake," wrote University of Oregon psychologist Paul Slovic in a notable study.[10]

What happens when trust dissipates? Washington's policy reversal seeded resentment and incentivized countries to resist America's sway and to "self-insure" against America's unreliability at great cost to them. Allies asked tough questions about the price of protection and whether close relations with Washington would be exploited against them. As tensions ratcheted up over Trump's threat to invade Greenland, Canadian Prime Minister Mark Carney directly addressed this dilemma, noting that "great powers" had begun using "economic integration as weapons, tariffs as leverage, financial infrastructure as coercion, supply chains as vulnerabilities to be exploited." He declared that "you cannot live within the lie of mutual benefit through integration, when integration becomes the source of your subordination."[11]

Brussels faced a particularly hard reckoning. During Biden's presidency, Ursula von der Leyen formed a close relationship with Washington. A watershed moment was her coordination with Biden in the lead-up to Russia's invasion of Ukraine. In 2021, US policymakers began issuing dark forecasts about Moscow's intentions toward Kyiv, but few in Europe were paying attention. Von der Leyen was an exception. After she attended an Oval Office meeting in November 2021, where Biden laid out in stark terms Russia's buildup of troops and weaponry, she sprang into action, directing her team to work with US counterparts on a package of sanctions that could be imposed if Moscow attacked. When Putin gave the go-ahead for war, the EU was ready. In the ensuing months, it rolled out a massive sanctions strategy in lockstep with the United States.[12]

Von der Leyen continued her tight cooperation with Biden through the end of his term. She was quick to recognize the growing threat from China and worked closely with the United States to shift Europe toward a more confrontational posture against Beijing. When hiccups in the transatlantic relationship emerged, such as concerns that the Inflation Reduction Act would put European companies at a disadvantage, "it was von der Leyen who worked with the White House to minimize the impact on the European economy." In fact, her "perceived closeness" to Washington ruffled some feathers in Europe. Former European Council President Charles Michel questioned whether it was wise for Europe to be "hitching its wagon too firmly to the United States." Emmanuel Macron warned against Europe becoming a "follower" of the United States. Yet these tensions were marginal; the core alliance remained solid.[13]

But Trump's second presidency once again destabilized the relationship, forcing von der Leyen to pivot. Prior to the start of Trump's term, the EU anticipated major upheaval and reached out to other countries about forging new trade deals and closer economic ties. (In January 2026, for instance, the EU signed a major trade pact with four South American countries, creating a common free trade market for 700 million people.) As relations deteriorated, von der Leyen delivered a stark message about the challenges facing Europe. "Battlelines for a new world order based on power are being drawn right now," she warned. "Europe must fight. For its place in the world, in which many major powers are either ambivalent or openly hostile to Europe."[14]

Can America Still Prevail in the Tech War?

America's tech war is in trouble. US allies are hedging their bets, overseas regulators are considering crackdowns on American platforms, and countries are seeking alternatives to US products and services. As the White House pursues a series of reckless policies aimed at taming universities and dismantling public research funding, China's innovation agenda forges ahead. Just how much will Trump weaken America's position? What are the prospects that America can still prevail in its tech rivalry against China? Four factors will be determinative: whether the United States maintains its alliances, the extent to which US tech companies remain global innovation leaders, whether Trump follows through on his plans to gut scientific research and destroy America's innovation advantage, and whether China can capitalize on US missteps.

Can the United States Keep Its Allies?

Will Trump sever long-standing alliances and pursue a "go it alone" policy, or will he find a way to coexist with America's strategic partners? Under Biden, keeping Western democracies aligned against China was a top priority. To that end, he invested considerable political capital into bolstering "minilateral" regional groupings, promoting "latticework" security arrangements, and coordinating export controls. His strategy proved effective. Europe began to distance itself from Beijing, walking away from a comprehensive investment agreement in 2021 and enacting a package of economic security measures designed to tighten export controls and prevent R&D from leaking to China. Japan, South Korea, and the United States signed a milestone accord at Camp David in 2024 to strengthen their security cooperation. In short, Biden took Eric Hirschhorn's warning to heart and valiantly endeavored to dam the entire river, not just half of it.

But Trump's approach differed markedly. For starters, his commitment to pursuing a tech war against China was suspect. Far too often, the signals coming from the White House caused head-scratching. For months, officials prepared for the release of a major policy statement on AI that would clarify Trump's approach to securing US dominance in the field. When the policy came out in July 2025, one of its main pillars was strengthening AI export controls, cautioning that "denying our foreign adversaries access to this resource, then, is a matter of both geostrategic competition and national security." Yet, before year's end, Trump would dramatically shift course, giving Nvidia the green light to export its H200 AI chips to China (the H200 is the company's second-best processor—superior to anything offered by Huawei and six times more powerful than Nvidia's previous flagship chip approved for export to China, the H20). Foreign policy experts were aghast. Jake Sullivan called the decision "crazy," telling *The New York Times* that "We are literally handing away our advantage. China's leaders can't believe their luck."[15]

America's allies struggled to make sense of this shift. Were export controls still a priority? Or was the real message that for the right deal, the rules would go out the window?

Trump also abandoned another national security priority when he suspended a key export regulation that would ban subsidiaries of Chinese firms from accessing advanced technology. Experts were aghast, calling the decision a breakthrough for Beijing. "Export controls have now become a tradable item in the relationship," said Christopher Padilla, a top Commerce official under President George W. Bush. "You've discarded decades of precedent." As Trump's commitment to the tech war wavered, his lieutenants were loath to expend precious capital placating US allies. Rather, they heaped scorn on their partners, dismissing them as freeloaders and pressuring them to spend more for their security. (Said J. D. Vance in a leaked Signal chat: "I just hate bailing Europe out again.")[16]

Trump's 2025 national security strategy, released in December, sent an even more chilling message to America's allies. In the document, he soft-pedaled the United States' rivalry with China, backtracking from the steely assertions of his 2017 national security strategy that "China and Russia want to shape a world antithetical to U.S. values and interests." Instead, its focus was on how America could leverage its power to make money. It proposed that Latin America should provide "sole source contracts" for US companies. Taiwan's significance to the United States was reduced to semiconductors and shipping routes. Washington would stop "hectoring" wealthy Gulf states about human rights and prioritize selling them technology. Most notable, was the document's attack against Europe. It declared that the bloc faced "civilizational erasure" and risked becoming "unrecognizable" in two decades due to immigration. It warned that America couldn't afford to write off the continent; rather, it must "cultivate resistance to Europe's trajectory" by enhancing commercial ties, political collaboration, and weapons sales to nations that shared Trump's far-right ideology. As Henry Farrell commented, "This is, quite straightforwardly, a program for regime change in Europe, aimed at turning it into an illiberal polity."[17]

The document "landed like a grenade" in Brussels. Leaders were already hesitant to strike deals with Washington or to stake their nations' prospects on America's promises. Only weeks later, Trump escalated his efforts to pry Greenland from Denmark, drawing stark warnings from European leaders about the potential breakup of NATO. Europe would accelerate its recalibration. Proposals were fast-tracked to replace US defense contractors with domestic suppliers for military equipment. Trade officials contemplated counter-tariffs and reducing intellectual property protections for American companies. Even if Trump's true aim was to win better deals rather than sink America's alliances, his rhetoric and negotiating style brought real costs. He inflamed allies, alienated potential partners, and painted the United States as untrustworthy and unreliable.[18]

These actions were especially unfortunate in light of China's growing advantages of scale. Historically, countries with larger populations, greater geographic expanse, and bigger economies have prevailed over their rivals. In the twentieth century, it was the United States that took advantage of scale, leveraging its land mass and rising population to surpass the United Kingdom, Germany, Japan, and eventually the Soviet Union. "Today, it is China that benefits from that dynamic and the United States is at risk of being overtaken technologically, deindustrialized economically, and defeated militarily by a rival with far greater size and productive capacity," wrote Kurt M. Campbell and Rush Doshi in *Foreign Affairs*.[19] Given these factors, the only viable way for the United States to outcompete China was to partner with allies and revitalize its coalitions. Unfortunately, Trump pushed America in the opposite direction, putting its tech containment strategy at risk and jeopardizing the United States' standing.

Will US Tech Companies Hold Their Advantage?

The second issue is whether US tech companies will remain globally predominant. As previous chapters discussed, the urgency of the Cold War caused Congress to commit billions of dollars toward scientific research. "The United States wanted to build the most powerful bombs and cure the worst diseases. It wanted to be first to explore the outer edges of the solar system. It wanted to grow more efficient crops," wrote Jeremy W. Peters and Andrea Fuller for *The New York Times*. "And so, it offered millions, and then billions, to researchers at universities across the country—in Cambridge, Mass., and Berkeley, Calif., but also in Minnesota, Indiana and Mississippi." This funding turned universities like Stanford into academic powerhouses; a major spillover effect was the creation of a vibrant tech ecosystem that laid the groundwork for Silicon Valley. Thanks to the easy availability of venture capital funding, scrappy startups eventually transformed into billion-dollar corporations that became household names. Today, US tech firms lead the world on everything from cloud computing to AI development. The only country with a reasonable prospect of catching up to America's technology lead is China, and it is only recently that its companies have successfully competed head-on with American firms.[20]

But warning signs are emerging. For one, there is a growing international backlash against Big Tech. This pushback has been building over many years, embodying deep concerns about America's market-driven model, its arrogant approach to personal data, the perceived unaccountability of its platforms, and the massive profits extracted from users directly into the hands of California's billionaire tycoons. Already, Europe had passed a slew of regulations, from AI rules to privacy and digital platform regulations, designed to cut Big Tech's inordinate influence down to size. But Trump's rise to power introduced additional complications.

After his election, US tech companies gambled that staying close to Trump would pay dividends. To that end, they lavished him with financial support, hosted black-tie parties and fundraisers in his honor, and allowed him to claim credit for multibillion-dollar tech investments. At Trump's inauguration, some of the most exclusive seats were reserved for figures like Meta CEO Mark Zuckerberg, Amazon head Jeff Bezos, Google CEO Sundar Pichai, and of course Elon Musk. Trump was quick to tout this realignment, commenting that tech executives supported Democrats until they realized Biden "didn't know what the hell he was talking about." He added, "They were all with him, every one of them, and now they are all with me."[21]

While this show of support for Trump might have looked like smart politics to Big Tech, global reactions told another story. Europeans were dismayed by Silicon Valley's embrace of MAGA and showed their displeasure with their pocketbooks. Unsurprisingly, Musk's companies were the first to pay the price. Tesla sales plummeted, dropping by about 40 percent across the continent by the middle of 2025. Europeans also started moving away from Starlink, with the European Commission probing how it could support domestic alternatives. A looming question is whether the brand damage experienced by Tesla and SpaceX will spread to other US companies. If Europe decides to play hardball, "the casualties could be enormous," writes analyst William Echikson for DC-based think tank the Center for European Policy Analysis. "European regulators could pull the plug on Elon Musk's X, restrict intellectual property rights, and prevent US tech companies from bidding on public contracts or even from investing in the EU."[22]

The cloud services industry is a good example. At present, Google and Microsoft control about two thirds of the European market for cloud computing. But European politicians, academics, and entrepreneurs are increasingly coalescing around the idea that they must develop sovereign cloud infrastructure to reduce their dependence on US technology. Recently, the European Commission appointed its first-ever "technology sovereignty chief," Henna Virkkunen. Calls to create a EuroStack to prioritize using locally developed technologies across digital infrastructure, cloud platforms, and applications are intensifying. And it's not just Europe. Around the world, other countries like Brazil, India, Saudi Arabia, and the UAE are exploring how they can develop digital sovereignty, cultivate indigenous tech champions, and break their dependencies on Silicon Valley. If these efforts gain steam—and are backed by real resources—US tech corporations could find their market positions eroding.[23]

Will Trump Destroy Science and Technology Research?

A third issue is whether Trump will continue his assault against science and technology research—killing the goose that laid America's innovation egg. American universities are highly dependent on government funding for their R&D; in 2023 alone, they received $60 billion in federal money, a thirty-fold increase from the 1950s, when the research system was just getting off the ground. Absent this long-term investment, America's current technological superiority and its development of globally competitive companies would be far from assured.[24]

Yet today, Trump and his conservative allies seem intent on dismantling this system. The White House has unleashed an unprecedented campaign of retribution against universities and scientific funding. It has slashed federal research institutions, frozen billions in academic funding, and sent extortionary letters to university presidents demanding radical changes.

Trump's team has slowed down the pipeline of talented foreign STEM students, revoking thousands of visas with little logic or public explanation. They have even threatened to block certain universities, such as Harvard, from enrolling international students at all. The big beneficiary from these moves is China, which has struggled to retain its own STEM graduates, causing Xi to declare that Beijing "must win the global battle to cultivate 'human capital.'" (Fittingly, just as Trump put in place new restrictions for America's H1-B visas for highly skilled foreign workers, including assessing employers a mandatory $100,000 fee, Xi moved in the opposite direction and established a special "K visa" with loosened requirements for applicants.)[25]

Considerable damage has already occurred. Commentators like economist Paul Krugman have begun making comparisons to the Chinese Cultural Revolution (where Mao Zedong used his influence as a revolutionary leader to wage war on China's bureaucracy and its elites): "Once you've seen the parallel between what MAGA is trying to do and China's Cultural Revolution, the similarities are everywhere."[26] As Trump wields his power to overturn America's governmental and cultural elite, one is forced to ask: How far will this campaign extend? How much harm will accrue?

His confrontational strategy arrives during a period of considerable economic and scientific transition. For centuries, power was derived from tangible resources that a government could control—such as populations that could be conscripted, territory that could be conquered, or even goods that could be circulated or withheld. But circumstances have changed. Today, as Stanford scholar Amy Zegart argues, "countries increasingly derive power from intangible resources—the knowledge and technologies such as AI that are super-charging economic growth, scientific discovery, and military potential." Even before Trump's election, the United States was in a precarious position. America's K–12 educational outcomes had fallen, US universities were struggling to compete globally for academic talent, and "chronic federal underinvestment" in basic research was damaging the country's long-term prospects for innovation breakthroughs.[27]

In contrast, China had dramatically accelerated its investments in scientific advancements—increasing basic research funding at six times the US rate (it is now on track to overtake American spending within the decade). If left unchecked, Trump's policies will hasten these declines. Critical innovations in fields such as battery technologies, nuclear and renewable energy, and robotics won't die off; instead, they will migrate to more supportive ecosystems. "China, Europe, and other regions with rational research funding models will happily fill the void," writes Andrew Dessler, a professor of atmospheric sciences at Texas A&M University. "Ultimately, in the name of saving 0.006 percent of the federal budget, we're surrendering America's technological leadership." Rather than shore up a rickety system in need of reform, there is a very real prospect that the White House's actions will fracture the country's innovation backbone, yielding advantages to Beijing that the United States will struggle to regain.[28]

Can China Capitalize on Trump's Blunders?

Finally, in light of Trump's missteps, can China take advantage of the resulting vacuum? One key question involves whether China can maintain its momentum in cutting-edge technologies and effectively challenge American competitors. Will its scientific R&D efforts bear fruit? Will it continue to achieve breakthroughs in emerging fields? In this regard, the signs are auspicious. As chapter 14 explained, Xi has made notable progress transforming China into a science and technology superpower. Its flagship initiatives, such as Made in China 2025, have reaped major dividends. Even in areas where Chinese firms lag, such as generative AI and leading-edge semiconductors, recent developments show that its engineers are catching up. Of course, these advancements are no guarantee of future success. China's highly personalistic system, centered around the interests of a single leader, is a major vulnerability. As Xi ages and questions arise about his successor, there is no way to predict how domestic politics could interfere with the country's modernization push, making China's long-term prospects uncertain.

A second consideration is diplomatic: Can China exploit Trump's chaos to build an alternative set of international alliances? As Trump alienates allies and infuriates trading partners, Beijing has eagerly stepped into the void. Xi Jinping has mounted a "diplomatic full-court press" to persuade other countries not to submit to Trump's tariff war and to lock arms in solidarity with China. Chinese ministers have held video calls with European counterparts pushing for cooperation. Diplomats have contacted officials in Tokyo and Seoul lobbying for closer ties. Soon after Trump's April tariff announcement, Xi traveled to Vietnam and Malaysia, where he was greeted by "crowds of supporters." His message was clear: China was the leading defender of the global order and the United States had become unreliable. In Hanoi, Xi urged Vietnam to join China against "unilateral bullying." In Malaysia, he called on Southeast Asian governments to "reject decoupling, supply disruption," and "tariff abuse."[29]

Yet it won't be easy for China to capitalize on Trump's mistakes. Other countries are well aware of Beijing's tendency to use economic bullying to get what it wants. When South Korea installed THAAD missile batteries from the United States in 2017, Beijing retaliated by kneecapping Korean conglomerate Lotte, forcing it to close scores of supermarkets and to eventually leave the country completely. Or in 2019, when Mesut Özil, who played for the English premier league football team Arsenal, criticized China's treatment of its Uyghur Muslim population, Chinese state-run television canceled a broadcast of an Arsenal match and removed Özil from the Chinese version of a popular video game. Even his 6 million followers on Chinese social media app Weibo disappeared overnight (along with fifty thousand members of his Chinese fan club).[30]

More recently, China attempted to strong-arm Lithuania—a small Baltic country with a population of less than 3 million. In November 2021, Lithuanian officials allowed for the opening of a "Taiwan Representative Office" in its capital city of Vilnius. Beijing was furious. Chinese officials viewed this office—particularly the name choice of "Taiwan" versus "Taipei"—as a direct affront to the One China principle. Lithuania's decision sparked an immediate reaction. Almost immediately, the country disappeared from China's customs system, "effectively blocking all bilateral exports and imports." Because the two nations had only a small trading relationship, Beijing escalated with secondary sanctions. One month later, its officials warned that international companies that sourced products from Lithuania could also face penalties. Soon thereafter, reports circulated that Continental, a major German auto parts firm with close ties to Lithuanian suppliers, was unable to clear customs in China. Beijing continued to mount the pressure, but Lithuania stood firm, leaving Xi fuming and showing the limits of his coercion strategy.[31]

Now, as China tries to turn the tables by labeling Washington a coercive power, it is struggling to gain traction. A big problem is that Beijing is using coercion to fight coercion; CCP officials have threatened to retaliate against countries striking deals with the United States, asserting that it would "resolutely take countermeasures" against them. Unsurprisingly, these moves have failed to resonate with leaders in Europe and elsewhere. Countries like Australia that have weathered antagonistic relations with China in recent years have flatly rejected calls to "join hands" with China against Trump.[32]

Where Beijing finds a more receptive audience is among nonaligned or hedging nations, including the rising tech powers. In these capitals, where relationships tend to be more transactional, and where the United States has a more checkered history, China has enticing cards to play. If American officials continue threatening other governments—imposing tariffs, disparaging their priorities, and withdrawing soft power inducements (such as terminating foreign aid or ending international exchange programs)—China is well positioned as a more dependable and stable partner.

In short, Trump has already set back America's strategy and undercut its positioning in the global tech war in serious ways. It is premature to conclude that his policies will hand China an enduring or decisive advantage, but momentum has clearly shifted. Trump's devaluing of US alliances is rash and counterproductive. The White House's moves to slash government personnel, shut down agencies, and push deep research cuts are bewildering. Xi's message to other countries is that China is a force for stability and an antidote to Trump's recklessness. In phone calls and remarks, Chinese diplomats suggest that Beijing wants to team up to "uphold the post-WWII world order" now under threat from the United States.[33] Over time, these divergences will become more and more meaningful.

These trends raise a chilling question: Can the United States and China find a path through their deepening disagreements, or are we witnessing the prelude to something far more dangerous than a mere technological rivalry?

26

Managing the Future

Around midnight on March 1, 2023, a piercing alarm from my iPhone jolted me awake in my hotel room in the center of Kyiv. As I struggled to get my bearings, the voice of Mark Hamill, Luke Skywalker from *Star Wars*, rang out: "Air raid alert. Proceed to the nearest shelter," he warned. "Don't be careless. Your overconfidence is your weakness." Bleary eyed, I threw on some workout clothes and stumbled to the elevator. I descended to the hotel's nearly empty lobby, turned in to a narrow corridor, and walked down two flights of stairs into an underground garage. The area had been repurposed into an improvised shelter. There were several makeshift beds scattered in the cavernous room and rigid plastic chairs to sit on. Over in the corner, the hotel had placed a basic stationary bike. One of the hotel guests was working up a sweat, burning midnight calories to distract himself from potential drone strikes overhead. It was unclear how long we would have to wait. As the minutes ticked by, more hotel guests trickled in. The shelter was chilly, and I found myself wishing I had brought my winter coat.

I glanced down at my phone and scrolled through the Air Alert app that had roused me. Its digital interface displayed a map of Ukraine; areas shaded in bright red signified that residents were at risk from incoming Russian missiles, artillery, and exploding drones. The app was developed by a home security company named Ajax Systems on the second day of the war. Municipal air raid sirens built during the Soviet Union were proving to be inadequate; they were "very old-style pieces of shit" that people frequently couldn't hear, said Valentine Hrytsenko, the company's chief marketing officer. Ukrainians were starting to cobble together ad hoc warning systems, but they were undependable and insecure. Ajax partnered with Ukraine's digital ministry to come up with a better solution; within a week they had developed a beta version linked to Ukraine's air defense system. As for Mark Hamill's involvement, he admired Ukraine's plucky resilience, remarking that their resistance to Russia reminded him of *Star*

Wars. Giving voice to the air raid app was his way of contributing. "A fairy tale about good versus evil is resonant with what's going on in Ukraine," said Hamill in an interview. "The Ukrainian people rallying to the cause and responding so heroically . . . It's impossible not to be inspired by how they've weathered this storm." The app proved immensely valuable and was downloaded more than 14 million times in the invasion's first year.[1]

Despite the cold, I began to doze off. After nearly three hours, Hamill's voice rang out once more: "The air alert is over." He signed off with a stirring cry: "May the Force be with you." As I climbed back up the stairs to my hotel room, I thought about the uniqueness of this situation and how much digital technology had shifted the contours of modern war. Here we were in the midst of active hostilities, taking cues from our smartphones about whether to shelter from impending drone attacks, listening to urgent alerts read by a Hollywood actor.

▪ ▪ ▪

My goal for this book is to underscore an essential point: advanced technology has become a powerful tool of competition and a primary focal point of geopolitical rivalry. To be clear, my argument is not that technology itself is responsible for driving conflict; I believe other factors are more significant catalysts of rivalry, such as ideology, territorial ambition, and economic gains. But the race to develop and acquire cutting-edge innovations is playing a major role in shaping how countries compete and wage war, and these dynamics will only intensify in the years ahead. This trend should not come as a surprise; adversaries have vied for technological supremacy since the dawn of history. From Egypt's mastery of the chariot and horse-drawn war technology in the sixteenth century BCE to England's deployment of the longbow in the fifteenth century, this pattern of competition has carried on without pause.

Today, the United States continues to pursue a strategy of tech containment to maintain its leadership and keep its rivals at bay. Its theory of the case is that denying critical technologies to its adversaries will preserve Washington's advantages and hinder its competitors from catching up. But as this book lays out, things aren't so simple. It has proven to be exceedingly difficult to slow China's technological progress. Even in areas of relative US advantage, Chinese scientists, researchers, and engineers have found ways to catch up to America's advances. And they aren't the only ones. Following closely behind are a slew of middle-power countries—both advanced industrial economies and emerging states—that are investing major resources into developing their own capabilities.

American and Chinese leaders face a dilemma. Both sides are committed to leveraging state power and private industry to enhance their capabilities and stop their rivals from progressing. Taken to their logical extremes, these actions will bring the two countries to the brink of confrontation. Yet neither side desires war. How can Washington and Beijing get closer to a managed or peaceful rivalry while preserving their respective strategies, interests, and objectives?

Getting to first steps is hard. "Each country believes it has the upper hand, and each rejects the other's key concepts for international order," write scholars Robert Daly and Robert Litwak. In the nuclear deterrence field, strategists speak about the concept of "escalation dominance." This term was first coined by RAND analysts in 1995 to describe America's ability to deter its adversaries by having the option to retaliate up or down the escalation ladder at its choosing. In the present contest for global power playing out across multiple domains—economic, technological, military, reputational—both sides are searching for new tools that will give them escalation dominance against the other side. But neither China nor the United States has sufficient resources to attain a decisive victory over the other. While there are few limits regarding the extent of destruction either party could inflict on the other, doing so will not get Washington or Beijing closer to actual victory.[2]

Unfortunately, neither side seems prepared to recognize this reality; their views remain shaped by historical narratives. The United States sees itself as a global guardian, victorious in two world wars and then the Cold War. "This legacy of righteous victory, founded on wealth and love of freedom, justifies Washington's continued assertion of global leadership in the United States and beyond," note Daly and Litwak. China's experience over the same period is anchored in victimhood and triumph over adversity. It faced humiliation and defeat from Japan, then civil and cultural upheaval, and finally success as a "benighted nation that rose to rival the resident superpower."[3] These dueling narratives make compromise difficult. Washington is loath to make accommodations, lest it reinforce the notion that America is a declining power giving way to a Sino-centric world order. Conversely, Beijing seeks to bolster the perception that it is a country on the rise, poised to overtake the United States and seize the future.

Fear also plays a role. Both sides are feeding a cycle of mutual distrust. China's rapidly expanding military capabilities and its growing technological progress underscore American fears about the country's mounting ambitions. From Chinese spy balloons floating over the US mainland, ostensibly surveilling sensitive military installations, to warnings about Beijing's cyber intrusion into America's critical infrastructure, US decisionmakers are in a state of high alarm. Likewise, Washington's actions to isolate China from the international trading system, ban semiconductor exports, and force third countries like Panama to divest Chinese companies from their port holdings feed into Beijing's paranoia about US intentions.[4]

As geopolitical tensions simmer, both countries are pushing hard to develop powerful new technologies that they believe will give them crucial advantages. Fueled by distrust of their competitors, governments and companies are racing to build ever more formidable systems with little thought to making sure that they are stable or secure. "If other companies and countries accelerate their development of AI while we are trying to slow it down and make it safer, they will develop superintelligence first and dominate the world," author Yuval Noah Harari told *Wired*. "So we have no choice but to move forward as fast as possible to stay ahead of the unreliable competition." This dynamic sets up a moral dilemma, where the "mere perception" of an AI arms race incentivizes companies and governments to take shortcuts and forgo critical safety controls for the sake of staying ahead. Or even worse, if one side perceived that its competitor was gaining crucial advantages, it might be tempted to take reckless gambles to keep pace, such as tolerating significant loss-of-control risks (i.e., the potential for an advanced AI system to act autonomously in ways that diverge from or clash with human goals or values). For a technology whose safety is dependent on gradual, measured, and cooperative development, such an AI arms race "may be catastrophically dangerous."[5]

Can a solution be found?

In periods of innovation and disruption, it is typical for a sequence of "hope, fear, and resignation" to play out. "States hoped that they could build new war-fighting tools that would ensure victory over their rivals in new domains, and that they could turn those victories into lasting political gains," writes American University scholar Joshua Rovner. But hope quickly turns to fear as leaders consider the consequences of their rivals making better investments and developing more effective operations. The period of fear can be especially hazardous if rivals engage in escalatory behavior out of concern they will fall permanently behind their competitors. In time, however, states settle into "a kind of resignation" about the limits of these innovations and their effect on military operations. Practical and political issues temper early enthusiasm for these technologies, and expectations are lowered.[6]

As Rovner reminds us, this is hardly the first time that major powers have grappled with the destructive potential of new technologies. Just as international leaders in the post–World War II era agreed to create new treaties and institutions, such as the Geneva Conventions and United Nations, to mitigate the grievous effects of modern war, our current age demands a similar effort—developing shared understandings and common rules to govern competition in the age of AI. A good starting point would be for Washington and Beijing to initiate a formal dialogue about the risks from AI and emerging technologies. These discussions could foster mutual trust between both parties while serving as a safety valve to prevent a collision course to war.

Near the end of Biden's presidency, top administration officials began laying the groundwork for exactly this. In June 2024, Secretary of Defense Lloyd Austin emphasized in public remarks that "war with China is neither imminent nor unavoidable" and that both sides had an interest in making sure "there are no misperceptions and no miscalculations . . . that can spiral out of control."[7] Five months later, Biden met with Xi for two hours in Lima, Peru, where they agreed to avoid giving AI systems control over nuclear weapons. The deal came as a surprise. Throughout Biden's term, Chinese officials had rebuffed American attempts to hold talks on nuclear safety or proliferation. One year earlier, following a four-hour sit-down with Xi, Biden called the Chinese president a "dictator" at a press conference. Things did not improve in the following months, with Beijing even canceling a working-group meeting on nuclear arms control to protest an impending US weapons sale to Taiwan. But the looming transition back to Trump, as well as Biden's determination to "emergency-proof" the US-China relationship, paved the way for this small but significant breakthrough. As the two leaders departed, Biden emphasized that such in-person conversations "prevent miscalculations, and they ensure the competition between our two countries will not veer into conflict."[8]

Now it is up to Trump whether America will build on Biden's late-term progress or shirk diplomacy for saber-rattling. Thus far, the signals are mixed. He initially ratcheted up tensions with Beijing and closed off diplomatic channels, but more recently he has pivoted and met with Xi face-to-face. What happens next remains unclear. Washington and Beijing increasingly lack shared priorities, perspectives, and frameworks; their leaders face domestic political pressure that enhances the prospects of a conflict that neither side wants. Perhaps it will take a catalyzing incident, akin to the Cuban missile crisis, to force both sides to the table.

Nonetheless, in this area, "patient, determined diplomacy" could make a big difference.[9]

In addition to holding regular talks, China and the United States could borrow a page from the Cold War and establish basic guidelines to head off errors in judgment and communication breakdowns. "In the 1970s, U.S. and Soviet leaders calmed rising tensions between their navies by setting rules for unplanned encounters on the high seas," writes Lauren Kahn in *Foreign Affairs*. Xi and Trump could adopt a similar path for AI systems, agreeing on "basic guidelines now, along with protocols to maximize transparency and minimize the risk of fatal miscalculation and miscommunication."[10]

This lesson isn't limited to managing the US-China rivalry; it has global implications, particularly in the military sphere, where the gap between AI advances and regulation is especially pronounced. Armed forces are competing to develop autonomous weapons and integrate AI systems into their operational frameworks. The proliferation risks are heightened due to an essential asymmetry: while building advanced models demands significant investment and specialized knowledge, deploying them requires far fewer resources. This dynamic makes AI "much easier to proliferate than nuclear technology," noted Paul Scharre, executive vice president at the Center for a New American Security; the implication is that an AI arms race could arise with few controls in place to limit inadvertent escalation or other unintended consequences.[11]

All nations would benefit from establishing a "floor for conduct" for how militaries will use emerging technologies in war. The current landscape is essentially a free-for-all, with states pursuing security technologies with minimal constraints. This behavior risks a race to the bottom, where regimes deploy ever more lethal combinations of tools for damaging impact. Take Israel's AI targeting systems or Ukraine and Russia's autonomous drones. There are few rules governing how much human control is required and what level of accountability is necessary. In such a vacuum, there is little incentive for tech firms or their military clients to prioritize safeguards. The consequence is that the urgency of war supersedes the dictates of law.[12]

Israel's assassination of Ibrahim Biari, a senior Hamas commander based in the northern Gaza Strip, illustrates the risks. Biari helped coordinate the October 7, 2023, massacres, representing a legitimate military target. Israeli intelligence scoured Gaza, searching for his location without success. Most likely, he was hiding somewhere in Hamas's network of tunnels. Out of frustration, the IDF turned to a new technology "infused with artificial intelligence" to help flush him out. Israel's Unit 8200 had developed a tracking tool ten years earlier but had never before used it in battle. As the pressure to find Biari intensified, Unit 8200 upgraded this system with AI capabilities. Soon, officials picked up Biari's calls and deployed the AI tool, "which gave an approximate location for where he was making his calls." Armed with that information, on October 31, the IDF called in an air strike on Biari's suspected whereabouts, unleashing a barrage of missiles that killed him. What the tool did not indicate—and what Israel either didn't know or chose to overlook—was the presence of hundreds of civilians in the target area. It soon came to light that along with the target, over 125 civilians also died in the strike.[13]

One could argue that the circumstances behind Biari's assassination were highly anomalous, that the AI tool merely provided targeting coordinates and had no role in facilitating the strike itself. But that assumes a static future; that the military kill chain—from target identification to the dispatching of forces and the initiation of attacks on the target—won't incorporate increasing levels of autonomy. Inevitably, it is only a matter of time before automated operations become a standard part of war fighting. And in the absence of clear guidelines about AI's role in warfare, military organizations will likely pursue extreme measures, especially those operating with fewer constraints than US forces.

Take it from none other than Ukrainian President Zelensky. In a major speech at the UN in the fall of 2025, he warned the assembly that "We are now living through the most destructive arms race in human history—because this time, it includes artificial intelligence." He noted that a decade ago, war looked very different. "No one imagined that cheap drones could create 'dead zones,' areas stretching for dozens of kilometers where nothing moves, no vehicles, no life. People used to imagine that only after a nuclear strike—now it's drone reality." His conclusion? "We need global rules—now—for how AI can be used in weapons. And this is just as urgent as preventing the spread of nuclear weapons." Fortunately, countries do not have to start from scratch to remedy this. They can build off existing policies, such as the United States' declaration on the responsible military use of AI and autonomy (voluntary guidelines released in 2023 and endorsed by nearly sixty countries) as a first step toward developing shared norms for these powerful weapons.[14]

■ ■ ■

We're living in an era full of surprises. Innovations that seemed far-fetched just a few years ago are now mainstream. We have entered the machine age, a world replete with self-driving cars, AI agents, robotic dark factories, and autonomous drones. Close to Shanghai in the industrial hub of Ningbo, Chinese EV maker Zeekr's factory is humming. Over eight hundred robots zoom around the plant's floor, "cheerfully trilling Kenny G tunes" to warn people that they are approaching. Robots move containers of aluminum blocks to self-operating lifts, which bring the metal to a smelting unit atop a forty-foot-tall mechanism. The molten aluminum is then shaped into assorted vehicle panels and diverse components, which are transported to the neighboring warehouse by a fleet of automated carts, supplemented by occasional human forklift operators. At the assembly line, "hundreds of robotic arms, working in teams of up to 16," weld the panels into car bodies. This manufacturing work takes place in a "dark factory" area where robots operate without human workers and the lights are turned off, save for brief periods when engineers come by to clean the machines and make sure the equipment is operating properly. Zeekr's facility is not an outlier. Today, dark factories operate all over China.[15]

Viewed in a certain light, this future appears dystopian. Human labor is being supplanted by machines. Workers are relegated to checking on the robots' progress but otherwise keep out of the way. "Most of our colleagues' jobs involve sitting in front of a computer monitor," Pinky Wu, a Zeekr worker, told *The New York Times*. Is this the future we want? But Zeekr's factory also shows possibility. Engineers have been freed up to think more creatively, leaving design drudgery to AI software. "I have more spare time to open my mind and explore for myself which kinds of fashion trends to include in the cars' interior," said Carrie Li, a Zeekr designer based in Shanghai.[16] Technology can inspire both apprehension and creative expression simultaneously.

Pax Americana is giving way to an emerging global framework whose contours remain unclear. Will it resemble the Concert of Europe in the nineteenth century, a system designed to maintain the balance of power between a handful of major powers in order to spare Europe from the scourge of another conflict? Are we entering a Chinese century, dominated by Beijing's material power, manufacturing scale, and centralized autocracy? Or could we see an American resurgence, with the United States rekindling its domestic innovation and inspiring others to follow suit? This story remains yet unwritten, shaped by the collective choices of countless individuals—innovators, risk-takers, technologists, political leaders—whose actions will redefine the world for generations.

For Selçuk Bayraktar, the brash MIT-trained inventor of Türkiye's TB2 drones, cutting-edge innovations are no longer the exclusive purview of large powers. He spent years honing his drone designs, testing prototypes in the mountains of southeastern Türkiye by tracking Kurdish rebels. His UAVs have been embraced by militaries around the world due to their reliability, technological innovation, and cost-effectiveness. And they have proven to be a valuable bargaining chip for Turkish President Erdoğan, enabling him to press for political advantages and extract meaningful diplomatic concessions.

In the Emirates, Sheikh Tahnoon bin Zayed, the country's formidable national security adviser, is staking his country's wealth on transitioning from its hydrocarbon past into an AI future. He wants to turn the Emirates into an AI powerhouse and has committed billions to new chip factories, massive data centers, and AI startups. But he has found that navigating the waters between China's and America's interests can be tricky. US officials have pushed him to sever ties with Beijing—to rip out Huawei infrastructure and partner exclusively with American firms. Tahnoon is cautiously orienting toward the United States, but he isn't ready to let go of his Chinese relationships, either. As the tech war between the two superpowers becomes more inflamed, it remains to be seen how long Tahnoon will be able to maintain this diplomatic balancing act.

DeepSeek head Liang Wenfeng shocked the world when he unveiled a powerful AI reasoning model in early 2025 that measured up to America's leading products—at a fraction of the cost. He came to the industry by developing an AI algorithm to place bets on the Chinese stock exchange, then pivoted to scientific research and developing large language models with astonishing success. In interviews, Liang claimed that he wasn't in it for the money, but wanted to "develop AI for the public good" and to break the monopoly of big tech firms. Indeed, he has since made his revolutionary code open-source, allowing anyone to access it. Now he has become the darling of Beijing's political class and is helming the ship of China's buzziest company through seriously choppy geopolitical waters.[17]

Palmer Luckey, Anduril's controversial founder, loves big cars and military gear; he drives a customized tan Humvee outfitted with machine gun mounts and orange toy guns. In 2014, he was casting about for his next venture after parting ways with Facebook. At a venture capital retreat in British Columbia, he started brainstorming with Trae Stephens and Peter Thiel and landed on a big idea: a startup that would challenge the dominance of the big defense primes and capture a share of the Pentagon's budget. Anduril's selling point would be its technology: it would create cutting-edge software to "rebuild the arsenal of democracy" and bring the

Pentagon's battle management and weaponry into the machine age. So far, their bet has paid off. Anduril has been on a roll with the company now valued at $60 billion. In early 2025, the firm broke ground on a $1 billion "hyperscale" factory in Columbus, Ohio, that will bring four thousand jobs and produce tens of thousands of drones and autonomous systems.[18]

As bombs and missiles rained down on Kyiv in February 2022, Mykhailo Fedorov, then Ukraine's minister of digital transformation, swiftly worked his Rolodex, calling on CEOs and politicians worldwide to help Ukraine's cause. He wasn't responsible for military strategy or leading Ukraine's diplomatic efforts, but he was in charge of something just as important: safeguarding all of Ukraine's data and spearheading the country's digital response to Russia's invasion. His tireless efforts paid off. Firms such as Apple, Spotify, PayPal, and Oracle immediately halted their sales in Russia and offered in-kind support to Kyiv. After Fedorov posted on X pleading with Elon Musk to send Ukraine Starlink terminals, Musk began shipping the devices over. Fedorov kept running at full speed—setting up an IT army to hit back at Moscow and raising hundreds of millions online to bankroll Ukraine's drone campaign. He is "running Ukraine's war like a startup," showing how creativity and quick thinking can keep a country in the fight.[19]

Jensen Huang, Nvidia's rock star CEO, controls one of the most valuable commodities in the world: AI chips that power the world's most advanced computing systems. His products are in such high demand that even tech titans are desperate to acquire more of his GPUs. Oracle CEO Larry Ellison recalled to *Fortune* about a dinner he attended with Huang and Elon Musk in Palo Alto. "I would describe the dinner as Oracle—me and Elon—begging Jensen for GPUs," Ellison said. "Please take our money. Please take our money. By the way, I got dinner. No, no, take more of it. We need you to take more of our money, please."[20] Obtaining Nvidia chips is serious business. Chinese companies are working through middlemen and black marketeers to circumvent export restrictions. Anyone with real AI ambitions knows that getting their hands on Nvidia's coveted microchips is essential, which has made Huang tech's biggest power broker.

We look for clean narratives to make sense of a complicated world. It is no different with tech. Whether advanced innovations are harnessed for destructive or productive uses is not preordained. Ultimately, tech is a reflection of our troubled times. Should leaders and governments choose to resolve disputes through war and aggression—abandoning generations of painstaking efforts to maintain collective peace and security—technology will serve as a catalyst for these fractures. Conversely, if humanity chooses a more enlightened path and focuses its competitive urges toward economic, political, and creative progress—inventing pioneering medical applications, dazzling works of artistry, and more inclusive democracies—then we have a much better shot at a flourishing future.

Appendix I: Rising Tech Powers Analysis

Country	Top 50 GDP	G7/EU/LrgPower/ 5Eyes/Treaty	Top 75 AI	Top 75 WIPO	Top 60 mil exp	Arms exports	Special cases	Rising State
Brazil	x		x	x	x	x		x
Iran	x		x	X	x	x		x
Israel	x		x	X	x	x		x
Saudi Arabia	x		x	X	x	x		x
Singapore	x		x	X	x	x		x
South Africa	x			x	x	x	x	x
Taiwan	x		x	NA	x	x		x
Türkiye	x		x	x	x	x		x
United Arab Emir-ates	x		x	x	x	x		x
Australia	x	x	x	x	x	x		
Austria	x	x	x	x	x	x		
Belgium	x	x	x	x	x	x		

Country	Top 50 GDP	G7/EU/LrgPower/ 5Eyes/Treaty	Top 75 AI	Top 75 WIPO	Top 60 mil exp	Arms exports	Special cases	Rising State
Canada	x	x	x	x	x	x		
China	x	x	x	x	x	x		
Czechia	x	x	x	x	x	x		
Denmark	x	x	x	x	x	x		
Finland	x	x	x	x	x	x		
France	x	x	x	x	x	x		
Germany	x	x	x	x	x	x		
India	x	x	x	x	x	x	x	
Ireland	x	x	x	x				
Italy	x	x	x	x	x	x		
Japan	x	x	x	x	x	x		
Nether-lands	x	x	x	x	x	x		
Norway	x	x	x	x	x	x		
Poland	x	x	x	x	x	x		
Portugal	x	x	x	x	x	x		
Romania	x	x	x	x	x	x		
Russia	x	x	x	x	x	x		

Country	Top 50 GDP	G7/EU/LrgPower/ 5Eyes/Treaty	Top 75 AI	Top 75 WIPO	Top 60 mil exp	Arms exports	Special cases	Rising State
South Korea	x	x	x	x	x	x		
Spain	x	x	x	x	x	x		
Sweden	x	x	x	x	x	x		
Switzer-land	x	x	x	x	x	x		
United Kingdom	x	x	x	x	x	x		
United States	x	x	x	x	x	x		
Argentina	x		x		x			
Bangla-desh	x		x					
Chile	x		x	x	x			
Colombia	x		x	X	x			
Egypt	x		x		x	x		
Indonesia	x		x	X	x			
Kazakh-stan	x							
Malaysia	x		x	X	x			
Mexico	x		x	x	x			
Nigeria	x		X					
Pakistan	x				x	x		

Country	Top 50 GDP	G7/EU/LrgPower/ 5Eyes/Treaty	Top 75 AI	Top 75 WIPO	Top 60 mil exp	Arms exports	Special cases	Rising State
Peru	x		x					
Philip-pines	x		x	x	x			
Thailand	x		x	x	x			
Vietnam	x		x	x	x			

Appendix II: China's Progress Overcoming Tech Choke Points

China's most acute choke points (2018)[1]	China's progress (as of December 2025)
Photolithography for microchips	Companies have just started making DUV machines; reportedly Chinese scientists have built an EUV prototype, but it has yet to produce working chips.[2]
Underwater network connectors	China's ZTT claims to have developed its own wet connectors.[3]
Vacuum evaporators	
High-end bearing steel	
High-end RF components	Chinese firms have shown "significant progress" in engineering RF chips.[4]
Microspheres	Suzhou NanoMicro is catching up for pharmaceutical-use microspheres but still has to import foreign raw materials and reaction kettles to produce microspheres.[5]
Heavy-duty gas turbines	China Energy Investment Corp. commissioned the Anji Power Plant using an advanced gas turbine that's been locally manufactured for the first time.[6]
Aviation design software	Chinese aerospace engineers have a revolutionary software design they say allows them to overcome a major barrier to stealth aircraft development.[7]
Main bearings for tunnel boring machines (TBMs)	Has successfully made its first giant main bearing for TBMs, "a major engineering breakthrough."[8]
High-strength stainless steel for rocket engines	"Chinese scientists have found a way to overcome metal fatigue in a process that transforms common stainless steel into a material strong enough for aerospace and other high-grade applications," according to a study in *Science*.[9]

China's most acute choke points (2018)[1]	China's progress (as of December 2025)
Aviation-grade steel	"A team of Chinese scientists has developed a technique to make aluminum alloys dramatically more heat resistant" overcoming a big obstacle that has limited applications of light-weight metal that is crucial for aerospace and transport.[10]
Transmission electron microscopes	Chinese firm has built the first transmission electron microscope, called the TH-F120.[11]
Operating systems for mobile phones and desktops	Huawei's HarmonyOS for smartphones has overtaken AppleIOS with a 17 percent market share in China.[12]
Lithium battery separators for new energy vehicles	Semcorp may have surpassed Asahei Kasei in separator production. Unclear if manufacturing core equipment.[13]

Appendix III: Data Tables for Drone Strikes and Fatalities

Global Incidences of Air/Drone Events and Associated Fatalities (2020–2025)[1]

Year	Air/Drone Events	Air/Drone Fatalities
2020	6,380	11,379
2021	6,645	13,696
2022	8,524	11,862
2023	15,495	31,881
2024	51,197	39,489
2025	58,272	32,769

Top Conflicts with Drone Strikes and Related Fatalities (2020–2025)[2]

Country	2020–2025 Air/Drone Events	2020–2025 Air/Drone Fatalities
Ukraine	54,265	16,874
Russia	23,683	1,641
Palestine	13,426	53,438
Iraq	12,108	2,147
Lebanon	10,259	4,215
Myanmar	8,567	7,134
Yemen	7,536	7,647
Syria	5,628	4,876
Sudan	2,688	7,153

Country	2020–2025 Air/Drone Events	2020–2025 Air/Drone Fatalities
Afghanistan	1,819	12,742
Burkina Faso	1,195	3,632
Somalia	815	4,098
Nigeria	718	7,793
Global total	146,789	141,170

Acknowledgments

The idea for this book crystalized during an unexpected visit to Kyiv in the spring of 2023, one year into the war Ukraine has been fighting against its much larger Russian foe. I am grateful for my Carnegie companions on that trip, Arthur Nelson and Gavin Wilde, along with Chris Beall, who helped me process the events on that trip and were welcome companions on two very long overnight train rides into and out of Kyiv.

The Carnegie Endowment for International Peace has been my professional home for over five years. First and foremost, I am deeply grateful to Tom Carothers for his mentorship, advice, and intellectual prodding. He has made my writing sharper and more incisive and challenged my thinking when my ideas required more rigor and thought. Various past and present colleagues in the Democracy, Conflict, and Governance program have my gratitude as well. Special thanks to Fiona Brauer, McKenzie Carrier, Shreya Joshi, Brian (Chun Hey) Kot, and Judy Lee for their research assistance at different stages in the book writing process, from uncovering critical information, offering editing suggestions, and helping with citations.

Other Carnegie colleagues who read early excerpts or who were valuable sounding boards for the book include: Jon Bateman, Sophia Besch, Nicole Grajewski, Mike Kofman, Ariel (Eli) Levite, Eugene Rumer, and Scott Singer. Many thanks to Dan Baer, Frances Brown, and Mariano-Florentino (Tino) Cuéllar for your exceptional leadership and support.

The following people generously took time out of their busy schedules to participate in a book workshop organized by American University: Nate Allen, Samantha Bradshaw (who hosted the workshop), Samuel Bresnick, Geoffrey Gertz, Trey Herr, Jonathan Hillman, Isaac Kardon, Jennifer Kavanagh, Jacklyn Kerr, Abraham Newman, Joshua Rovner, and Nicholas Wright (who flew all the way from London to attend). Special thanks to Kate Munro for organizing the workshop logistics. Others whose guidance I benefited from in this process include Matthew Ford and Seva Gunitsky.

Without the support and guidance of my agent, Bridget Matzie, from

Aevitas Creative Management, this book wouldn't have been possible. She's been with me every step of the way, honing the initial idea into something that would be appealing to editors, and providing crucial guidance as I navigated the publishing terrain. I'm very lucky to have her as my agent. Many thanks as well to Vanessa Kerr for her support for the UK market.

Kevin Reilly and his team at St. Martin's Press saw the potential for this book and took a chance on the idea, for which I'm grateful. His guidance has been instrumental in making the book tighter, livelier, and more compelling. Much appreciation to Gabrielle Gantz, Martin Quinn, Laura Clark, Lizz Blaise, Rob Grom, Jonathan Bennett, Lena Shekhter, Ryan T. Jenkins, and Greg Villepique for their assistance. At Monoray thanks to Jake Lingwood for guidance about how to make this a truly global story, as well as to Charlotte Sanders and Chloë Johnson-Hill. Many thanks to Liz Wetzel for her help promoting the book.

I owe a debt to the journalists, scholars, activists, and writers whose research and intrepid investigations I have drawn upon to write this book. Without their curiosity and relentless pursuit of the truth, the world would be far more opaque.

None of this would be possible without the love and support from friends and family. To my parents, Ronald and Wing-Yee Liu, who opened my eyes to the wider world and encouraged me to explore beyond my own horizons, as well as to my brother, David, for his insights and support. To Cheryl, David, Brad, Aaron, and Jacqui, who are a constant sounding board for ideas and stories that have informed the project. To my children, Caleb and Louisa, who keep me grounded and are a daily reminder of the joys of life in the face of serious and weighty topics. Most importantly, I want to express deep gratitude and love to my wife, Elizabeth, for her steadfast belief in the book and in me. I am eternally grateful for her patience and for standing by me every step of this journey.

Notes

Introduction

1. Steven Lee Myers and Zoe Mou, "'New Chapter' in Space Exploration as China Reaches Far Side of the Moon," *The New York Times,* January 2, 2019, https://www.nytimes.com/2019/01/02/world/asia/china-change-4-moon.html.
2. Stu Woo, Clarence Leong, and Micah Maidenberg, "Historic Moon Mission Moves China Ahead in Space Race with U.S.," *The Wall Street Journal,* June 25, 2024, https://www.wsj.com/world/asia/historic-moon-mission-gets-china-one-small-step-ahead-in-space-race-41894d41.
3. Kenneth Chang, "Japan Becomes the Latest Country to Land on the Moon," *The New York Times,* January 19, 2024, https://www.nytimes.com/2024/01/12/science/japan-slim-moon-landing.html; Christian Shepherd and James Kynge, "China's Ambitions in Space: National Pride or Taking on the Americans?," *Financial Times,* May 8, 2021, https://www.ft.com/content/8a6bb0c0–9a6f-46c0–8438–48984c5e32dc.
4. Shepherd and Kynge, "China's Ambitions in Space"; Office of the Director of National Intelligence, "Annual Threat Assessment of the U.S. Intelligence Community," 2024, https://www.dni.gov/files/ODNI/documents/assessments/ATA-2024-Unclassified-Report.pdf.
5. James Marson et al., "Inside the Ukrainian Drone Operation That Devastated Russia's Bomber Fleet," *The Wall Street Journal,* June 3, 2025, https://www.wsj.com/world/ukraine-russia-drone-attack-bombers-cc77e534.
6. David L. Stern, "Ukraine Intelligence Chief Gives New Details on Spiderweb Drone Operation," *The Washington Post,* June 23, 2025, https://www.washingtonpost.com/world/2025/06/23/ukraine-spiderweb-drones-russia-airfields/.
7. Kateryna Zakharchenko, "'117 Drones and 2,200 kg of Explosives': SBU Chief Unveils Ukraine's Secret Ops," *Kyiv Post,* June 23, 2025, https://www.kyivpost.com/post/55032; Stern, "Ukraine Intelligence Chief Gives New Details on Spiderweb Drone Operation."
8. James Marson, Jane Lytvynenko, and Serhii Bosak, "Inside Ukraine's Daring 'Operation Spiderweb' Attack on Russia," *The Wall Street Journal,* December 9, 2025, https://www.wsj.com/world/europe/ukraine-russia-drone-attack-operation-spiderweb-24d821ab.
9. Marson et al., "Inside Ukraine's Daring 'Operation Spiderweb' Attack on Russia."
10. Marson et al., "Inside the Ukrainian Drone Operation That Devastated Russia's Bomber Fleet"; Zakharchenko, "'117 Drones and 2,200 kg of Explosives'"; Kateryna Bondar, "How Ukraine's Operation 'Spider's Web' Redefines Asymmetric Warfare," Center for Strategic and International Studies, June 2, 2025, https://www.csis.org/analysis/how-ukraines-spider-web-operation-redefines-asymmetric-warfare.
11. Stern, "Ukraine Intelligence Chief Gives New Details on Spiderweb Drone Operation"; Francesca Ebel et al., "Ukraine Attacks Russian Air Bases in Far-Reaching Drone Strikes," *The Washington Post,* June 1, 2025, https://www.washingtonpost.com/world/2025/06/01/ukraine-russia-war-drone-attack-siberia/; Natalie Kitroeff, "Inside Operation Spider's Web," *The Daily,* podcast

transcript, June 4, 2025, https://www.nytimes.com/2025/06/04/podcasts/the-daily/ukraine-russia-drone-attack.html?showTranscript=1; Julien de Troullioud de Lanversin, "Ukrainian Attack on Russian Bombers Shows How Cheap Drones Could Upset Global Security," *Bulletin of the Atomic Scientists*, June 5, 2025, https://thebulletin.org/2025/06/ukrainian-attack-on-russian-bombers-shows-how-cheap-drones-could-upset-global-security/.

12. Anastasia Stognei, Fabrice Deprez, and Christopher Miller, "'An Epic Failure': Russia Reels from Surprise Ukrainian Attack on Bomber Fleet," *Financial Times*, June 3, 2025, https://www.ft.com/content/132e4327-11da-4412-b36b-7363604879e6; Marson et al., "Inside the Ukrainian Drone Operation That Devastated Russia's Bomber Fleet"; Bondar, "How Ukraine's Operation 'Spider's Web' Redefines Asymmetric Warfare"; Daniel Michaels and Jane Lytvynenko, "Ukraine's Stunning Assault Roils Russia's Global Military Strategy," *The Wall Street Journal*, June 2, 2025, https://www.wsj.com/world/ukraines-stunning-assault-upends-russias-global-military-strategy-094f8c1c. See also Yaroslav Trofimov, Drew Hinshaw, and Joe Parkinson, "Modern Tech and Old-School Spycraft Are Redefining War," *The Wall Street Journal*, June 8, 2025, https://www.wsj.com/world/modern-technology-spycraft-war-eabda84f?mod=world_lead_pos2.

13. Max Boot, "Ukraine Just Rewrote the Rules of War," *The Washington Post*, June 1, 2025, https://www.washingtonpost.com/opinions/2025/06/01/ukraine-drone-attack-russia-bombers/.

14. Adam Entous, "The Partnership: The Secret History of the War in Ukraine," *The New York Times*, March 29, 2025, https://www.nytimes.com/interactive/2025/03/29/world/europe/us-ukraine-military-war-wiesbaden.html.

15. Edward Luce, "The New Washington Consensus," *Financial Times*, April 19, 2023, https://www.ft.com/content/42922712-cd33-4de0-8763-1cc271331a32.

16. Science and Security Board, "Closer than Ever Before: It Is Now 85 Seconds to Midnight," *Bulletin of the Atomic Scientists*, January 27, 2026, https://thebulletin.org/doomsday-clock/2026-statement/; Sophie Tanno, "EU Urges Citizens to Stockpile 72 Hours' Worth of Supplies amid War Risk," CNN, March 26, 2025, https://www.cnn.com/2025/03/26/europe/european-union-stockpile-member-states-intl-latam?cid=ios_app.

17. Thomas J. Wright, *All Measures Short of War: The Contest for the Twenty-First Century and the Future of American Power* (New Haven: Yale University Press, 2017), 2–3.

18. Rosa Brooks, *How Everything Became War and the Military Became Everything: Tales from the Pentagon* (New York: Simon and Schuster, 2016).

1. The End of American Unipolarity

1. Francis Fukuyama, "The End of History?," *The National Interest* 16 (Summer 1989); Francis Fukuyama, *The End of History and the Last Man* (New York: Free Press, 1992); Thomas J. Wright, *All Measures Short of War: The Contest for the Twenty-First Century and the Future of American Power* (New Haven: Yale University Press, 2017).

2. Wright, *All Measures Short of War*, 3.

3. G. John Ikenberry, *Liberal Leviathan* (Princeton, NJ: Princeton University Press, 2011), 269.

4. Vladimir Putin, "Putin's Prepared Remarks at 43rd Munich Conference on Security Policy," speech, Munich Conference on Security Policy, February 12, 2007, https://www.washingtonpost.com/wp-dyn/content/article/2007/02/12/AR2007021200555.html; Daniel Fried and Kurt Volker, "The Speech In Which Putin Told Us Who He Was," *Politico*, February 18, 2022, https://www.politico.com/news/magazine/2022/02/18/putin-speech-wake-up-call-post-cold-war-order-liberal-2007-00009918.

5. The White House, "Memorandum of Conversation. Subject: Meeting with Boris Yeltsin, President of Russia," transcript, February 1, 1992, National Security Archive, https://nsarchive

.gwu.edu/document/29670-document-8-memorandum-conversation-subject-meeting-boris-yeltsin-president-russia.

6. William J. Burns, "How the U.S.-Russian Relationship Went Bad," Carnegie Endowment, March 8, 2019, https://carnegieendowment.org/posts/2019/03/how-the-us-russian-relationship-went-bad?lang=en.

7. Burns, "How the U.S.-Russian Relationship Went Bad."

8. Eugene Rumer and Richard Sokolsky, "Thirty Years of U.S. Policy Toward Russia: Can the Vicious Circle Be Broken?," Carnegie Endowment, June 20, 2019, https://carnegieendowment.org/research/2019/06/thirty-years-of-us-policy-toward-russia-can-the-vicious-circle-be-broken?lang=en.

9. Susan L. Shirk, *Overreach: How China Derailed Its Peaceful Rise* (New York: Oxford University Press, 2022), 19; Rober Zoellick, "Wither China? From Membership to Responsibility: Remarks to the National Committee on U.S.-China Relations," speech, New York City, September 21, 2005, https://www.ncuscr.org/wp-content/uploads/2020/04/migration_Zoellick_remarks_notes06_winter_spring.pdf.

10. Shirk, *Overreach*, 18.

11. Shirk, *Overreach*, 17.

12. Pei, Minxin. "China in Xi's 'New Era': A Play for Global Leadership," *Journal of Democracy* 29, no. 2 (2018): 38; Wright, 26.

13. Charles Krauthammer, "The Unipolar Moment," *Foreign Affairs*, January 1, 1990, https://www.foreignaffairs.com/articles/1990-01-01/unipolar-moment; Francis Fukuyama, "More Proof That This Really Is the End of History," *The Atlantic*, October 17, 2022, https://www.theatlantic.com/ideas/archive/2022/10/francis-fukuyama-still-end-history/671761/.

14. See Michael Horowitz, *The Diffusion of Military Power* (Princeton, NJ: Princeton University Press, 2010), 4; James D. Fearon, "Domestic Political Audiences and the Escalation of International Disputes," *American Political Science Review* 88, no. 3 (1994): 587.

15. Siri Aas Rustad, "Conflict Trends: A Global Overview, 1946–2024," *PRIO Paper* (Oslo: PRIO, 2025).

16. Data collected from Rustad, "Conflict Trends: A Global Overview, 1946–2024;" Shawn Pettersson Davies et al., "Organized Violence 1989–2024, and the Challenges of Identifying Civilian Victims," *Journal of Peace Research* 62, no. 4, 2025; Nils Petter Gleditsch et al., "Armed Conflict 1946–2001: A New Dataset," *Journal of Peace Research* 39, no. 5, 2002.

17. G. John Ikenberry, "Three Worlds: The West, East and South and the Competition to Shape Global Order," *International Affairs* 100, no. 1 (January 2024): 121–38, https://academic.oup.com/ia/article/100/1/121/7506681?rss=1.

18. Hal Brands, "The Next Global War," *Foreign Affairs*, January 26, 2024, https://foreignaffairs.com/united-states/next-global-war.

19. See Johannes Hoogeveen, Johan A. Mistiaen, and Haoyu Wu, "Accelerating Poverty Reduction in Sub-Saharan Africa Requires Stability," World Bank Blogs, March 5, 2024, https://blogs.worldbank.org/en/africacan/accelerating-poverty-reduction-sub-saharan-africa-requires-stability; U.S. Agency for International Development, "Vision for Ending Extreme Poverty," 2015, https://www.circleofblue.org/waternews/wp-content/uploads/2015/09/US-AID_Ending_Extreme_Poverty_Sept_2015.pdf.

20. See Franklin Foer, "The New Washington Consensus," *The Atlantic*, May 9, 2023, https://www.theatlantic.com/ideas/archive/2023/05/biden-economics-industrial-policy-trump-nationalism/673988/.

21. Luce, "The New Washington Consensus;" Peter D. Feaver, "How Trump Will Change the World," *Foreign Affairs*, November 6, 2024, https://www.foreignaffairs.com/united-states/how-trump-will-change-world.

22. Richard Baldwin, "China Is the World's Sole Manufacturing Superpower: A Line Sketch of the Rise," *VoxEU*, January 17, 2024, https://cepr.org/voxeu/columns/china-worlds-sole-manufacturing-superpower-line-sketch-rise.

23. See Dale C. Copeland, *A World Safe for Commerce: American Foreign Policy from the Revolution to the Rise of China* (Princeton, NJ: Princeton University Press, 2024), 355.

24. Data compiled from Jonathan D. Moyer, Alanna Markle, and Adam Szymanski-Burgos, "Relative National Power Codebook," Diplometrics, Frederick S. Pardee Institute for International Futures, Josef Korbel School of International Studies, University of Denver, 2024.

25. Copeland, *A World Safe for Commerce*, 356–57.

26. Nicholas Welch and Jordan Schneider, interview with Rahm Emmanuel, *ChinaTalk*, podcast audio, April 2, 2024, https://www.chinatalk.media/p/rahm-on-china-colorful-diplomacy?utm_source=substack&utm_medium=email.

27. Jeffrey Ding, *Technology and the Rise of Great Powers: How Diffusion Shapes Economic Competition* (Princeton, NJ: Princeton University Press, 2024), 181–83.

28. Barry R. Posen, "Command of the Commons: the Military Foundation of US Hegemony," *International Security* 28, no. 1 (2003): 10.

29. William Burns, "Spycraft and Statecraft: Transforming the CIA for an Age of Competition," *Foreign Affairs*, January 30, 2024, https://www.foreignaffairs.com/united-states/cia-spycraft-and-statecraft-william-burns.

30. Alexander Gabuev and Oliver Stuenkel, "The Battle for the BRICS," *Foreign Affairs*, September 24, 2024, https://www.foreignaffairs.com/russia/battle-brics; Eric Bazail-Eimil, Robbie Gramer, and Suzanne Lynch, "A Fresh Push to Hobble Russia's War Economy," *Politico National Security Daily*, October 24, 2024, https://www.politico.com/newsletters/national-security-daily/2024/10/24/a-fresh-push-to-hobble-russias-war-economy-00183727; Timofey Bordachev, "BRICS Summit: Marching Towards a New World Order," RT, October 23, 2024, https://www.rt.com/russia/606271-brics-before-kazan-summit/; Lydia Polgreen, "We've Just Had a Glimpse of the World to Come," *The New York Times*, October 29, 2024, https://www.nytimes.com/2024/10/29/opinion/brics-summit-russia-west.html.

31. Posen, "Command of the Commons."

2. Rival Powers Compete for Technology

1. Divya Siddarth, "Reimagining Democracy's Defense," *Journal of Democracy* 34, no. 3 (2023): 173–77, https://dx.doi.org/10.1353/jod.2023.a900443; Gerard DiPippo, Ilaria Mazzocco, Scott Kennedy, and Matthew P. Goodman, "Red Ink: Estimating Chinese Industrial Policy Spending in Comparative Perspective," Center for Strategic and International Studies, May 2022, https://www.csis.org/analysis/red-ink-estimating-chinese-industrial-policy-spending-comparative-perspective; Michael J. Mazarr, Bryan Frederick, and Yvonne K. Crane, *Understanding a New Era of Strategic Competition*, RAND, 2022, https://www.rand.org/content/dam/rand/pubs/research_reports/RRA200/RRA290-4/RAND_RRA290-4.pdf.

2. Jennifer Wong Leung, Stephan Robin, and Linus Cohen, "ASPI's Critical Technology Tracker: 2025 updates and 10 new technologies," Australian Strategic Policy Institute, December 1, 2025, https://www.aspi.org.au/strategist-posts/aspis-critical-technology-tracker-2025-updates-and-10-new-technologies/; National Security Commission on Artificial Intelligence, "Final Report," 2021, https://reports.nscai.gov/final-report/.

3. See Elsa B. Kania and Lorand Laskai, "Myths and Realities of China's Military-Civil Fusion Strategy," Center for a New American Security, January 28, 2021, https://www.cnas.org/publications/reports/myths-and-realities-of-chinas-military-civil-fusion-strategy; Koichiro Takagi, "New Tech, New Concepts: China's Plans for AI and Cognitive Warfare," *War on the Rocks*, April 13, 2022, https://warontherocks.com/2022/04/new-tech-new-concepts-chinas

-plans-for-ai-and-cognitive-warfare/; David Brown, "Why China Could Win the New Global Arms Race," BBC, July 28, 2022, https://www.bbc.com/news/world-asia-china-59600475; Lindsay Maizland, "China's Modernizing Military," Council on Foreign Relations, February 5, 2020, https://www.cfr.org/backgrounder/chinas-modernizing-military.

4. The White House, "Winning the Race: America's AI Action Plan," July 2025, https://www.whitehouse.gov/wp-content/uploads/2025/07/Americas-AI-Action-Plan.pdf.

5. Max Boot, *War Made New: Technology, Warfare, and the Course of History, 1500 to Today* (New York: Penguin, 2006), 23.

6. Boot, *War Made New*, 23.

7. Paul Kennedy, *The Rise and Fall of the Great Powers: Economic Change and Military Conflict from 1500 to 2000* (New York: Vintage, 1987).

8. Julian Borger, "The Drone Operators Who Halted Russian Convoy Headed for Kyiv," *The Guardian*, March 28, 2022, https://www.theguardian.com/world/2022/mar/28/the-drone-operators-who-halted-the-russian-armoured-vehicles-heading-for-kyiv; Claire Press and Svitlana Libet, "How Russia's 35-Mile Armoured Convoy Ended in Failure," BBC, February 22, 2023, https://www.bbc.com/news/world-europe-64664944.

9. Luke Mogelson, "Battling Under a Canopy of Drones," *The New Yorker*, April 8, 2024, https://www.newyorker.com/magazine/2024/04/15/battling-under-a-canopy-of-drones.

10. Audrey Kurth Cronin, *Power to the People: How Open Technological Innovation Is Arming Tomorrow's Terrorists* (New York: Oxford University Press, 2019), 12, 19, 33–34; Audrey Kurth Cronin, "How Private Tech Companies Are Reshaping Great Power Competition," Henry A. Kissinger Center for Global Affairs, Johns Hopkins School of Advanced International Studies, https://sais.jhu.edu/kissinger/programs-and-projects/kissinger-center-papers/how-private-tech-companies-are-reshaping-great-power-competition#_ednref11.

11. Michael C. Horowitz, *The Diffusion of Military Power: Causes and Consequences for International Politics* (Princeton, NJ: Princeton University Press, 2010), 3–4.

12. Stephen Biddle, *Military Power: Explaining Victory and Defeat in Modern Battle* (Princeton, NJ: 2004), 24, 35, 46.

13. Jeffrey Ding, "The Diffusion Deficit in Scientific and Technological Power: Re-Assessing China's Rise," *Review of International Political Economy* 31, no.1 (2023):174, https://doi.org/10.1080/09692290.2023.2173633.

14. Lennart Heim, Markus Anderljung, and Haydn Belfield, "To Govern AI, We Must Govern Compute," *Lawfare*, March 28, 2024, https://www.lawfaremedia.org/article/to-govern-ai-we-must-govern-compute; Jaime Sevilla, et al., "Compute Trends Across Three Eras of Machine Learning," in *International Joint Conference on Neural Networks*, 2022, https://arxiv.org/abs/2202.05924.

15. Henry A. Kissinger, Eric Schmidt, and Daniel Huttenlocher, *The Age of AI: And Our Human Future* (New York: Little, Brown and Company, 2021), 107–109; John Thornhill, "The Age of AI and Our Human Future—a Technological Triple Threat," *Financial Times*, October 27, 2021, https://www.ft.com/content/5aa09af9-cd6c-46b6-a1be-e37ad1e33758.

16. Colin H. Kahl, "America Is Winning the Race for Global AI Primacy—for Now," *Foreign Affairs*, January 17, 2025, https://www.foreignaffairs.com/united-states/america-winning-race-global-ai-primacy-now; Lauren Leffer, "In the Race to Artificial General Intelligence, Where's the Finish Line?," *Scientific American*, June 25, 2024, https://www.scientificamerican.com/article/what-does-artificial-general-intelligence-actually-mean/; Jim Mitre and Joel B. Predd, "Artificial General Intelligence's Five Hard National Security Problems," RAND, February 10, 2025, https://www.rand.org/pubs/perspectives/PEA3691-4.html.

17. Mitre and Predd, "Artificial General Intelligence's Five Hard National Security Problems."

18. Mitre and Predd, "Artificial General Intelligence's Five Hard National Security Problems."

19. Steven Feldstein, *The Rise of Digital Repression* (New York: Oxford University Press, 2021), 39–40.

20. Steven Feldstein, "The Consequences of Generative AI for Democracy, Governance and War," in International Institute for Strategic Studies (IISS), ed., *Survival* (October–November 2023): 117–42, Routledge, 2023.

3. Rising Tech Powers

1. John F. Antal, *7 Seconds to Die: A Military Analysis of the Second Nagorno-Karabakh War and the Future of Warfighting* (Havertown, PA: Casemate Publishers, 2022), 11–12.

2. Antal, *7 Seconds to Die*, 48; Isabel Debre, "Israeli Arms Quietly Helped Azerbaijan Retake Nagorno-Karabakh, to the Dismay of Region's Armenians," Associated Press, October 5, 2023, https://apnews.com/article/armenia-azerbaijan-nagorno-karabakh-weapons-israel-6814437bcd744ac1c4df0409a74406c.

3. Jakub Janovsky et al., "The Fight for Nagorno-Karabakh: Documenting Losses on the Sides of Armenia And Azerbaijan," *Oryx*, September 27, 2020, https://www.oryxspioenkop.com/2020/09/the-fight-for-nagorno-karabakh.html; Antal, *7 Seconds to Die*, 22.

4. Andrew Higgins and Ivan Nechepurenko, "A Stunningly Sudden End to a Long, Bloody Conflict in the Caucasus," *The New York Times*, September 27, 2023, https://www.nytimes.com/2023/09/27/world/europe/nagorno-karabakh-armenia-azerbaijan.html.

5. Higgins and Nechepurenko, "A Stunningly Sudden End to a Long, Bloody Conflict in the Caucasus."

6. Robyn Dixon and Francesca Ebel, "Russia Failed to Keep Peace in Nagorno-Karabakh, Pivoting Away from Armenia," *The Washington Post*, September 30, 2023, https://www.washingtonpost.com/world/2023/09/30/russia-nagorno-karabakh-peacekeepers-failure/.

7. Dan Sabbagh, "UK Wants New Drones in Wake of Azerbaijan Military Success," *The Guardian*, December 29, 2020, https://www.theguardian.com/world/2020/dec/29/uk-defence-secretary-hails-azerbaijans-use-of-drones-in-conflict; Eric Schmidt, "Innovation Power: Why Technology Will Define the Future of Geopolitics," *Foreign Affairs*, March/April 2023, https://www.foreignaffairs.com/united-states/eric-schmidt-innovation-power-technology-geopolitics.

8. Baldur Thorhallsson and Sverrir Steinsson, "Small State Foreign Policy," *Oxford Research Encyclopedia of Politics*, May 2017, https://doi.org/10.1093/acrefore/9780190228637.013.484.

9. Nan Tian et al., "Trends in World Military Expenditure, 2023," Stockholm International Peace Research Institute, April 2024, https://www.sipri.org/sites/default/files/2024-04/2404_fs_milex_2023.pdf; "Global Innovation Index 2025," World Intellectual Property Organization, accessed December 8, 2025, https://www.wipo.int/web-publications/global-innovation-index-2025/en/gii-2025-results.html.

10. "Global Trends 2024," US National Intelligence Council, March 2021, https://www.dni.gov/files/ODNI/documents/assessments/GlobalTrends_2040.pdf.

11. "GDP, Current Prices," International Monetary Fund, 2025, accessed April 28, 2025, https://www.imf.org/external/datamapper/NGDPD@WEO/OEMDC/ADVEC/WEOWORLD.

12. "Chapter One: Defence and Military Analysis," *The Military Balance* 125, no. 1 (2025): 6–11, https://doi.org/10.1080/04597222.2025.2445473.

13. "Global Innovation Index 2025."

14. "The Global AI Index," Tortoise Media, September 19, 2024, https://www.tortoisemedia.com/intelligence/global-ai#rankings.

15. Frederic Wehrey and Andrew Bonney, "The Middle East's AI Warfare Laboratory," *War on the Rocks*, April 28, 2025, https://warontherocks.com/2025/04/the-middle-easts-ai-warfare-laboratory/.

16. Roger Connor, "The Predator, a Drone That Transformed Military Combat," National Air and Space Museum, March 9, 2018, https://airandspace.si.edu/stories/editorial/predator-drone-transformed-military-combat; Arms Transfers Database, Stockholm International Peace Research Institute, accessed January 11, 2025, https://armstransfers.sipri.org/ArmsTransfer/TransferData.

17. Steven Feldstein, "How Global Demand for Military Drones Is Transforming International Security and Geopolitics," *Georgetown Journal of International Affairs* 24, no. 2 (Fall 2023): 146–155, https://doi.org/10.1353/gia.2023.a913640.

18. Samuel Granados, et al., "How Hamas Breached Israel's 'Iron Wall,'" *The Washington Post*, October 10, 2023, https://www.washingtonpost.com/world/2023/10/10/how-hamas-entered-israel/.

19. Emily O. Goldman and Andrew L. Ross, "Conclusion: The Diffusion of Military Technology and Ideas—Theory and Practice," in *The Diffusion of Military Technology and Ideas*, eds. Emily O. Goldman and Leslie C. Eliason (Stanford: Stanford University Press, 2003), 388; Javed Ali, "Chemical Weapons and the Iran-Iraq War: A Case Study in Noncompliance," *The Nonproliferation Review* 8, no. 1 (2001): 43–58, https://www.nonproliferation.org/wp-content/uploads/npr/81ali.pdf.

20. Horowitz, *The Diffusion of Military Power: Causes and Consequences for International Politics*, 26; Emil Avdaliani, "Iran and Russia Enter A New Level of Military Cooperation," Stimson Center, March 6, 2024, https://www.stimson.org/2024/iran-and-russia-enter-a-new-level-of-military-cooperation/; Joby Warrick and Souad Mekhennet, "Iran Seeks China's Help with Surveillance Satellites, Officials Say," *The Washington Post*, August 16, 2024, https://www.washingtonpost.com/national-security/2024/08/16/iran-space-china-satellites-military/; Steve Stecklow and Babak Dehghanpisheh, "Exclusive: Huawei Hid Business Operation in Iran After Reuters Reported Links to CFO," Reuters, July 14, 2020, https://www.reuters.com/article/business/exclusive-huawei-hid-business-operation-in-iran-after-reuters-reported-links-to-idUSKBN23A19B/.

21. Cheng-Chwee Kuik, "Getting Hedging Right: A Small-State Perspective," *China International Strategy Review* 3 (2021): 300–315, https://doi.org/10.1007/s42533-021-00089-5.

22. Lisandra Paraguassu, "Lula to Seek Chinese Semiconductor Technology, Investment in Beijing," Reuters, March 24, 2023, https://www.reuters.com/technology/lula-seek-chinese-semiconductor-technology-investment-beijing-2023-03-24/; Christopher S. Chivvis and Beatrix Geaghan-Breiner, "Brazil in the Emerging World Order," Carnegie Endowment, December 18, 2023, https://carnegieendowment.org/research/2023/12/brazil-in-the-emerging-world-order?lang=en; "Huge Crowds Rally for Brazil's Bolsonaro Ahead of Verdict in Coup Trial," Al Jazeera, September 8, 2025, https://www.aljazeera.com/news/2025/9/8/huge-crowds-rally-in-support-of-brazils-bolsonaro-ahead-of-coup-verdict.

23. Hussein Kalout, "Trump's Collision Course with Brazil," *Foreign Affairs*, September 2, 2025, https://www.foreignaffairs.com/united-states/trumps-collision-course-brazil.

24. The framework builds from insights in Goldman and Ross, "Conclusion: The Diffusion of Military Technology and Ideas—Theory and Practice," 387–90; and Horowitz, *The Diffusion of Military Power*, 25–29.

25. Nahal Toosi, "Trump Is Turning Out to Be a Very Pro-China President," *Politico*, February 5, 2025, https://www.politico.com/news/magazine/2025/02/05/trump-pro-china-moves-00202500.

26. Audrey Kurth Cronin, *Power to the People: How Open Technological Innovation Is Arming Tomorrow's Terrorists* (New York: Oxford University Press, 2020), 34.

4. Snowball Units and Starlink Terminals

1. Anton Troianovski and Jason Horowitz, "The World Awaits Putin's Word on Ukraine, but on That, He Is Silent," *The New York Times*, January 26, 2022, https://www.nytimes.com/2022/01/26/world/europe/putin-ukraine.html; Michael Schwirtz, Scott Reinhard, and Josh Holder, "How Russia Has Increased Its Military Buildup," *The New York Times*, January 27, 2022, https://www.nytimes.com/interactive/2022/01/27/world/europe/russia-forces.html.

2. Peter Guest, "Mykhailo Fedorov Is Running Ukraine's War Like a Startup," *Wired*, July 25, 2023, https://www.wired.com/story/ukraine-runs-war-startup/.

3. "Defending Ukraine: Early Lessons from the Cyber War," Microsoft blog, June 22, 2022, https://query.prod.cms.rt.microsoft.com/cms/api/am/binary/RE50KOK; Greg Ip, "Big Tech Is the West's Surprise Weapon in Competition with Russia, China," *The Wall Street Journal*, August 03, 2022, https://www.wsj.com/articles/big-tech-is-the-wests-surprise-weapon-in-competition-with-russia-china-11659535233; Russ Mitchell, "How Amazon Put Ukraine's 'Government in a Box'—and Saved Its Economy from Russia," *Los Angeles Times*, December 15, 2022, https://www.latimes.com/business/story/2022-12-15/amazon-ukraine-war-cloud-data.

4. Mitchell, "How Amazon Put Ukraine's 'Government in a Box'"; Ip, "Big Tech Is the West's Surprise Weapon."

5. David E. Sanger, Julian E. Barnes, and Kate Conger, "As Tanks Rolled into Ukraine, So Did Malware. Then Microsoft Entered the War," *The New York Times*, February 28, 2022, https://www.nytimes.com/2022/02/28/us/politics/ukraine-russia-microsoft.html.

6. Steven Feldstein, "Why Catching Up to Starlink Is a Priority for Beijing," Carnegie Endowment for International Peace, September 3, 2024, https://carnegieendowment.org/emissary/2024/08/china-starlink-qianfan-satellite-internet-launch-priority?lang=en; Jonathan McDowell, "Jonathan's Space Pages-Starlink Statistics," last updated March 14, 2026, https://planet4589.org/space/con/star/stats.html.

7. Mykhailo Fedorov (@FedorovMykhailo), "@elonmusk, while you try to colonize Mars—Russia try to occupy Ukraine! While your rockets successfully land from space—Russian rockets attack Ukrainian civil people! We ask you to provide Ukraine with Starlink stations and to address sane Russians to stand," X, February 26, 2022, 7:06 a.m., https://x.com/FedorovMykhailo/status/1497543633293266944.

8. Dr. Jeffrey Lewis (@ArmsControlWonk), "According @googlemaps, there is a 'traffic jam' at 3:15 in the morning on the road from Belgorod, Russia to the Ukrainian border. It starts *exactly* where we saw a Russian formation of armor and IFV/APCs show up yesterday. Someone's on the move," X, February 23, 2022, 8:26 p.m., https://x.com/ArmsControlWonk/status/1496657816740036616.

9. Omar Abdel-Baqui, "Google Disables Live Traffic Data Access in Ukraine," *The Wall Street Journal*, February 28, 2022, https://www.wsj.com/livecoverage/russia-ukraine-latest-news-2022-02-28/card/L6NIeQg1OPgB9zzcEEky; Cat Zakrzewski, "4,000 Letters and Four Hours of Sleep: Ukrainian Leader Wages Digital War," *The Washington Post*, March 30, 2022, https://www.washingtonpost.com/technology/2022/03/30/mykhailo-fedorov-ukraine-digital-front/.

10. Audrey Kurth Cronin, "How Private Tech Companies Are Reshaping Great Power Competition," *The Kissinger Center Papers*, August 2023, https://sais.jhu.edu/kissinger/programs-and-projects/kissinger-center-papers/how-private-tech-companies-are-reshaping-great-power-competition#_ednref11; Kaushik Roy, "The Hybrid Military Establishment of the East India Company in South Asia: 1750–1849," *Journal of Global History* 6, no. 2 (2011): 195–218.

11. Garrett Fisk, "Henry Ford and the World Wars," *Military History of the Upper Great Lakes*, October 12, 2015, https://ss.sites.mtu.edu/mhugl/2015/10/12/henry-ford/; Kathryn Kemp, "House of Morgan," *International Encyclopedia of the First World War*, last updated March 5,

2015, https://encyclopedia.1914-1918-online.net/article/house-of-morgan/; Christine H. Fox and Emelia S. Probasco, "Big Tech Goes to War," *Foreign Affairs*, October 19, 2022, https://www.foreignaffairs.com/ukraine/big-tech-goes-war.

12. Shirin Ghaffary and Alex Kantrowitz, "'Don't Be Evil' Isn't a Normal Company Value. But Google Isn't a Normal Company," *Vox*, February 16, 2021, https://archive.ph/jIuI2; Alan Z. Rozenshtein, "Surveillance Intermediaries," *Stanford Law Review* 70 (2018): 118.

13. Henry Farrell and Abraham Newman, *Underground Empire: How America Weaponized the World Economy* (New York: Henry Holt and Company, 2023), 157–58.

14. "Microsoft Suspends New Sales in Russia," Microsoft blog, March 04, 2022, https://blogs.microsoft.com/on-the-issues/2022/03/04/microsoft-suspends-russia-sales-ukraine-conflict/; "Defending Ukraine: Early Lessons from the Cyber War," Microsoft blog; Farrell and Newman, *Underground Empire*, 159–60; "How Tech Is Supporting Ukraine," US Chamber of Commerce Technology Engagement Center, accessed February 6, 2025, https://americaninnovators.com/news/how-tech-is-supporting-ukraine/.

15. Milan Babic, Jan Fichtner, and Eelke M. Heemskerk, "States Versus Corporations: Rethinking the Power of Business in International Politics," *The International Spectator* 52, no. 4 (2018): 20–43, https://www.tandfonline.com/doi/full/10.1080/03932729.2017.1389151; Geoffrey Gertz and Miles M. Evers, "Geoeconomic Competition: Will State Capitalism Win?," *The Washington Quarterly* 43, no. 2 (2020): 117–36, https://doi.org/10.1080/0163660X.2020.1770962.

16. Joscha Abels, "Private Infrastructure in Geopolitical Conflicts: The Case of Starlink and the War in Ukraine," *European Journal of International Relations* 30, no. 4 (2024): 842–66, https://doi.org/10.1177/13540661241260653; Gertz and Evers, "Geoeconomic Competition: Will State Capitalism Win?"

17. Interview with a former senior Biden official, Washington, DC, May 14, 2024.

18. "GDP, Current Prices," International Monetary Fund, accessed January 30, 2025, https://www.imf.org/external/datamapper/NGDPD@WEO/OEMDC/ADVEC/WEOWORLD.

19. "Companies Ranked by Revenue," CompaniesMarketcap.com, accessed April 30, 2025, https://companiesmarketcap.com/largest-companies-by-revenue/.

20. Interview with a former senior Biden official, Washington, DC, May 14, 2024.

21. Abishur Prakash, "How Technology Companies Are Shaping the Ukraine Conflict," *Scientific American*, October 28, 2022, https://www.scientificamerican.com/article/how-technology-companies-are-shaping-the-ukraine-conflict/.

22. Silke Wünsch, "How the World Has United in Song via #StandWithUkraine," DW, March 09, 2022, https://www.dw.com/en/how-the-world-has-united-in-song-via-standwithukraine/a-61062544; Cat Zakrzewski, "Pressure Mounts on Major Tech Companies to Take Tougher Line Against Russia," *The Washington Post*, February 26, 2022, https://www.washingtonpost.com/technology/2022/02/26/apple-youtube-ukraine-russia-pressure/.

23. "Defending Ukraine: Early Lessons from the Cyber War," Microsoft blog; Bureau of Economic Analysis, "Direct Investment by Country and Industry, 2023," July 23, 2024, https://www.bea.gov/sites/default/files/2024–07/dici0724.pdf; "Information Technology (IT) Market Revenue in Russia from 2013 to 2023," Statista, November 16, 2024, https://www.statista.com/statistics/517157/russia-it-market-size/.

24. Adam Satariano et al., "Elon Musk's Unmatched Power in the Stars," *The New York Times*, July 28, 2023, https://www.nytimes.com/interactive/2023/07/28/business/starlink.html.

25. Sam Altman, "Transcript for Sam Altman: OpenAI, GPT-5, Sora, Board Saga, Elon Musk, Ilya, Power & AGI," *Lex Fridman Podcast* 419, March 17, 2024, https://lexfridman.com/sam-altman

-2-transcript#chapter9__7_trillion_of_compute; Bob Work, "Remarks by Deputy Secretary Work on Third Offset Strategy," Brussels, Belgium, April 28, 2016, https://www.defense.gov/News/Speeches/Speech/Article/753482/remarks-by-deputy-secretary-work-on-third-offset-strategy/.

26. Jake Sullivan, "Remarks by APNSA Jake Sullivan on AI and National Security," Washington, DC, October 24, 2024, archived November 11, 2024, Internet Archive, https://web.archive.org/web/20250116072220/https://www.whitehouse.gov/briefing-room/speeches-remarks/2024/10/24/remarks-by-apnsa-jake-sullivan-on-ai-and-national-security/.

27. Ryan McMorrow and Sun Yu, "The Vanishing Billionaire: How Jack Ma Fell Foul of Xi Jinping," *Financial Times*, April 15, 2021, https://www.ft.com/content/1fe0559f-de6d-490e-b312-abba0181da1f.

28. Lavender Au, "Jack Ma Isn't Back," *Wired*, June 15, 2024, https://www.wired.com/story/jack-ma-isnt-back/; McMorrow and Yu, "The Vanishing Billionaire"; Steven Feldstein, "Is the Reign of Tech Titans Coming to an End?," *Bulletin of the Atomic Scientists*, November 5, 2024, https://thebulletin.org/2024/11/is-the-reign-of-tech-titans-coming-to-an-end/; Zijing Wu, "How Jack Ma's Pivot to AI Rehabilitated Alibaba," *Financial Times*, March 16, 2025, https://www.ft.com/content/df2bccee-1730–402f-bb92–9d743018324f.

29. Ronan Farrow, "Elon Musk's Shadow Rule," *The New Yorker*, August 21, 2023, https://www.newyorker.com/magazine/2023/08/28/elon-musks-shadow-rule; Satariano et al., "Elon Musk's Unmatched Power in the Stars"; Roula Khalaf, "Elon Musk: 'Aren't You Entertained?,'" *Financial Times*, October 07, 2022, https://www.ft.com/content/5ef14997–982e-4f03–8548-b5d67202623a; Thomas Grove et al., "Elon Musk's Secret Conversations with Vladimir Putin," *The Wall Street Journal*, October 25, 2024, https://www.wsj.com/world/russia/musk-putin-secret-conversations-37e1c187; Rohan Goswami, "China's CCP Warns Elon Musk Against Sharing Wuhan Lab Leak Report," CNBC, February 28, 2023, https://www.cnbc.com/2023/02/28/chinas-ccp-warns-elon-musk-against-sharing-wuhan-lab-leak-report.html; See also Feldstein, "Is the Reign of Tech Titans Coming to an End?"

30. Eric Lipton, "Elon Musk Dominates Space Launch. Rivals Are Calling Foul," *The New York Times*, May 28, 2024, https://www.nytimes.com/2024/05/28/us/politics/elon-musk-space-launch-competition.html.

31. Paul Lipscombe, "Taiwan in Talks with Amazon over Kuiper Satellite Internet Comms," Data Center Dynamics, December 20, 2024, https://www.datacenterdynamics.com/en/news/taiwan-in-talks-with-amazon-over-kuiper-satellite-internet-comms/; Igor Mikolic-Torreira et al., "The National Security Memorandum on Artificial Intelligence—CSET Experts React," Center for Security and Emerging Technology, October 24, 2024, https://cset.georgetown.edu/article/the-national-security-memorandum-on-artificial-intelligence-cset-experts-react/; Kevin Klyman et al., "Expanding Academia's Role in Public Sector AI," Stanford University Human-Centered Artificial Intelligence, December 2024, https://hai.stanford.edu/expanding-academias-role-public-sector-ai.

32. Daniella Cheslow, "Uncle Sam Takes a Cut of American Tech Profits," *Politico: Digital Future Daily* newsletter, August 11, 2025, https://www.politico.com/newsletters/digital-future-daily/2025/08/11/uncle-sam-takes-a-cut-of-american-tech-profits-00503838; Demetri Sevastopulo and Michael Acton, "Nvidia and AMD to Pay 15% of China Chip Sale Revenues to US Government," *Financial Times*, August 10, 2025, https://www.ft.com/content/cd1a0729-a8ab-41e1-a4d2–8907f4c01cac.

33. Josh Rogin, "National Security for Sale: Inside Trump's U-Turn on Technology Export Restrictions for China," *WP Intelligence*, August 13, 2025, https://www.washingtonpost.com/documents/2e56ab4a-e6f0–4dfe-bce7-a3588c108f38.pdf; Tripp Mickle, Lauren Hirsch, and Ana Swanson, "Intel Agrees to Sell U.S. a 10% Stake in Its Business," *The New York Times*, August 22, 2025, https://www.nytimes.com/2025/08/22/technology/trump-intel-stake.html.

34. Steven Feldstein, "The Pentagon Feuding with an AI Company Is a Very Bad Sign," *Foreign Policy*, February 25, 2026, https://foreignpolicy.com/2026/02/25/anthropic-pentagon-feud-ai/; Sheera Frenkel, Cade Metz and Julian Barnes, "How Talks Between Anthropic and the Defense Dept. Fell Apart," *The New York Times*, March 2, 2026, https://www.nytimes.com/2026/03/01/technology/anthropic-defense-dept-openai-talks.html; Elizabeth Dwoskin, Ian Duncan, Tara Copp, and Faiz Siddiqui, "Pentagon Assault on Anthropic Sends Shock Waves Across Silicon Valley," *The Washington Post*, February 28, 2026, https://www.washingtonpost.com/technology/2026/02/28/pentagon-anthropic-fight-silicon-valley/.

35. Tripp Mickle, "Trump Has Made Himself Commander in Chief of the Chip Industry," *The New York Times*, August 13, 2025, https://www.nytimes.com/2025/08/13/technology/trump-chips-nvidia-intel.html; Tony Romm and Ana Swanson, "After U.S. Takes Stake in Intel, Trump Pledges 'Many More' Deals," *The New York Times*, August 25, 2025, https://www.nytimes.com/2025/08/25/us/politics/trump-intel-economy-strategy.html.

36. Ian Bremmer, "The Technopolar Paradox: How the Fusion of Technology and State Power Is Reshaping the World," *Foreign Affairs*, November/December 2021, https://www.foreignaffairs.com/united-states/technopolar-paradox-ian-bremmer-fusion-tech-state-power.

5. From the Last Supper to the "Pro-Military Arm of Silicon Valley"

1. Jonathan Chang and Meghna Chakrabarti, "'The Last Supper': How a 1993 Pentagon Dinner Reshaped the Defense Industry," WBUR, March 01, 2023, https://www.wbur.org/onpoint/2023/03/01/the-last-supper-how-a-1993-pentagon-dinner-reshaped-the-defense-industry.

2. Mark Bowden, "The Crumbling Foundation of America's Military," *The Atlantic*, December 17, 2024, https://www.theatlantic.com/politics/archive/2024/12/weapons-production-munitions-shortfall-ukraine-democracy/680867/.

3. Margaret O'Mara, *The Code: Silicon Valley and the Remaking of America* (New York: Penguin Press, 2019), 383–84; Raj Shah, interview conducted by the author, Burlingame, California, April 23, 2024.

4. Sam Winter-Levy, "Silicon Valley Hasn't Revolutionized Warfare—Yet," *Foreign Policy*, September 20, 2024, https://foreignpolicy.com/2024/09/20/silicon-valley-pentagon-us-military-technology-warfare/; Barton Gellman and Ashkan Soltani, "NSA Infiltrates Links to Yahoo, Google Data Centers Worldwide, Snowden Documents Say," *The Washington Post*, October 30, 2013, https://www.washingtonpost.com/world/national-security/nsa-infiltrates-links-to-yahoo-google-data-centers-worldwide-snowden-documents-say/2013/10/30/e51d661e-4166–11e3–8b74-d89d714ca4dd_story.html.

5. Scott Shane, Cade Metz, and Daisuke Wakabayashi, "How a Pentagon Contract Became an Identity Crisis for Google," *The New York Times*, May 30, 2018, https://www.nytimes.com/2018/05/30/technology/google-project-maven-pentagon.html.

6. Aaron Gregg, "Google Bows Out of Pentagon's $10 Billion Cloud-Computing Race," *The Washington Post*, October 9, 2018, https://www.washingtonpost.com/business/2018/10/09/google-bows-out-out-pentagons-billion-cloud-computing-race/; Diane Greene, "Incorporating Google's AI Principles into Google Cloud," *The Keyword*, June 7, 2018, https://blog.google/products/google-cloud/incorporating-googles-ai-principles-google-cloud/.

7. Maureen Dowd, "Alex Karp Has Money and Power. So What Does He Want?," *The New York Times*, August 17, 2024, https://www.nytimes.com/2024/08/17/style/alex-karp-palantir.html; Sharon Weinberger, "Techie Software Soldier Spy: Palantir, Big Data's Scariest, Most Secretive Unicorn, Is Going Public. But Is Its Crystal Ball Just Smoke And Mirrors?," *Intelligencer*, September 28, 2020, https://nymag.com/intelligencer/2020/09/inside-palantir-technologies-peter-thiel-alex-karp.html; Raj Shah and Christopher Kirchhoff, *Unit X: How*

the Pentagon and Silicon Valley Are Transforming the Future of War (New York: Scribner, 2024), 125.

8. "Palantir Might Be the Most Overvalued Firm of All Time," *The Economist*, August 12, 2025, https://www.economist.com/finance-and-economics/2025/08/12/palantir-might-be-the-most-overvalued-firm-of-all-time; Weinberger, "Techie Software Soldier Spy"; "Palantir Valued at $20 Billion in Choppy Stock Exchange Debut," Reuters, September 30, 2020, https://www.reuters.com/article/business/palantir-valued-at-20-billion-in-choppy-stock-exchange-debut-idUSKBN26L3DQ/.

9. Elizabeth Dwoskin et al., "In Trump's Washington, Palantir Is Winning Big," *The Washington Post*, August 1, 2025, https://www.washingtonpost.com/technology/2025/08/01/palantir-trump-defense-tech-ai-software/; Reinhardt Krause, "Palantir Wins $250M U.S. Army Services Contract for AI," *Investor's Business Daily*, September 27, 2023, https://www.investors.com/news/technology/pltr-stock-palantir-wins-250-million-army-ai-services-contract/; Katrina Manson, "Palantir Wins $100 Million US Contract for AI Targeting Tech," Bloomberg, September 19, 2024, https://www.bloomberg.com/news/articles/2024-09-19/palantir-wins-100-million-us-contract-for-ai-targeting-tech?sref=QmOxnLFz.

10. Colin Demarest, "Axios Future of Defense," *Axios* newsletter, December 10, 2025, https://www.axios.com/newsletters/axios-future-of-defense-a6753010-d059-11f0-97b3-a39cce533403.html.

11. Tom Simonite, "3 Years After the Project Maven Uproar, Google Cozies to the Pentagon," *Wired*, November 18, 2021, https://www.wired.com/story/3-years-maven-uproar-google-warms-pentagon/; "Pentagon Splits $9 Billion Cloud Contract Among Google, Amazon, Oracle and Microsoft," Reuters, December 08, 2022, https://www.reuters.com/technology/pentagon-awards-9-bln-cloud-contracts-each-google-amazon-oracle-microsoft-2022-12-07/; see also "Solutions for Department of Defense," Google Cloud blog, https://cloud.google.com/gov/federal-defense-and-intel.

12. Jason French, Josh Chin, Jemal R. Brinson, and Liza Lin, "How American and Chinese Drone Arsenals Stack Up," *The Wall Street Journal*, November 14, 2025, https://www.wsj.com/politics/national-security/china-america-military-drones-compared-7c56be7a; Charles Clover et al., "Russia, Ukraine and the Race for Chinese Drone Components," *Financial Times*, January 21, 2026, https://www.ft.com/content/58416bf8-97d0-4d99-8152-cff9ec2dd3ef.

13. Michael Brown, "A Plan to Revitalize the Arsenal of Democracy," *War on the Rocks*, May 10, 2024, https://warontherocks.com/2024/05/a-plan-to-revitalize-the-arsenal-of-democracy/.

6. "A Feeling of a Thousand Snipers in the Sky"

1. Julian E. Barnes et al., "Ukrainians Were Likely Behind Kremlin Drone Attack, U.S. Officials Say," *The New York Times*, May 24, 2023, https://www.nytimes.com/2023/05/24/us/politics/ukraine-kremlin-drone-attack.html.

2. Sabrina Tavernise, "Inside Ukraine's Drone Attacks on Russia," *The Daily* podcast, *The New York Times*, September 19, 2023, https://www.nytimes.com/2023/09/19/podcasts/the-daily/ukraine-drones.html?showTranscript=1; Lizzie Johnson and Kamila Hrabchuk, "These Drones Are Hitting Russia's Energy Sector. Moscow Wants Them Stopped," *The Washington Post*, March 19, 2025, https://www.washingtonpost.com/world/2025/03/19/ukraine-russia-long-range-drone-strikes/.

3. Michael C. Horowitz, "Battles of Precise Mass," *Foreign Affairs*, October 22, 2024, https://www.foreignaffairs.com/world/battles-precise-mass-technology-war-horowitz.

4. "How Ukraine Uses Cheap AI-Guided Drones to Deadly Effect Against Russia," *The Economist*, December 2, 2024, https://www.economist.com/europe/2024/12/02/how-ukraine-uses-cheap-ai-guided-drones-to-deadly-effect-against-russia; Horowitz, "Battles of Precise Mass"; Mark Bowden, "The Crumbling Foundation of America's Military," *The Atlantic*, December 17, 2024, https://www.theatlantic.com/politics/archive/2024/12/weapons-production-munitions

-shortfall-ukraine-democracy/680867/; Mark Santora et al., "A Thousand Snipers in the Sky: The New War in Ukraine," *The New York Times*, March 3, 2025, https://www.nytimes.com/interactive/2025/03/03/world/europe/ukraine-russia-war-drones-deaths.html?smid=nytcore-ios-share&referringSource=articleShare.

5. Charles Clover et al., "Russia, Ukraine and the Race for Chinese Drone Components"; Nataliya Gumenyuk, "Ukraine's New Way of War," *The Atlantic*, May 27, 2025, https://www.theatlantic.com/international/archive/2025/05/ukraine-troops-front-lines/682910/; David Hambling, "Flying Aircraft Carriers: Ukraine Refines FPV Mothership Tactics," *Forbes*, April 9, 2025, https://www.forbes.com/sites/davidhambling/2025/04/09/flying-aircraft-carriers-ukraine-refines-fpv-mothership-tactics/; C. J. Chivers, "A Harrowing Escape From the Drone-Infested Hellscape of Ukraine's Front Lines," *The New York Times*, November 5, 2025, https://www.nytimes.com/2025/11/05/magazine/ukraine-russia-war-drones.html; Santora et al., "A Thousand Snipers in the Sky."

6. Jack Watling and Nick Reynolds, "Tactical Developments During the Third Year of the Russo-Ukrainian War," Royal United Services Institute, February 2025, https://static.rusi.org/tactical-developments-third-year-russo-ukrainian-war-february-2205.pdf.

7. Alistair MacDonald, "AI-Powered Drone Swarms Have Now Entered the Battlefield," *The Wall Street Journal*, September 2, 2025, https://www.wsj.com/world/ai-powered-drone-swarms-have-now-entered-the-battlefield-2cab0f05; David Hambling, "Ukrainian FPV Operator Reviews AI-Enabled 'Lock On Target' Drones," *Forbes*, March 5, 2025; Jack Watling, "Emergent approaches to combined arms manoeuvre in Ukraine," Royal United Services Institute, October 2025, https://static.rusi.org/emergent-approaches-combined-arms-manoeuvre-ukraine-oct-25.pdf?utm_id=2122081; Rachel Rees, "Ukraine's 'drone war' hastens development of autonomous weapons," *Financial Times*, May 27, 2025, https://www.ft.com/content/165272fb-832f-4299-a0d2–1be8efcf5758.

8. Caitlin Welsh, Joseph Glauber, and Emma Dodd, "Why Is Russia Blocking Ukraine's Food Exports?," Center for Strategic and International Studies, September 15, 2023, https://www.csis.org/analysis/why-russia-blocking-ukraines-food-exports.

9. James Marson, "How Ukraine's Naval Drones Turned the Tide in the Battle of the Black Sea," *The Wall Street Journal*, June 25, 2024, https://www.wsj.com/world/naval-drones-innovation-warfare-ukraine-russia-ce35adfa?mod=mhp; Joseph Ataman, Frederik Pleitgen, and Daria Tarasova-Markina, "A Ukrainian Pilot Outlines How Drones Powered by Jet Skis Sunk a Russian Warship," CNN, February 5, 2024, https://www.cnn.com/2024/02/05/europe/ukraine-drone-jet-skis-russian-warship-intl.

10. Constant Méheut, "Ukraine's Seaborne Grain Exports Bounce Back to Near Prewar Levels," *The New York Times*, May 12, 2024, https://www.nytimes.com/2024/05/12/world/europe/ukraine-black-sea-grain-exports.html.

11. Marson, "How Ukraine's Naval Drones Turned the Tide in the Battle of the Black Sea"; Ian Lovett, "Ukraine Claims Sinking of Russian Submarine in Missile Strike," *The Wall Street Journal*, August 3, 2024, https://www.wsj.com/world/europe/ukraine-claims-sinking-of-russian-submarine-in-missile-strike-2dd5984d; Victoria Butenko, Christian Edwards, and Alex Stambaugh, "Ukraine Says It Has Sunk Another Warship, Disabling a Third of Russia's Black Sea Fleet," CNN, February 14, 2024, https://www.cnn.com/2024/02/14/europe/ukraine-says-russian-warship-caesar-kunikov-destroyed-black-sea-intl-hnk?cid=ios_app.

12. Adam Entous, "The Partnership: The Secret History of the War in Ukraine," *The New York Times*, March 29, 2025, https://www.nytimes.com/interactive/2025/03/29/world/europe/us-ukraine-military-war-wiesbaden.html; Marson, "How Ukraine's Naval Drones Turned the Tide in the Battle of the Black Sea"; Alisa Orlova and Kateryna Zakharchenko, "Ukraine-Built MAGURA V5 Naval Drones Sink Russian Patrol Ship," *Kyiv Post*, March 5, 2024, https://www.kyivpost.com/post/29016.

13. Nitasha Tiku and Elizabeth Dwoskin, "How Silicon Valley Learned to Love America, Drones and Glory," *The Washington Post*, February 17, 2024, https://www.washingtonpost.com/technology/2024/02/17/silicon-valley-military-tech-defense-contractors/.

7. Defense Tech Gold Rush

1. James O'Donnell, "We Saw a Demo of the New AI System Powering Anduril's Vision for War," *MIT Technology Review*, December 10, 2024, https://www.technologyreview.com/2024/12/10/1108354/we-saw-a-demo-of-the-new-ai-system-powering-andurils-vision-for-war/.
2. O'Donnell, "We Saw a Demo of the New AI System Powering Anduril's Vision for War."
3. Mikayla Easley, "Anduril, General Atomics Move into Next Phase of Air Force CCA Drone Program," DefenseScoop, April 24, 2024, https://defensescoop.com/2024/04/24/anduril-general-atomics-air-force-cca-program/; Sheera Frenkel, "Silicon Valley Bet on War. The Bets Are Paying Off." *The New York Times*, March 18, 2026, https://www.ny times.com/2026/03/18/technology/silicon-valley-war-defense-tech.html.
4. Kirsten Grind and Keach Hagey, "Why Did Facebook Fire a Top Executive? Hint: It Had Something to Do with Trump," *The Wall Street Journal*, November 11, 2018, https://www.wsj.com/articles/why-did-facebook-fire-a-top-executive-hint-it-had-something-to-do-with-trump-1541965245.
5. Sharon Weinberger and Heather Somerville, "Tech Bros Are Betting They Can Help Win a War with China," *The Wall Street Journal*, August 9, 2024, https://www.wsj.com/tech/anduril-drones-palmer-luckey-china-ukraine-china-951494ec.
6. Raj Shah and Christopher Kirchhoff, *Unit X: How the Pentagon and Silicon Valley Are Transforming the Future of War* (New York: Scribner, 2024), 125; Anduril Technologies, "Rebooting the Arsenal of Democracy: Anduril Mission Document," *Medium*, June 6, 2022, https://blog.anduril.com/rebooting-the-arsenal-of-democracy-anduril-mission-document-67fdbf442799; Ed Ludlow, "Kushner's Thrive, Andreesen Lead $4 Billion Anduril Round," Bloomberg, March 3, 2026, https://www.bloomberg.com/news/articles/2026-03-03/thrive-capital-andreessen-to-lead-4-billion-anduril-round.
7. Greg Williams, "F-35 Testing Report Reveals Problems with Production Decisions," Project on Government Oversight, November 21, 2024, https://www.pogo.org/analysis/f-35-testing-report-reveals-problems-with-production-decisions; "The F-35 Will Now Exceed $2 Trillion As the Military Plans to Fly It Less," US Government Accountability Office, May 16, 2024, https://www.gao.gov/blog/f-35-will-now-exceed-2-trillion-military-plans-fly-it-less.
8. Eva Dou and Aaron Gregg, "Elon Musk's Martian dreams are a boon to the U.S. military," *The Washington Post*, December 7, 2024, https://www.washingtonpost.com/technology/2024/12/07/musk-mars-technology-us-national-security/.
9. Tiku and Dwoskin, "How Silicon Valley Learned to Love America, Drones and Glory"; Eric Lipton, "New Spin on a Revolving Door: Pentagon Officials Turned Venture Capitalists," *The New York Times*, December 30, 2023, https://www.nytimes.com/2023/12/30/us/politics/pentagon-venture-capitalists.html; Weinberger and Somerville, "Tech Bros Are Betting They Can Help Win a War With China"; Raj Shah, interview conducted by the author, Burlingame, California, April 23, 2024.
10. Data from: "Vertical Snapshot: Defense Tech," PitchBook, August 05, 2025, https://pitch book.com/news/reports/2025-vertical-snapshot-defense-tech.
11. Lipton, "New Spin on a Revolving Door."
12. Lizette Chapman, "The 10 Defense Tech Startups to Watch in 2025," Bloomberg, January 16, 2025, https://www.bloomberg.com/features/2025-tech-defense-startups-to-watch/ Janakiram MSV, "Meta Invests $1.4 Billion in Scale AI to Strengthen Model Training," *Forbes*, June 23, 2025, https://www.forbes.com/sites/janakirammsv/2025/06/23/meta

-invests-14-billion-in-scale-ai-to-strengthen-model-training/; Utkarsh Shetti, "Applied Intuition Valued at $15 Billion in Latest Fund Raise," Reuters, June 17, 2025, https://www.reuters.com/business/autos-transportation/applied-intuition-valued-15-billion-latest-fund-raise-2025-06-17/; Rebecca Torrance and Paula Sambo, "Shield AI in Talks to Raise $1 Billion at $12 Billion Valuation," Bloomberg, February 13, 2026, https://www.bloomberg.com/news/articles/2026-02-13/shield-ai-in-talks-to-raise-1-billion-at-12-billion-valuation; "Hermeus IPO," Forge, accessed March 18, 2026, https://forgeglobal.com/hermeus_ipo."Hermeus IPO," Forge, accessed March 18, 2026, https://forgeglobal.com/hermeus_ipo.

13. "Could a German Startup Disrupt Europe's Arms Industry?," *The Economist*, February 13, 2025, https://www.economist.com/business/2025/02/13/could-a-german-startup-disrupt-europes-arms-industry.

14. Christina Kyriasoglou, Jake Rudnitsky, and Jenny Leonard, "Ukraine Holds Off on New Helsing Drone Orders After Setbacks," Bloomberg, January 19, 2026, https://www.bloomberg.com/news/articles/2026-01-19/ukraine-holds-off-on-new-helsing-drone-orders-after-setbacks.

15. Harry Davies and Yuval Abraham, "'A Million Calls an Hour': Israel Relying on Microsoft Cloud for Expansive Surveillance of Palestinians," *The Guardian*, August 6, 2025, https://www.theguardian.com/world/2025/aug/06/microsoft-israeli-military-palestinian-phone-calls-cloud; Michael Biesecker, Sam Mednick, and Garance Burke, "As Israel Uses US-made AI Models in War, Concerns Arise About Tech's Role in Who Lives and Who Dies," AP News, February 18, 2025, https://apnews.com/article/israel-palestinians-ai-technology-737bc17af7b03e98c29cec4e15d0f108.

16. Biesecker et al., "As Israel Uses US-Made AI Models in War, Concerns Arise About Tech's Role in Who Lives and Who Dies."

17. Davies and Abraham, "'A Million Calls an hour.'"

18. Davies and Abraham, "'A Million Calls an Hour.'"

19. Harry Davies and Yuval Abraham, "Microsoft Blocks Israel's Use of Its Technology in Mass Surveillance of Palestinians," *The Guardian*, September 25, 2025, https://www.theguardian.com/world/2025/sep/25/microsoft-blocks-israels-use-of-its-technology-in-mass-surveillance-of-palestinians.

20. Cory Weinberg, "OpenAI Projections Imply Losses Tripling to $14 Billion in 2026," *The Information*, October 09, 2024, https://www.theinformation.com/articles/openai-projections-imply-losses-tripling-to-14-billion-in-2026; Heather Somerville and Deepa Seetharaman, "OpenAI Enters Silicon Valley's Hot New Business: War," *The Wall Street Journal*, December 4, 2024, https://www.wsj.com/tech/ai/openai-enters-silicon-valleys-hot-new-business-war-7beccf6e; Gerrit De Vynck, "OpenAI Employees Question the Ethics of Military Deal with Startup Anduril," *The Washington Post*, December 06, 2024, https://www.washingtonpost.com/technology/2024/12/06/openai-anduril-employee-military-ai/.

21. Nitasha Tiku and Gerrit De Vynck, "Google's Drops Pledge Not to Use AI for Weapons or Surveillance," *The Washington Post*, February 4, 2025, https://www.washingtonpost.com/technology/2025/02/04/google-ai-policies-weapons-harm/; Casey Newton, "Google Is Ready to Work on AI Weapons Now," *Platformer*, February 4, 2025, https://www.platformer.news/google-ai-principles-military-safety/?ref=platformer-newsletter; "Chinese Military: Google's Removal of the 'AI Weapons Ban' Is an Extremely Dangerous Signal," *Geopolitechs*, February 6, 2025, https://www.geopolitechs.org/p/the-official-voice-of-the-chinese.

22. Josh Rogin, "Disrupting Defense: How Silicon Valley Plans to Take Over the War Business," *WP Intelligence*, June 20, 2025, https://www.washingtonpost.com/documents/9eb3cb47-c4a4–4270–9c5f-6eda551c0dc1.pdf.

23. Heather Somerville, "Investors Are Betting on Defense Startups. The Pentagon Isn't," *The Wall*

Street Journal, January 25, 2024, https://www.wsj.com/tech/defense-startups-risk-becoming-failed-experiment-without-more-pentagon-dollars-dc9e663a.

24. Charles Grassley, "These Toilet Seat Lids Aren't Gold-Plated, but They Cost $14,000," *The New York Times*, December 19, 2018, https://www.nytimes.com/2018/12/19/opinion/pentagon-budget-military-spending-waste.html; Heather Somerville and Brett Forrest, "How American Drones Failed to Turn the Tide in Ukraine," *The Wall Street Journal*, April 10, 2024, https://www.wsj.com/world/how-american-drones-failed-to-turn-the-tide-in-ukraine-b0ebbac3; "Western Drones Are Underwhelming on the Ukrainian Battlefield," *The Economist*, October 23, 2025, https://www.economist.com/europe/2025/10/23/western-drones-are-underwhelming-on-the-ukrainian-battlefield.

25. Somerville, "Investors Are Betting on Defense Startups."

26. Valerie Insinna, "LaPlante Pokes Silicon Valley 'Tech Bros,' Calls for Increased Munitions Production for Ukraine," Breaking Defense, November 8, 2022, https://breakingdefense.com/2022/11/laplante-pokes-silicon-valley-tech-bros-calls-for-increased-munitions-production-for-ukraine/; Weinberger and Somerville, "Tech Bros Are Betting They Can Help Win a War with China."

27. Marc Santora, "On E-Scooters and ATVs, Russian Forces Swarm Ukrainian Positions in the East," *The New York Times*, December 31, 2024, https://www.nytimes.com/2024/12/31/world/europe/ukraine-russia-attacks-eastern-front.html.

28. Antoine J. Bousquet, *The Scientific Way of Warfare: Order and Chaos on the Battlefields of Modernity* (New York: Oxford University Press, 2022), 16.

29. Thomas Powers, "The Whiz Kid vs. the Old Boys," *The New York Times*, December 3, 2000, https://archive.nytimes.com/www.nytimes.com/library/magazine/home/20001203mag-powers.html; Ian Reynolds, "Seeing, Knowing, and Deciding: The Technological Command Dream That Never Dies?," *War on the Rocks*, July 13, 2022, https://warontherocks.com/2022/07/seeing-knowing-and-deciding-the-technological-command-dream-that-never-dies/.

30. Stephen Biddle, "Back in the Trenches," *Foreign Affairs*, August 10, 2023, https://www.foreignaffairs.com/ukraine/back-trenches-technology-warfare; Gavin Wilde, "Technology Alone Won't Break the Stalemate in Ukraine," *Foreign Policy*, March 19, 2024, https://foreignpolicy.com/2024/03/19/technology-ai-drones-stalemate-ukraine-russia-manpower/.

31. T. X. Hammes, "Game-Changers: Implications of the Russo-Ukraine War for the Future of Ground Warfare," Atlantic Council, 2022, https://www.atlanticcouncil.org/wp-content/uploads/2023/04/Game-Changers-or-Little-Change-Lessons-for-Land-War-in-Ukraine-.pdf.

32. Paul Lushenko and Keith Carter, "A New Military-Industrial Complex: How Tech Bros Are Hyping AI's Role in War," *Bulletin of the Atomic Scientists*, October 7, 2024, https://thebulletin.org/2024/10/a-new-military-industrial-complex-how-tech-bros-are-hyping-ais-role-in-war/.

33. Alexander C. Karp, "Our Oppenheimer Moment: The Creation of A.I. Weapons," *The New York Times*, July 25, 2023, https://www.nytimes.com/2023/07/25/opinion/karp-palantir-artificial-intelligence.html; Calder McHugh, "The MAGA-fication of a Tech CEO," *Politico: Digital Future Daily* newsletter, November 6, 2025, https://www.politico.com/newsletters/digital-future-daily/2025/11/06/the-maga-evolution-of-palantirs-ceo-00639832.

34. Olena Bilousova et al., "Rethinking European Security in the Face of the Russian Threat," KSE Institute, June 2025, https://kse.ua/wp-content/uploads/2025/06/Rethinking_European_Security_in_the_Face_of_the_Russian_Threat.pdf.

35. Jack Shanahan, interview conducted by the author, June 24, 2024.

36. Michael Brown, interview conducted by the author, June 20, 2024.

37. Ben Smith, "The scramble to remake America's way of war," Semafor, November 3, 2025, https://www.semafor.com/article/11/03/2025/its-pe-vs-vc-in-the-scramble-to-remake-americas-way-of-war; David Jeans, "Exclusive: The US Navy Is Building a Drone Fleet to Take on China. It's Not

Going Well." Reuters, August 20, 2025, https://www.reuters.com/business/aerospace-defense/us-navy-is-building-drone-fleet-take-china-its-not-going-well-2025-08-20/.

8. Rise of Economic Warfare

1. Danielle Kurtzleben, "Trump Unveils Sweeping 10% Tariff and 'Reciprocal' Tariffs on Dozens of Nations," NPR, April 2, 2025, https://www.npr.org/2025/04/02/nx-s1-5345802/trump-tariffs-liberation-day.
2. Jon Bateman, "U.S.-China Technological 'Decoupling,'" Carnegie Endowment, 2022, https://carnegie-production-assets.s3.amazonaws.com/static/files/Bateman_US-China_Decoupling_final.pdf.
3. Ana Swanson, "A Rising Star in the Biden Administration Faces a $100 Billion Test," *The New York Times*, November 26, 2022, http://nytimes.com/2022/11/26/business/economy/gina-raimondo-chips-biden.html.
4. Jonathan Swan, "The Decline and Fall of Wilbur Ross," *Axios*, January 21, 2018, https://www.axios.com/2018/01/21/the-decline-and-fall-of-wilbur-ross-1516568275.
5. Yuka Hayashi, "Everyone Wants to Talk to Gina Raimondo—Even China," *The Wall Street Journal*, August 26, 2023, https://www.wsj.com/politics/policy/everyone-wants-to-talk-to-gina-raimondoeven-china-38629492.
6. Katrina Northrop, "Commerce at the Center," *The Wire China*, May 28, 2023, https://www.thewirechina.com/2023/05/28/department-of-commerce-at-the-center/.
7. Northrop, "Commerce at the Center."
8. Jake Sullivan, "Remarks by National Security Advisor Jake Sullivan at the Special Competitive Studies Project Global Emerging Technologies Summit," Washington, DC, September 16, 2022, archived January 19, 2025, Internet Archive, https://web.archive.org/web/20250119092002/https://www.whitehouse.gov/briefing-room/speeches-remarks/2022/09/16/remarks-by-national-security-advisor-jake-sullivan-at-the-special-competitive-studies-project-global-emerging-technologies-summit/.
9. Alex W. Palmer, "'An Act of War': Inside America's Silicon Blockade Against China," *The New York Times*, July 12, 2023, https://www.nytimes.com/2023/07/12/magazine/semiconductor-chips-us-china.html.
10. Gabriel Felbermayr, Aleksandra Kirilakha, Constantinos Syropoulos, Erdal Yalcin, and Yoto V. Yotov, "The Global Sanctions Data Base," *European Economic Review* 129 (2020): 103561; Nicholas Mulder, *The Economic Weapon: The Rise of Sanctions as a Tool of Modern War* (New Haven: Yale University Press, 2022), 14.
11. Albert O. Hirschman, *National Power and the Structure of Foreign Trade* (Berkeley and Los Angeles: University of California Press, 1945), 15; Daniel W. Drezner, "Global Economic Sanctions," *Annual Review of Political Science* 27 (2024): 9–24, https://www.annualreviews.org/content/journals/10.1146/annurev-polisci-041322-032240.
12. Jeff Stein and Federica Cocco, "How Four U.S. Presidents Unleashed Economic Warfare Across the Globe," *The Washington Post*, July 25, 2024, https://www.washingtonpost.com/business/interactive/2024/us-sanction-countries-work/; Nicholas Mulder, "How America Learned to Love (Ineffective) Sanctions," *Foreign Policy*, January 30, 2022, https://foreignpolicy.com/2022/01/30/us-sanctions-reliance-results/.
13. Data compiled from Constantinos Syropoulos, Gabriel Felbermayr, Aleksandra Kirilakha, Erdal Yalcin, and Yoto V. Yotov, "The Global Sanctions Data Base–Release 3: COVID-19, Russia, and Multilateral Sanctions," *Review of International Economics* 32, no. 1 (2023): 12:48, https://onlinelibrary.wiley.com/doi/abs/10.1111/roie.12691.
14. Stein and Cocco, "How Four U.S. Presidents Unleashed Economic Warfare Across the Globe."

15. Agathe Demarais, *Backfire: How Sanctions Reshape the World Against US Interests* (New York: Columbia University Press, 2022), x, 43; Jacob J. Lew, "U.S. Treasury Secretary Jacob J. Lew on the Evolution of Sanctions and Lessons for the Future," March 30, 2016, 40 min., 28 sec., https://carnegieendowment.org/events/2016/03/us-treasury-secretary-jacob-j-lew-on-the-evolution-of-sanctions-and-lessons-for-the-future?lang=en.

16. Interview with a senior Biden administration official, Washington, DC, May 14, 2024.

17. David Lin et al., "Welcome to the Arena: Who's Ahead, Who's Behind, and Where We Are Headed Next in the U.S.-China Technology Competition," Special Competitive Studies Project, January 2025, https://www.scsp.ai/wp-content/uploads/2025/01/Gaps-Analysis-2025-Report.pdf.

9. What Is a Strategic Technology?

1. "Drones by the Numbers," Federal Aviation Administration, October 1, 2024, https://www.faa.gov/node/54496.

2. U.S. Department of War, "Under Secretary of War for Research and Engineering Emil Michael Announces Six Critical Technology Areas for the War Department," November 17, 2025, https://www.war.gov/News/Releases/Release/Article/4333074/under-secretary-of-war-for-research-and-engineering-emil-michael-announces-six/.

3. Jeffrey Ding and Allan Dafoe, "The Logic of Strategic Assets: From Oil to AI," *Security Studies* 30, no. 2 (2021): 184.

4. "Benefit-Cost Analysis Guidance for Rail Projects," Federal Railroad Administration, June 2016, https://railroads.dot.gov/sites/fra.dot.gov/files/fra_net/16837/Benefit%20Cost%20Analysis%20Guidance%20for%20Rail%20Projects_Dec_2016_Clean.pdf.

5. Desmond Butler et al., "Elon Musk's Business Empire is Built on $38 Billion in Government Funding," *The Washington Post*, February 26, 2025, https://www.washingtonpost.com/technology/interactive/2025/elon-musk-business-government-contracts-funding/.

6. My analytic framework builds off insights from Ding and Dafoe, "The Logic of Strategic Assets," and Daniel W. Drezner, "Technological Change and International Relations," *International Relations* 33, no. 2 (2019): 286–303.

7. The following passage reflects ideas from a previously published article—Steve Feldstein, "The TikTok Deal Is America's White Flag in the Tech War With China," *Foreign Policy*, September 29, 2025, https://foreignpolicy.com/2025/09/29/tik-tok-deal-tech-war-china/.

8. Arvind Narayanan, "TikTok's Secret Sauce," Knight First Amendment Institute at Columbia University, December 15, 2022, https://knightcolumbia.org/blog/tiktoks-secret-sauce; Lauren Hirsch and Andrew Duehren, "TikTok Investors Set to Pay $10 Billion Fee to Trump Administration," *The New York Times*, March 13, 2026, https://www.nytimes.com/2026/03/13/business/trump-tiktok-10-billion-fee.html.

9. Drew Harwell and Eva Dou, "Trump Turns Biden's TikTok Law into a Big Win," *The Washington Post*, September 23, 2025, https://archive.ph/h1z1u#selection-263.0-263.45.

10. Benedict J. Clements and Ian Parry, "Subsidies: Some Work, Others Don't," *IMF Back to Basics*, September 2018, https://www.imf.org/en/Publications/fandd/issues/2018/09/what-are-subsidies-basics; Elizabeth Van Heuvelen, "Subsidy Wars: Cooperation and Common Understanding Could Dial Back Trade Tensions," *IMF Back to Basics*, June 2023, https://www.imf.org/en/Publications/fandd/issues/2023/06/B2B-subsidy-wars-elizabeth-van-heuvelen.

11. Tim Colton and LaVar Huntzinger, "A Brief History of Shipbuilding in Recent Times," Center for Naval Analyses, September 2002, https://www.cna.org/archive/CNA_Files/pdf/d0006988.a1.pdf.

12. John Frittelli, "U.S. Commercial Shipbuilding in a Global Context," Congressional Research Service, November 15, 2023, https://www.congress.gov/crs-product/IF12534; Jerry Hendrix, "The Age of American Naval Dominance Is Over," *The Atlantic*, March 13, 2023, https://www.theatlantic.com/magazine/archive/2023/04/us-navy-oceanic-trade-impact-russia-china/673090/; Jeffrey L. Seavy, "The United States Must Improve Its Shipbuilding Capacity," U.S. Naval Institute, February 2024, https://www.usni.org/magazines/proceedings/2024/february/united-states-must-improve-its-shipbuilding-capacity; Niharika Mandhana, "China's Shipyards Are Ready for a Protracted War. America's Aren't," *The Wall Street Journal*, February 13, 2024, https://www.wsj.com/world/china/chinas-shipyards-are-ready-for-a-protracted-war-americas-arent-d6f004dd; Max Boot, "The U.S. Is Unprepared for a Major War. Can Pete Hegseth Fix That?," *The Washington Post*, January 13, 2025, https://www.washingtonpost.com/opinions/2025/01/13/pete-hegseth-defense-secretary-confirmation-hearing/.

10. "We Cannot Let China Get These Chips. Period."

1. Jake Sullivan, "The Biden Administration's International Economic Agenda," speech, Brookings Institution, April 27, 2023, https://www.brookings.edu/events/the-biden-administrations-international-economic-agenda-a-conversation-with-national-security-advisor-jake-sullivan/.
2. "How to Define Artificial General Intelligence," *The Economist*, March 28, 2024, https://www.economist.com/the-economist-explains/2024/03/28/how-to-define-artificial-general-intelligence.
3. Nestor Maslej et al., "Artificial Intelligence Index Report 2025," Institute for Human-Centered AI, April 2025, https://hai-production.s3.amazonaws.com/files/hai_ai_index_report_2025.pdf; Nathan Benaich, "State of AI Report," October 9, 2025, https://www.stateof.ai/.
4. Peter Martin, "Raimondo Says Commerce Needs More Money to Halt China Chip Drive," Bloomberg, December 2, 2023, https://www.bloomberg.com/news/articles/2023-12-02/raimondo-says-commerce-needs-more-money-to-halt-china-chip-drive; Jake Sullivan, "Remarks by National Security Advisor Jake Sullivan at the Special Competitive Studies Project Global Emerging Technologies Summit," Special Competitive Studies Project, September 16, 2022, https://bidenwhitehouse.archives.gov/briefing-room/speeches-remarks/2022/09/16/remarks-by-national-security-advisor-jake-sullivan-at-the-special-competitive-studies-project-global-emerging-technologies-summit/.
5. Cameron Cavanagh, "U.S. Economic Restrictions on China: Small Yard, High Fence?," *Georgetown Security Studies Review*, December 26, 2023, https://georgetownsecuritystudiesreview.org/2023/12/26/u-s-economic-restrictions-on-china-small-yard-high-fence/; "Executive Order on Addressing the Threat from Securities Investments that Finance Certain Companies of the People's Republic of China," The White House, June 3, 2021, archived January 16, 2025, https://web.archive.org/web/20250116071643/https://www.whitehouse.gov/briefing-room/presidential-actions/2021/06/03/executive-order-on-addressing-the-threat-from-securities-investments-that-finance-certain-companies-of-the-peoples-republic-of-china/; "Executive Order on Ensuring Robust Consideration of Evolving National Security Risks by the Committee on Foreign Investment in the United States," The White House, September 15, 2022, archived January 16, 2025, https://web.archive.org/web/20250116063010/https://www.whitehouse.gov/briefing-room/presidential-actions/2022/09/15/executive-order-on-ensuring-robust-consideration-of-evolving-national-security-risks-by-the-committee-on-foreign-investment-in-the-united-states/.
6. Kevin Wolf, interview conducted by author, Zoom, March 25, 2024.
7. Henry Farrell and Abraham L. Newman, "Weaponized Interdependence: How Global Economic Networks Shape State Coercion," *International Security* 44, no. 1 (2019): 42–79; Henry Farrell and Abraham Newman, *Underground Empire: How America Weaponized the World Economy* (New York: Henry Holt and Company, 2023).

8. Dmitri Alperovitch, "How the Right U.S. Chip Strategy Can Keep Taiwan Free," *The Washington Post*, April 29, 2024, https://www.washingtonpost.com/opinions/2024/04/29/china-us-computer-chip-strategy-breakout-free-taiwan/.

9. Cagan Koc, "ASML Says Chip Controls Will Push China to Create Own Technology," Bloomberg, January 25, 2023, https://www.bloomberg.com/news/articles/2023-01-25/asml-says-chip-controls-will-push-china-to-create-own-technology.

10. Paul Mozur, Aaron Krolik, and Adam Satariano, "Chinese Traders and Moroccan Ports: How Russia Flouts Global Tech Bans," *The New York Times*, December 19, 2023, https://www.nytimes.com/2023/12/19/technology/russia-flouts-global-tech-bans.html; Warren P. Strobel and Michael R. Gordon, "How Putin Rebuilt Russia's War Machine with Help from U.S. Adversaries," *The Wall Street Journal*, June 19, 2024, https://www.wsj.com/world/how-putin-rebuilt-russias-war-machine-with-help-from-u-s-adversaries-1ea6c2d1.

11. Ansgar Baums, "The 'Chokepoint' Fallacy of Tech Export Controls," Stimson Center, February 6, 2024, https://www.stimson.org/2024/the-chokepoint-fallacy-of-tech-export-controls.

12. Gregory Zuckerman and Raffaele Huang, "DeepSeek Chief's Journey from Math Geek to Global Disruptor," *The Wall Street Journal*, January 28, 2025, https://www.wsj.com/tech/ai/liang-wenfeng-deepseek-ai-8bba3bb3; Meaghan Tobin, Paul Mozur, and Alexandra Stevenson, "DeepSeek's Rise: How a Chinese Start-Up Went from Stock Trader to A.I. Star," *The New York Times*, January 28, 2025, https://www.nytimes.com/2025/01/28/business/deepseek-owner-china-ai.html.

13. Tobin, Mozur, and Stevenson, "DeepSeek's Rise: How a Chinese Start-Up Went from Stock Trader to A.I. Star"; Zuckerman and Huang, "DeepSeek Chief's Journey from Math Geek to Global Disruptor"; "Why Chinese AI Has Stunned the World," *The Economist*, January 23, 2025, https://www.economist.com/briefing/2025/01/23/chinas-ai-industry-has-almost-caught-up-with-americas.

14. "Why Chinese AI Has Stunned the World."

15. "Why Chinese AI Has Stunned the World."

16. Marc Andreesen (@pmarca), "Deepseek R1 is one of the most amazing and impressive breakthroughs I've ever seen—and as open source, a profound gift to the world," X, January 24, 2025, 4:19 a.m., https://x.com/pmarca/status/1882719769851474108.

17. "The Global Diffusion of AI Is Now Irreversible: A Chinese AI Engineer's Rebuttal to Dario Amodei," *Geopolitechs*, January 30, 2025, https://www.geopolitechs.org/p/the-global-diffusion-of-ai-is-now; Lingling Wei, "DeepSeek Is One More Example of China's 'Good-Enough' Strategy," *The Wall Street Journal*, February 4, 2025, https://wsjchina.cmail20.com/t/d-e-sjhhltl-djtkukjlit-r/.

18. Deepa Seetharaman and Fabiola Arámburo, "OpenAI Says China's DeepSeek Trained Its AI by Distilling US Models, Memo Shows," Reuters, February 12, 2026, https://www.reuters.com/world/china/openai-accuses-deepseek-distilling-us-models-gain-advantage-bloomberg-news-2026-02-12/.

19. Jack Clark, "DeepSeek Means AI Proliferation Is Guaranteed; Maritime Wardrones; and More Evidence of LLM capability overhangs," *Import AI*, January 27, 2025, https://importai.substack.com/p/import-ai-397-deepseek-means-ai-proliferation.

20. Christine Mui, "Welcome to the Global 'AI Sovereignty' Race," *Politico: Digital Future Daily* newsletter, September 18, 2024, https://www.politico.com/newsletters/digital-future-daily/2024/09/18/should-the-u-s-seek-ai-sovereignty-00179910.

21. Pablo Chavez, "The Rising Tide of Sovereign AI," *Silicon Angle*, February 10, 2024, https://siliconangle.com/2024/02/10/rising-tide-sovereign-ai/; Tripp Mickle and Paul Mozur, "Nvidia's Booming A.I. Business Collides with U.S.-Chinese Tensions," *The New York Times*,

December 19, 2024, https://www.nytimes.com/2024/12/19/technology/nvidia-chip-sales-us-china.html.

22. Brian Caulfield, "NVIDIA CEO: Every Country Needs Sovereign AI," Nvidia, February 12, 2024, https://blogs.nvidia.com/blog/world-governments-summit/; Mickle and Mozur, "Nvidia's Booming A.I. Business Collides with U.S.-Chinese Tensions."

23. David Sacks (@DavidSacks), "There was major news from China in the AI race this week: 1. Huawei introduced a new AI chip to compete with Nvidia. 2. The Chinese gov . . . ," X, September 18, 2025, 6:14 AM, https://x.com/DavidSacks/status/1968619522484543756; The White House, "Winning the Race: America's AI Action Plan," July, 2025, https://www.whitehouse.gov/wp-content/uploads/2025/07/Americas-AI-Action-Plan.pdf.

24. Demetri Sevastopulo, "US Weighs Tougher Restrictions on AI Chip Exports to China," *Financial Times,* June 28, 2023, https://www.ft.com/content/a6c6b769–349e-440f-a55b-c92371d00dd9.

25. Tripp Mickle, David McCabe, and Ana Swanson, "How the Big Chip Makers Are Pushing Back on Biden's China Agenda," *The New York Times,* October 5, 2023, https://www.nytimes.com/2023/10/05/technology/chip-makers-china-lobbying.html.

26. Matteo Crosignani et al., "The Anatomy of Export Controls," *Liberty Street Economics,* April 12, 2024, https://libertystreeteconomics.newyorkfed.org/2024/04/the-anatomy-of-export-controls/.

27. Interview with a senior Biden administration official, Washington, DC, March 29, 2024; interview with a senior Commerce Department official, Washington, DC, June 5, 2024.

28. Liza Lin, "Nvidia's Huang Calls U.S. Export Controls a Failure," *The Wall Street Journal,* May 21, 2025, https://www.wsj.com/world/asia/nvidias-huang-calls-u-s-export-controls-a-failure-praises-trump-060940b5; Ana Swanson and Tripp Mickle, "As Trump Courts a More Assertive Beijing, China Hawks Are Losing Out," *The New York Times,* July 20, 2025, https://www.nytimes.com/2025/07/20/us/politics/trump-china-technology.html.

29. Tripp Mickle and Ana Swanson, "How Trump and Nvidia's C.E.O. Became Partners on the International Stage," *The New York Times,* November 19, 2025, https://www.nytimes.com/2025/11/19/technology/trump-nvidia-jensen-huang.html.

30. Lingling Wei, Amrith Ramkumar, and Robbie Whelan, "Trump Officials Torpedoed Nvidia's Push to Export AI Chips to China," *The Wall Street Journal,* November 3, 2025, https://www.wsj.com/world/china/trump-nvidia-china-chip-exports-51e00415.

11. Supercharging American Innovation

1. Margaret O'Mara, *The Code: Silicon Valley and the Remaking of America* (New York: Penguin Press, 2019), 21–22.

2. Steve Blank, "Secret History of Silicon Valley," posted November 20, 2008, by Computer History Museum, YouTube, 62 min., 46 sec., https://www.youtube.com/watch?v=ZTC_RxWN_xo.

3. O'Mara, *The Code,* 28.

4. O'Mara, *The Code,* 29.

5. Blank, "Secret History."

6. O'Mara, *The Code,* 68.

7. Leslie Berlin, "The First Venture Capital Firm in Silicon Valley: Draper, Gaither & Anderson," in *Making the American Century: Essays on the Political Culture of Twentieth Century America,* ed. Bruce J. Schulman (Oxford: Oxford University Press, 2014), https://academic.oup.com/book/12778/chapter-abstract/162938942?redirectedFrom=fulltext; Blank, "Secret History."

8. "U.S. R&D Totaled $892 Billion in 2022; Estimate for 2023 Indicates Further Increase to $940 Billion," NCSES, February 27, 2025, https://ncses.nsf.gov/pubs/nsf25327.

9. Audrey Kurth Cronin, "How Private Tech Companies Are Reshaping Great Power Competition," Henry A. Kissinger Center for Global Affairs, August 2023, https://sais.jhu.edu/kissinger/programs-and-projects/kissinger-center-papers/how-private-tech-companies-are-reshaping-great-power-competition.

10. Abby Vesoulis, "Commerce Secretary Gina Raimondo on the Chip Shortage, Supply Chain Issues, and Russian Sanctions," *Time*, March 9, 2022, https://time.com/6156329/commerce-secretary-raimondo-chips/; Deepa Shivaram, "Biden Is Giving Intel $8.5 Billion for Big Semiconductor Projects in 4 States," NPR, March 20, 2024, https://www.npr.org/2024/03/20/1239533039/biden-chips-arizona-intel.

11. Sean Hollister, "Nvidia Reveals Blackwell B200 GPU, the 'World's Most Powerful Chip' for AI," *The Verge*, March 18, 2024, https://www.theverge.com/2024/3/18/24105157/nvidia-blackwell-gpu-b200-ai.

12. Chris Miller, *Chip War: The Fight for the World's Most Critical Technology* (New York: Simon and Schuster, 2022), 219.

13. Viola Zhou, "TSMC's Debacle in the American Desert," *Rest of the World*, April 23, 2024, https://restofworld.org/2024/tsmc-arizona-expansion/.

14. Zhou, "TSMC's Debacle in the American Desert."

15. Zhou, "TSMC's Debacle in the American Desert."

16. Mohar Chatterjee, "Sam Altman's Changing Pitch to Washington," *Politico: Digital Future Daily* newsletter, February 3, 2025, https://www.politico.com/newsletters/digital-future-daily/2025/02/03/sam-altmans-changing-pitch-to-washington-00202185.

17. Cade Metz and Tripp Mickle, "OpenAI's Ambitious Plan to Make A.I. Flow Like Electricity," *The New York Times*, September 25, 2024, https://www.nytimes.com/2024/09/25/business/openai-plan-electricity.html.

18. James O'Donnell, "What's Next in Chips," *MIT Technology Review*, May 13, 2024, https://www.technologyreview.com/2024/05/13/1092319/whats-next-in-chips/; "SIA Whitepaper: Taking Stock of China's Semiconductor Industry," Semiconductor Industry Association, July 2021, https://www.semiconductors.org/wp-content/uploads/2021/07/Taking-Stock-of-China%E2%80%99s-Semiconductor-Industry_final.pdf.

19. Interview with a senior CHIPS office official, Washington, DC, April 15, 2024.

12. The Pitfalls of "Damming Half the River"

1. Gregory C. Allen and Emily Benson, "Clues to the U.S.-Dutch-Japanese Semiconductor Export Controls Deal Are Hiding in Plain Sight," Center for Strategic and International Studies, March 1, 2023, https://www.csis.org/analysis/clues-us-dutch-japanese-semiconductor-export-controls-deal-are-hiding-plain-sight.

2. Sander Tordoir, "Chinese Exports Threaten Europe Even More Than the US," *Politico*, June 7, 2024, https://www.politico.eu/article/chinese-exports-europe-united-states-xi-jinping-joe-biden-tariffs-cleantech-electric-vehicles-derisk-trade/; Sander Tordoir and Brad Setser, "How German Industry Can Survive the Second China Shock," Center for European Reform, January 16, 2025, https://www.cer.eu/publications/archive/policy-brief/2025/how-german-industry-can-survive-second-china-shock; David H. Autor, David Dorn, and Gordon H. Hanson, "The China Shock: Learning from Labor-Market Adjustment to Large Changes in Trade," *Annual Review of Economics* 8 (October 2016): 205–40, https://www.annualreviews.org/content/journals/10.1146/annurev-economics-080315-015041.

3. "EU-China—A Strategic Outlook," European Commission High Representative of the

Union for Foreign Affairs and Security Policy, March 12, 2019, https://commission.europa.eu/system/files/2019-03/communication-eu-china-a-strategic-outlook.pdf.

4. Andrew Small, *No Limits: The Inside Story of China's War with the West* (United Kingdom: Melville House, 2022), 23.

5. Andreas Kluth, "How China is Losing Europe," Bloomberg, May 7, 2020, https://www.bloomberg.com/view/articles/2020–05–07/from-france-to-sweden-china-is-losing-europe; "The Geopolitical Implications of the COVID-19 Pandemic," European Parliament, September 2020, https://www.europarl.europa.eu/RegData/etudes/STUD/2020/603511/EXPO_STU(2020)603511_EN.pdf.

6. Interview with Gerard de Graff, San Francisco, April 24, 2024.

7. Steven Erlanger and Matina Stevis-Gridneff, "Even as U.S. Beckons, European Leaders Head to Beijing," *The New York Times*, March 29, 2023, https://www.nytimes.com/2023/03/29/world/europe/china-european-union-us.html; James V. Wertsch, "Possible Limits to Putin and Xi's No-Limits Friendship," Wilson Center, October 20, 2023, https://www.wilsoncenter.org/blog-post/possible-limits-putin-and-xis-no-limits-friendship.

8. Markus Garlauskas, Joseph Webster, and Emma C. Verges, "China's Support for Russia Has Been Hindering Ukraine's Counteroffensive," Atlantic Council, November 15, 2023, https://www.atlanticcouncil.org/blogs/new-atlanticist/chinas-support-for-russia-has-been-hindering-ukraines-counteroffensive/.

9. Author interview with Mark Leonard, March 28, 2024; Ursula von der Leyen, "Speech by President von der Leyen on EU-China Relations to the Mercator Institute for China Studies and the European Policy Centre," European Commission, March 29, 2023, https://ec.europa.eu/commission/presscorner/detail/en/speech_23_2063.

10. Sam Fleming, "Ursula von der Leyen Forged a Bolder, Stronger EU. Can She Keep It?," *Financial Times*, April 21, 2023, https://www.ft.com/content/d0bd82d1–8e7d-4379–9f59–630886c04293.

11. "European Chips Act," European Commission, https://commission.europa.eu/strategy-and-policy/priorities-2019–2024/europe-fit-digital-age/european-chips-act_en.

12. James Kynge, "Von der Leyen's 'De-risking' Is a Tougher Stance on China," *Financial Times*, March 30, 2023, https://www.ft.com/content/c89c9472–51d6–42fc-a305–22dba159cca2.

13. Noah Barkin, "Watching China in Europe—April 2024," German Marshall Fund newsletter, April 4, 2024, https://email.gmfus.org/rv/ff00c56614aaccf13cf4d8891e46bd046f043b03.

14. Barkin, "Watching China in Europe—April 2024"; Mackenzie Hawkins et al., "US Faces Pushback on More China Chip Curbs as Election Nears," Bloomberg, April 12, 2024, https://www.bloomberg.com/news/articles/2024–04–12/us-faces-pushback-on-more-china-chip-curbs-as-election-nears; Wo Chiang Lee, "German Chancellor Scholz's Visit to China Reveals Intention to Promote Trade," Institute for China and Contemporary Studies (ICCS), April 18, 2024, https://iccs.org.tw/en/NewsContent/176.

15. Arjun Neil Alim and Sam Jones, "German Companies' Dependence on China Will Last Decades, Warns Siemens," *Financial Times*, April 14, 2024, https://www.ft.com/content/e8634f5c-5a28–4741-a13d-024665456f43.

16. Author interview with Mark Leonard, March 28, 2024.

17. Josep Borrell, "Shangri-La Dialogue: Speech by High Representative Josep Borrell on Security in the Asia-Pacific Region," European External Action Service, June 1, 2024, https://www.eeas.europa.eu/eeas/shangri-la-dialogue-speech-high-representative-josep-borrell-security-asia-pacific-region_en; Elena Giordano, "Trump: The EU Is an 'Atrocity' on Trade," *Politico*, February 3, 2025, https://www.politico.eu/article/donald-trump-trade-tariffs-says-eu-is-an-atrocity/; Jeffrey Goldberg and Shane Harris, "Here Are the Attack Plans That Trump's Advisers Shared

on Signal," *The Atlantic,* March 26, 2025, https://www.theatlantic.com/politics/archive/2025/03/signal-group-chat-attack-plans-hegseth-goldberg/682176/.

18. Claude Malhuret, "Trump Is Nero While Washington Burns," *The Atlantic,* March 8, 2025, https://www.theatlantic.com/international/archive/2025/03/claude-malheuret-speech/681947/.

19. Susan B. Glasser, "Why Donald Trump Wants Greenland (and Everything Else)," *The New Yorker,* January 8, 2026, https://www.newyorker.com/news/letter-from-trumps-washington/why-donald-trump-wants-greenland-and-everything-else.

20. Anders Kongshaug, Claudia Ciobanu, and Stefanie Dazio, "A U.S. Takeover of Greenland Would Mark the End of NATO, Danish Prime Minister Says," *PBS News,* January 6, 2026, https://www.pbs.org/newshour/world/a-u-s-takeover-of-greenland-would-mark-the-end-of-nato-danish-prime-minister-says; Zolan Kanno-Youngs, "At Davos, a Clash Between Trump's World and the Old World," *The New York Times,* January 21, 2026, https://www.nytimes.com/2026/01/21/us/politics/davos-trump-speech-greenland-deal.html; Steve Kopack, "Trump pauses Greenland-linked tariffs on 8 European countries," *NBC News,* January 21, 2026, https://www.nbcnews.com/business/economy/trump-pauses-greenland-tariffs-rcna255270; Zoya Sheftalovich, "'Our American Dream Is Dead': EU Concedes Trump Is Not on Its Side," *Politico,* January 22, 2026, https://www.politico.eu/article/eu-leaders-donald-trump-greenland-summit-davos-relief/.

21. Barkin, "Watching China in Europe—April 2025"; Anthony Faiola, "Spurned by Trump, Europe and China Weigh Closer Economic Ties," *The Washington Post,* April 4, 2025, https://www.washingtonpost.com/world/2025/04/04/europe-china-trade-talks-trump-tariffs/; Natalie Allison et al., "Inside President Trump's Whirlwind Decision to Upend Global Trade," *The Washington Post,* April 4, 2025, https://www.washingtonpost.com/politics/2025/04/04/trump-tariffs-reason-advisers/.

22. Anu Bradford, "The Race to Regulate Artificial Intelligence," *Foreign Affairs,* June 27, 2023, https://www.foreignaffairs.com/united-states/race-regulate-artificial-intelligence-sam-altman-anu-bradford.

23. Mark Scott, "What You Need to Know About France's AI Action Summit," *Digital Politics* newsletter, February 5, 2025, https://www.digitalpolitics.co/newsletter029free/?ref=digital-politics-newsletter; Francesca Bria, "The Quest for European Technological Sovereignty: Building the Eurostack," *Tech Policy Press,* October 15, 2024, https://www.techpolicy.press/the-quest-for-european-technological-sovereignty-building-the-eurostack/.

24. Scott, "What You Need to Know About France's AI Action Summit."

25. Mario Draghi, "The Future of European Competitiveness," European Commission, September 2024, https://commission.europa.eu/document/download/97e481fd-2dc3–412d-be4c-f152a8232961_en.

26. Ben Cohen, "He Turned 55. Then He Started the World's Most Important Company," *The Wall Street Journal,* March 29, 2024, https://www.wsj.com/tech/tsmc-morris-chang-taiwan-semiconductor-chips-entrepreneurship-506fcbc4; "The Semiconductor Choke-Point," *The Economist,* June 13, 2024, https://www.economist.com/asia/2024/06/13/the-semiconductor-choke-point.

27. Tsai Ing-wen, "Taiwan and the Fight for Democracy," *Foreign Affairs,* October 5, 2021, https://www.foreignaffairs.com/articles/taiwan/taiwan-and-fight-democracy-tsai-ing-wen; Richard Cronin, "Semiconductors and Taiwan's 'Silicon Shield,'" Stimson, August 16, 2022, https://www.stimson.org/2022/semiconductors-and-taiwans-silicon-shield/; Sheila Chiang, "Taiwan's Strength in Semiconductors Could Be Its 'Achilles Heel,' Economist Says," CNBC, January 12, 2024, https://www.cnbc.com/2024/01/12/taiwans-strength-in-semiconductors-could-be-its-achilles-heel-economist-says.html.

28. Kathrin Hille and Demetri Sevastopulo, "US Warns Europe a Conflict over Taiwan Could Cause Global Economic Shock," *Financial Times,* November 10, 2022, https://www.ft.com/content/c0b815f3-fd3e-4807–8de7–6b5f72ea8ae5; Aidan Powers-Riggs, "Taipei Fears Wash-

ington Is Weakening Its Silicon Shield," *Foreign Policy*, February 17, 2023, https://foreignpolicy.com/2023/02/17/united-states-taiwan-china-semiconductors-silicon-shield-chips-act-biden/.

29. Interview with a senior Biden administration official, Washington, DC, June 20, 2024.

30. Elizabeth Frost and Kwangyin Liu, "TSMC Founder: 'In the Chip Sector, Globalization Is Dead,'" *CommonWealth Magazine*, March 19, 2023, https://english.cw.com.tw/article/article.action?id=3396; Powers-Riggs, "Taipei Fears Washington Is Weakening Its Silicon Shield"; "The Semiconductor Choke-Point"; Anton Shilov, "Taiwanese Govt Clears TSMC to Make 2nm Chips Abroad—Country Lowers Its 'Silicon Shield,'" Tom's Hardware, January 13, 2025, https://www.tomshardware.com/tech-industry/taiwanese-govt-clears-tsmc-to-make-2nm-chips-abroad-country-lowers-its-silicon-shield.

31. Joyu Wang and Nathaniel Taplin, "China Tried Using Economic Ties to Bring Taiwan Closer. It Isn't Working," *The Wall Street Journal*, November 23, 2023, https://www.wsj.com/world/asia/china-tried-using-economic-ties-to-bring-taiwan-closer-it-isnt-working-8a7d63dc; Karen M. Sutter, "U.S.-Taiwan Trade and Economic Relations," Congressional Research Service, August 18, 2025, https://www.congress.gov/crs-product/IF10256; "The Semiconductor Choke-Point"; "Overviews," Taiwan Semiconductor Industry Association, https://www.tsia.org.tw/EN/PublOverview?nodeID=60.

13. Punishing Russia's War Machine

1. Edward Wong and Michael Crowley, "Putin Insulated Russia's Economy. Will Biden's Sanctions Hold Him Back in Ukraine?," *The New York Times*, February 22, 2022, https://www.nytimes.com/2022/02/22/us/politics/russia-biden-sanctions-ukraine.html; Kimberly Donovan and Maia Nikoladze, "Russia Sanctions Database," Atlantic Council, last updated November 8, 2024, https://www.atlanticcouncil.org/blogs/econographics/russia-sanctions-database/; "On Second Anniversary of Russia's Further Invasion of Ukraine and Following the Death of Aleksey Navalny, Treasury Sanctions Hundreds of Targets in Russia and Globally," US Department of the Treasury, February 23, 2024, https://home.treasury.gov/news/press-releases/jy2117; "Five Russian Nationals, Including Suspected FSB Officer, and Two U.S. Nationals Charged with Helping the Russian Military and Intelligence Agencies Evade Sanctions," US Attorney's Office, Eastern District of New York, December 13, 2022, https://www.justice.gov/usao-edny/pr/five-russian-nationals-including-suspected-fsb-officer-and-two-us-nationals-charged; "U.S. Will Cut Russia Off from Tech, Resources If Putin Escalates—U.S. Official," Reuters, February 23, 2022, https://www.reuters.com/world/us-will-cut-russia-off-tech-resources-if-putin-escalates-us-official-2022–02–23/.

2. David E. Sanger, "The White House Threatens Sanctions Against Putin's 'Inner Circle' If Russia Invades Ukraine," *The New York Times*, January 31, 2022, https://www.nytimes.com/2022/01/31/world/europe/putin-sanctions-list.html; Julian Borger, "Biden Threatens Putin with Personal Sanctions If Russia Invades Ukraine," *The Guardian*, January 26, 2022, https://www.theguardian.com/world/2022/jan/26/biden-threatens-putin-with-personal-sanctions-if-russia-invades-ukraine; James Landale, "Ukraine: Boris Johnson Says Putin Threatened Him with Missile Strike," BBC, January 30, 2023, https://www.bbc.com/news/uk-64397745.

3. Jon Hovi, Robert Huseby, and Detlef F. Sprinz, "When Do (Imposed) Economic Sanctions Work?," *World Politics* 57, no. 4 (2005): 479–99.

4. Daniel W. Drezner, "Lose-Lose: The Economic Sanctions of the Russo-Ukrainian War," in *War in Ukraine: Conflict, Strategy and the Return of a Fractured World*, ed. Hal Brands (Baltimore: Johns Hopkins University Press, 2024), 273–87; Wilfried Jilge, "Russkiy Mir: 'Russian World,'" German Council on Foreign Relations, May 3, 2016, https://dgap.org/en/events/russkiy-mir-russian-world.

5. Victor Jack, "Sergey Lavrov Admits Russia Was Surprised by Scale of Western Sanctions,"

Politico, March 23, 2022, https://www.politico.eu/article/lavrov-admits-no-one-could-have-predicted-scale-of-western-sanctions/; Drezner, "Lose-Lose," 278.

6. "Russian Central Bank Hikes Rates to Fight Inflation Fuelled by Military Spending," Associated Press, September 13, 2024, https://apnews.com/article/russia-economy-central-bank-interest-rate-7aeaed4f26b8ce34eb91d16e46226d16; Kate Davidson, "Biden Turned the Ruble into Rubble. Then It Quickly Came Back," *Politico*, March 31, 2022, https://www.politico.com/news/2022/03/31/ruble-recovery-russia-biden-sanctions-00021850; Edward Wong and Michael Crowley, "With Sanctions, U.S. and Europe Aim to Punish Putin and Fuel Russian Unrest," *The New York Times*, March 4, 2022, https://www.nytimes.com/2022/03/04/us/politics/russia-sanctions-ukraine.html.

7. Niko Vorobyov, "What Happened to Russia's Anti-War Movement?," Al Jazeera, July 15, 2022, https://www.aljazeera.com/news/2022/7/15/what-happened-to-russias-anti-war-movement; "Impact of Sanctions on the Russian Economy," Council of the European Union, last updated October 12, 2023, https://www.consilium.europa.eu/en/infographics/impact-sanctions-russian-economy/.

8. "Peremptory Norms of General International Law (*Jus Cogens*)," in *Report of the International Law Commission* (New York: United Nations, 2019), https://legal.un.org/ilc/reports/2019/english/chp5.pdf.

9. "Common High Priority List," Bureau of Industry and Security, Department of Commerce, February 23, 2024, https://www.bis.doc.gov/index.php/russia-export-controls#:~:text=(February%2023%2C%202024),as%20new%20information%20becomes%20available.

10. Interview with a senior Commerce Department official, Washington, DC, April 30, 2024.

11. "Fact Sheet: United States and G7 Partners Impose Severe Costs for Putin's War Against Ukraine," The White House, May 08, 2022, archived July 26, 2024, Internet Archive, https://web.archive.org/web/20240726002839/https://www.whitehouse.gov/briefing-room/statements-releases/2022/05/08/fact-sheet-united-states-and-g7-partners-impose-severe-costs-for-putins-war-against-ukraine/; Jeanne Whalen, "Sanctions Forcing Russia to Use Appliance Parts in Military Gear, U.S. Says," *The Washington Post*, May 11, 2022, https://www.washingtonpost.com/technology/2022/05/11/russia-sanctions-effect-military/.

12. David Gauthiers et al., "As Russian Missiles Struck Ukraine, Western Tech Still Flowed," Reuters, August 8, 2022, https://www.reuters.com/investigates/special-report/ukraine-crisis-russia-missiles-chips/.

13. "The Impact of Sanctions and Export Controls on the Russian Federation," US Department of State, October 20, 2022, archived February 19, 2023, Internet Archive, https://web.archive.org/web/20230219023748/https://www.state.gov/the-impact-of-sanctions-and-export-controls-on-the-russian-federation/; Benjamin Katz and Georgi Kantchev, "Midair Fires and Malfunctions Surge on Russia's Planes as Sanctions Bite," *The Wall Street Journal*, February 4, 2024, https://www.wsj.com/business/airlines/midair-fires-and-malfunctions-surge-on-russias-planes-as-sanctions-bite-ba0df5df; Don Graves, "America's Hidden Tool Is Hobbling Russia's War Machine with Multilateral Impact," *War on the Rocks*, February 1, 2023, https://warontherocks.com/2023/02/americas-hidden-tool-is-hobbling-russias-war-machine-with-multilateral-impact/

14. Data from "Export Controls: A Key G-7 Tool to Halt Russia's War," The International Working Group on Russian Sanctions, June 12, 2024, https://fsi9-prod.s3.us-west-1.amazonaws.com/s3fs-public/2024-06/working_paper_20_export_controls_june_12_2024_final_update.pdf.

15. Olena Bilousova et al., "Challenges of Export Controls Enforcement: How Russia Continues to Import Components for Its Military Production," Kyiv School of Economics, January 2024, https://kse.ua/wp-content/uploads/2024/01/Challenges-of-Export-Controls-Enforcement.pdf.

16. Data provided by the KSE Institute, which calculated trade volumes for dual-use goods and CHPL items based on an analysis of statistics from the General Customs Administration of China. See "An Axis of Autocracy? China's Relations with Russia, Iran, and North Korea," 119th (2025), US-China Economic and Security Review Commission, Elina Ribakova, testimony, February 20, 2025, https://www.uscc.gov/sites/default/files/2025-02/Elina_Rib akova_Testimony.pdf?mc_cid=8f8c83e896.

17. "Components in the Aggressor's Weapon," Government of Ukraine-War Sanctions, accessed March 16, 2026, https://war-sanctions.gur.gov.ua/en/components; "Export Controls: A Key G-7 Tool to Halt Russia's War," The International Working Group on Russian Sanctions, June 12, 2024, https://fsi9-prod.s3.us-west-1.amazonaws.com/s3fs-public/2024-06/working_paper_20_export_controls_june_12_2024_final_update.pdf.

18. James Byrne et al., "The Orlan Complex: Tracking the Supply Chains of Russia's Most Successful UAV," Royal United Services Institute, December 15, 2022, https://rusi.org/explore-our-research/publications/special-resources/orlan-complex-tracking-supply-chains-russias-most-successful-uav.

19. Stephanie Baker and Daryna Krasnolutska, "How a Weapon Packed with US Technology Killed a 6-Year-Old Girl in Ukraine," Bloomberg, October 02, 2024, https://www.bloomberg.com/features/2024-russian-missiles-us-tech-ukraine/?sref=QmOxnLFz; Amy Mackinnon, "Russia's War Machine Runs on Western Parts," *Foreign Policy*, February 22, 2024, https://foreignpolicy.com/2024/02/22/russia-sanctions-weapons-ukraine-war-military-semiconductors/.

20. Baker and Krasnolutska, "How a Weapon Packed with US Technology Killed a 6-Year-Old Girl in Ukraine."

21. Baker and Krasnolutska, "How a Weapon Packed with US Technology Killed a 6-Year-Old Girl in Ukraine"; U.S. Senate, Committee on Homeland Security and Governmental Affairs, Majority Staff, "The U.S. Technology Fueling Russia's War in Ukraine," Majority Staff Report, September 10, 2024, https://www.hsgac.senate.gov/wp-content/uploads/09.10.2024-Majority-Staff-Report-The-U.S.-Technology-Fueling-Russias-War-in-Ukraine.pdf.

22. Baker and Krasnolutska, "How a Weapon Packed with US Technology Killed a 6-Year-Old Girl in Ukraine;" Gauthiers et al., "As Russian Missiles Struck Ukraine, Western Tech Still Flowed."

23. Ana Swanson and Matina Stevis-Gridneff, "Russia Is Importing Western Weapons Technology, Bypassing Sanctions," *The New York Times*, April 18, 2023, https://www.nytimes.com/2023/04/18/business/economy/us-russia-chips-sanctions.html; David Hambling, "Russia's Smartest Weapon May Have an American Brain," *Forbes*, March 28, 2023, https://www.forbes.com/sites/davidhambling/2023/03/28/does-russias-smartest-weapon-have-an-american-brain/.

24. Drezner, "Lose-Lose," 283.

25. Joseph Biden, "Remarks by President Biden on Russia's Unprovoked and Unjustified Attack on Ukraine," Washington, DC, February 24, 2022, archived February 26, 2022, Internet Archive, https://web.archive.org/web/20220226003825/https://www.whitehouse.gov/briefing-room/speeches-remarks/2022/02/24/remarks-by-president-biden-on-russias-unprovoked-and-unjustified-attack-on-ukraine/.

26. Kevin Wolf, interview conducted by author, Zoom, March 25, 2024.

27. Eric Bazail-Eimil, Robbie Gramer, and Suzanne Lynch, "A Fresh Push to Hobble Russia's War Economy," *Politico: National Security Daily* newsletter, October 24, 2024, https://www.politico.com/newsletters/national-security-daily/2024/10/24/a-fresh-push-to-hobble-russias-war-economy-00183727; Patricia Cohen, "Are Russian Sanctions Working? Debate Gains New Urgency with Trump," *The New York Times*, January 2, 2025, https://www.nytimes.com/2025/01/02/business/economy/russia-sanctions-ukraine.html.

14. China's Quest to Become a Tech Superpower

1. Elizabeth Economy, *The Third Revolution: Xi Jinping and the New Chinese State* (New York: Oxford University Press, 2018), 3, 5; Melissa De Witte, "China's Cultural Revolution was a Power Grab from Within the Government, Not from Without, Stanford Sociologist Finds," *Stanford Report*, October 29, 2019, https://news.stanford.edu/stories/2019/10/violence-unfolded-chinas-cultural-revolution.

2. Chen Jian, "The Man Who Almost Changed China," *Foreign Affairs*, January/February 2025, https://www.foreignaffairs.com/reviews/man-who-almost-changed-china; Graham Webster et al., "China's Plan to 'Lead' in AI: Purpose, Prospects, and Problems," *DigiChina*, August 1, 2017, https://digichina.stanford.edu/work/chinas-plan-to-lead-in-ai-purpose-prospects-and-problems/.

3. Kurt M. Campbell and Ely Ratner, "The China Reckoning," *Foreign Affairs*, February 13, 2018, https://www.foreignaffairs.com/articles/china/2018-02-13/china-reckoning; Susan L. Shirk, *Overreach: How China Derailed Its Peaceful Rise* (New York: Oxford University Press, 2023), 1.

4. Economy, *The Third Revolution*, 3, 5.

5. Tai Ming Cheung, *Innovate to Dominate: The Rise of the Chinese Techno-Security State* (Ithaca, NY: Cornell University Press, 2022), 304; Klaus Schwab, "The Fourth Industrial Revolution: What It Means, How to Respond," World Economic Forum, January 14, 2016, https://www.weforum.org/stories/2016/01/the-fourth-industrial-revolution-what-it-means-and-how-to-respond/.

6. Cited in Jeffrey Ding, *Technology and the Rise of Great Powers* (Princeton, NJ: Princeton University Press, 2024), 1.

7. Steve Tsang and Olivia Cheung, *The Political Thought of Xi Jinping* (New York: Oxford University Press, 2024), 121; Institute for Security and Development Policy, "Made in China 2025," June 2018, https://isdp.eu/wp-content/uploads/2018/06/Made-in-China-Backgrounder.pdf.

8. Xiaoying You, "Top Brass in China Reaffirm Goal to be World Leaders in Tech, AI," *Nature*, March 14, 2026, https://archive.ph/Kfvxn; Shruti Jargad, "Key Takeaways from China's Two Sessions," The Strategist—The Australian Strategic Policy Institute blog, March 19, 2026, https://www.aspistrategist.org.au/key-takeaways-from-chinas-new-five-year-plan/.

9. Ben Murphy, "Chokepoints: China's Self-Identified Strategic Technology Import Dependencies," Center for Security and Emerging Technology, May 2022, https://cset.georgetown.edu/publication/chokepoints; Yu Jie, "Breaking China's Technological Chokepoints," in "China's New Scientists," Chatham House, August 22, 2023, https://www.chathamhouse.org/2023/07/chinas-new-scientists/breaking-chinas-technological-chokepoints; Tsang and Cheung, *The Political Thought of Xi Jinping*, 121; Institute for Security and Development Policy, "Made in China 2025."

10. Tsang and Cheung, *The Political Thought of Xi Jinping*, 119–20; Cheung, *Innovate to Dominate*, 306.

11. Elsa B. Kania and Lorand Laskai, "Myths and Realities of China's Military-Civil Fusion Strategy," Center for a New American Security, January 28, 2021, https://www.cnas.org/publications/reports/myths-and-realities-of-chinas-military-civil-fusion-strategy; "How Dominant Is China in the Global Arms Trade?," *China Power*, updated May 27, 2021, https://chinapower.csis.org/china-global-arms-trade/; Seong Hyeon Choi, "China Cuts Arms Imports to Rely More on Its Own Weapons Tech but Russia Still Biggest Overseas Supplier: SIPRI," *South China Morning Post*, March 11, 2024, https://www.scmp.com/news/china/military/article/3254853/china-cuts-arms-imports-rely-more-its-own-weapons-tech-russia-still-biggest-overseas-supplier-sipri.

12. Zi Yang, "Opening Up While Closing Up," *Asia Policy* 16, no. 1 (2021): 2; Richard A. Bitzinger,

"China's Shift from Civil-Military Integration to Military-Civil Fusion," *Asia Policy* 16, no. 1 (2021): 5–24; Cheung, *Innovate to Dominate*, 140.

13. Josh Chin, "China Is Using the Private Sector to Advance Military AI," *The Wall Street Journal*, September 3, 2025, https://www.wsj.com/world/china/china-military-ai-partners-7836a2bc; Cole McFaul, Sam Bresnick, and Daniel Chou, "Pulling Back the Curtain on China's Military-Civil Fusion," Center for Security and Emerging Technology, September 2025, https://cset.georgetown.edu/publication/pulling-back-the-curtain-on-chinas-military-civil-fusion/.

14. Cheung, *Innovate to Dominate*, 139; Kania and Laskai, "Myths and Realities of China's Military-Civil Fusion Strategy."

15. Matt Pottinger, "Examining the Flow of U.S. Money into China's Military Might," testimony before the House Foreign Affairs Committee, 118th Congress, January 17, 2024, https://www.congress.gov/118/meeting/house/116738/witnesses/HHRG-118-FA00-Wstate-PottingerM-20240117.pdf; Ana Swanson, "Trump Bars Investment in Chinese Firms with Military Ties," *The New York Times*, November 12, 2020, https://www.nytimes.com/2020/11/12/business/economy/trump-china-investment-ban.html.

16. Rishi Iyengar, "The Technocrat," *Foreign Policy*, August 16, 2024, https://foreignpolicy.com/2024/08/16/gina-raimondo-us-china-tech-competition-chips-ai/.

17. Tsang and Cheung, *The Political Thought of Xi Jinping*, 123; Ngor Luong, Zachary Arnold, and Ben Murphy, "Understanding Chinese Government Guidance Funds: An Analysis of Chinese-Language Sources," Center for Security and Emerging Technology, March 2021, https://cset.georgetown.edu/publication/understanding-chinese-government-guidance-funds/.

18. Daniel Blaugher, Benton Gordon, and Matthew Dagher-Margosian, "Made in China 2025: Evaluating China's Performance," U.S.-China Economic and Security Review Commission--staff research report, November 14, 2025, https://www.uscc.gov/sites/default/files/2025-11/Made_in_China_2025--Evaluating_Chinas_Performance.pdf.

19. Liza Lin, "China Raises $48 Billion for Semiconductor Fund to Bolster Chip-Making Capabilities," *The Wall Street Journal*, May 27, 2024, https://www.wsj.com/tech/china-unveils-48-billion-fund-to-bolster-chip-industry-10d7f1ce.

20. Raymond Zhong and Cao Li, "With Money, and Waste, China Fights for Chip Independence," *The New York Times*, February 2, 2024, https://www.nytimes.com/2020/12/24/technology/china-semiconductors.html.

21. Eduardo Jaramillo, "After a Year of Corruption Scandals, China's National Chip Fund Forges Ahead," The China Project, January 4, 2023, https://thechinaproject.com/2023/01/04/after-a-year-of-corruption-scandals-chinas-national-chip-fund-forges-ahead/; Lizzi C. Lee, "China's Big Fund 3.0: Xi's Boldest Gamble Yet for Chip Supremacy," *The Diplomat*, June 6, 2024, https://thediplomat.com/2024/06/chinas-big-fund-3–0-xis-boldest-gamble-yet-for-chip-supremacy/; Luong, Arnold, and Murphy, "Understanding Chinese Government Guidance Funds."

22. Cheung, *Innovate to Dominate*, 280.

23. Cheung, *Innovate to Dominate*, 280.

24. Bob Davis and Lingling Wei, *Superpower Showdown: How the Battle Between Trump and Xi Threatens a New Cold War* (New York: Harper Business, 2020), 223.

25. Uri Friedman, "Donald Trump's Real Endgame with China," *The Atlantic*, October 4, 2018, https://www.theatlantic.com/international/archive/2018/10/trump-china-trade/572122/; Raymond Zhong, "Chinese Tech Giant on Brink of Collapse in New U.S. Cold War," *The New York Times*, May 9, 2018, https://www.nytimes.com/2018/05/09/technology/zte-china-us-trade-war.html.

26. Damian Paletta, Ellen Nakashima, and David J. Lynch, "Trump Administration Cracks Down on Giant Chinese Tech Firm, Escalating Clash with Beijing," *The Washington Post*, May 16,

2019, https://www.washingtonpost.com/world/national-security/trump-signs-order-to-protect-us-networks-from-foreign-espionage-a-move-that-appears-to-target-china/2019/05/15/d982ec50-7727-11e9-bd25-c989555e7766_story.html; Davis and Wei, *Superpower Showdown*, 222–25.

27. Zhong, "Chinese Tech Giant on Brink of Collapse in New U.S. Cold War"; Cheung, *Innovate to Dominate*, 300.

28. Gregory C. Allen, "China's New Strategy for Waging the Microchip Tech War," Center for Strategic and International Studies, May 3, 2023, https://www.csis.org/analysis/chinas-new-strategy-waging-microchip-tech-war.

29. Cheung, *Innovate to Dominate*, 304; Lingling Wei, "In Messaging Aimed at Trump, Xi Signals China's Lines in the Sand," *Wall Street Journal China Newsletter*, November 19, 2024, https://wsjchina.cmail20.com/t/d-e-sbduht-djtkukjlit-r/.

15. Seizing the "Crown Jewels of U.S. Innovation"

1. Jim Acosta and Matthew Hoye, "China vs. U.S. Media at Rare Beijing News Conference," CNN, November 12, 2024, https://www.cnn.com/2014/11/12/politics/us-china-summit/index.html.

2. "An Interactive Look at the U.S.-China Military Scorecard," RAND, 2015, https://www.rand.org/paf/projects/us-china-scorecard.html; *Wall Street Journal* editorial board, "China's Cyber-Theft Jet Fighter," *The Wall Street Journal*, November 12, 2014, https://www.wsj.com/articles/chinas-cyber-theft-jet-fighter-1415838777.

3. McGuirk, Rod, "Australian judge rejects US Marine pilot's appeal against extradition to US," Associated Press, April 16, 2026, https://apnews.com/article/australia-us-pilot-duggan-extradite-7c90c244a5c7a0c22968ff642f227a22

4. Alastair Gale, "China Recruited Western Pilots with Knowledge of Stealth Fighters and Aircraft Carriers," *The Wall Street Journal*, December 14, 2022, https://www.wsj.com/articles/china-recruited-western-pilots-with-knowledge-of-stealth-fighters-and-aircraft-carriers-11671038565; "American Ex–Fighter Pilot Accused of Illegally Training Chinese Aviators Can Be Extradited to U.S., Australian Judge Says," CBS News, May 24, 2024, https://www.cbsnews.com/news/daniel-duggan-accused-training-chinese-aviators-can-be-extradited-us-australia-judge/; Dan De Luce, "China Wants to Recruit Military Pilots from the U.S. and Its Allies to Strengthen Chinese Air Power, Bulletin Warns," NBC News, June 5, 2024, https://www.nbcnews.com/politics/national-security/china-wants-recruit-military-pilots-us-allies-train-chinese-pilots-rcna155680.

5. Elsa Kania and Peter Wood, "The People's Liberation Army and Foreign Technology," in *China's Quest for Foreign Technology*, eds. William C. Hannas and Didi Kirsten Tatlow (New York: Routledge, 2021), 226.

6. Ethan Chiu, "Failures, Successes, and Challenges in Advanced Semiconductor Manufacturing: Lessons from the Soviet Union, China, and Taiwan," *The Yale Review of International Studies*, December 3, 2023, https://yris.yira.org/column/failures-successes-and-challenges-in-advanced-semiconductor-manufacturing-lessons-from-the-soviet-union-china-and-taiwan/#857c3258-e7be-4bdf-b842-624fc14dda3c; Steven T. Usdin, "Zelenograd, the Soviet Silicon Valley, 1962–1965," in *Engineering Communism: How Two Americans Spied for Stalin and Founded the Soviet Silicon Valley* (New Haven: Yale University Press, 2005), https://academic.oup.com/yale-scholarship-online/book/17688/chapter-abstract/175382192?redirectedFrom=fulltext.

7. Chiu, "Failures, Successes, and Challenges in Advanced Semiconductor Manufacturing"; Usdin, "Zelenograd, the Soviet Silicon Valley, 1962–1965."

8. William C. Hannas and Huey-Meei Chang, "Chinese Technology Transfer," in *China's Quest for Foreign Technology: Beyond Espionage*, eds. William Hannas and Didi Kirsten Tatlow (New York: Routledge, 2021), 3–8; "The Theft of American Intellectual Property: Reassessments

of the Challenges and the United States Policy," Commission on the Theft of American Intellectual Property, 2017, https://www.nbr.org/wp-content/uploads/pdfs/publications/IP_Commission_Report_Update.pdf.

9. James Andrew Lewis, "Rethinking Technology Transfer Policy Toward China," Center for Strategic and International Studies, November 17, 2023, https://www.csis.org/analysis/rethinking-technology-transfer-policy-toward-china#:~:text=Foreign%20direct%20investment%20and%20joint,to%20gain%20access%20to%20technology; "National Security Strategy," The White House, February 2015, https://history.defense.gov/Portals/70/Documents/nss/NSS2015.pdf?ver=TJJ2QfM0McCqL-pNtKHtVQ%3d%3d; David B. Larter, "White House Tells the Pentagon to Quit Talking About 'Competition' with China," *Navy Times*, September 26, 2016, https://www.navytimes.com/news/your-navy/2016/09/26/white-house-tells-the-pentagon-to-quit-talking-about-competition-with-china/.

10. Rush Doshi, *The Long Game: China's Grand Strategy to Displace American Order* (Oxford: Oxford University Press, 2021), 261–63.

11. "National Security Strategy," The White House, December 2017, https://history.defense.gov/Portals/70/Documents/nss/NSS2017.pdf?ver=CnFwURrw09pJ0q5EogFpwg%3d%3d.

12. Michael Brown and Pavneet Singh, "China's Technology Transfer Strategy: How Chinese Investments in Emerging Technology Enable a Strategic Competitor to Access the Crown Jewels of U.S. Innovation," Defense Innovation Unit Experimental, January 2018, https://admin.govexec.com/media/diux_chinatechnologytransferstudy_jan_2018_(1).pdf.

13. Michael Brown, interview conducted by the author, Washington, DC, June 20, 2024.

14. Paul Mozur and Jane Perlez, "China Bets on Sensitive U.S. Start-Ups, Worrying the Pentagon," *The New York Times*, March 22, 2017, https://www.nytimes.com/2017/03/22/technology/china-defense-start-ups.html?partner=rss&emc=rss&_r=0.

15. Office of the United States Trade Representative, "Section 301 Report," May 22, 2018, https://ustr.gov/sites/default/files/Section%20301%20FINAL.PDF, 19; Sean O'Connor, "How Chinese Companies Facilitate Technology Transfer from the United States," U.S.-China Economic and Security Review Commission, May 6, 2019, https://www.uscc.gov/research/how-chinese-companies-facilitate-technology-transfer-united-states.

16. Kate O'Keeffe and Brian Spegele, "How a Big U.S. Chip Maker Gave China the 'Keys to the Kingdom,'" *The Wall Street Journal*, June 27, 2019, https://www.wsj.com/articles/u-s-tried-to-stop-china-acquiring-world-class-chips-china-got-them-anyway-11561646798.

17. O'Keeffe and Spegele, "How a Big U.S. Chip Maker Gave China the 'Keys to the Kingdom.'"

18. Chris Miller, *Chip War: The Fight For The World's Most Critical Technology* (New York: Simon and Schuster, 2022), 259.

19. O'Connor, "How Chinese Companies Facilitate Technology Transfer from the United States."

20. Trefor Moss and Tim Higgins, "Tesla to Build Auto Factory in Shanghai," *The Wall Street Journal*, July 10, 2018, https://www.wsj.com/articles/tesla-to-build-auto-factory-in-shanghai-1531230784; Brody Mullins and Susan Pulliam, "Elon Musk's Business Ties to China Create Unease in Washington," *The Wall Street Journal*, March 20, 2022, https://www.wsj.com/politics/policy/elon-musks-business-ties-to-china-create-unease-in-washington-11647768780; Sherisse Pham, "Tesla Is Accusing a Former Employee of Stealing Self-Driving Tech and Giving It to a Chinese Rival," CNN, March 22, 2019, https://www.cnn.com/2019/03/22/tech/tesla-xiaopeng-motors-lawsuit/index.html.

21. Yudhijit Bhattacharjee, "A New Kind of Spy," *The New Yorker*, April 28, 2014, https://www.newyorker.com/magazine/2014/05/05/a-new-kind-of-spy.

22. Bhattacharjee, "A New Kind of Spy."

23. Bhattacharjee, "A New Kind of Spy"; Charles Riley, "China's Long History of Spying on Business," CNN, May 20, 2014, https://money.cnn.com/2014/05/20/news/china-espionage-business/.

24. John Moolenaar and Bobbi Jo Foxx, "Moolenaar, Foxx Uncover American University Research That Aids Chinese Military," press release, Select Committee on the Chinese Communist Party, US House of Representatives, September 23, 2024, https://selectcommitteeontheccp.house.gov/media/press-releases/moolenaar-foxx-uncover-american-university-research-aids-chinese-military; Jessica Chen Weiss, "The Case Against the China Consensus," *Foreign Affairs*, September 16, 2024, https://www.foreignaffairs.com/united-states/case-against-china-consensus; Ana Swanson, "U.S. Research Aided Chinese Military Technology, House Republicans Say," *The New York Times*, September 23, 2024, https://www.nytimes.com/2024/09/23/us/politics/us-china-research-military.html.

25. Hannas and Chang, "Chinese Technology Transfer," 7; Brown and Singh, "China's Technology Transfer Strategy."

26. Mozur and Perlez, "China Bets on Sensitive U.S. Start-Ups, Worrying the Pentagon"; Daniel Griswold and Donald J. Boudreaux, "How the United States Should Respond to China's Intellectual Property Practices," Mercatus Center, April 4, 2019, https://www.mercatus.org/research/policy-briefs/how-united-states-should-respond-chinas-intellectual-property-practices; Shah and Kirchhoff, *Unit X*, 149–50.

16. "We Must Be Prepared for Worst-Case and Extreme Scenarios"

1. Raymond Zhong, "Chinese Tech Giant on Brink of Collapse in New U.S. Cold War," *The New York Times*, May 9, 2018, https://www.nytimes.com/2018/05/09/technology/zte-china-us-trade-war.html.

2. Shira Ovide, "The Strange Saga of Huawei," *The New York Times*, November 2, 2021, https://www.nytimes.com/2021/11/02/technology/huawei-us-policy.html.

3. Alex W. Palmer, "'An Act of War': Inside America's Silicon Blockade Against China," *The New York Times*, July 12, 2023, https://www.nytimes.com/2023/07/12/magazine/semiconductor-chips-us-china.html; Dan Strumpf, "U.S. Set Out to Hobble China's Huawei, and So It Has," *The Wall Street Journal*, October 7, 2021, https://www.wsj.com/articles/u-s-set-out-to-hobble-chinas-huawei-and-so-it-has-11633617478; Liza Lin, Stu Woo, and Raffaele Huang, "The U.S. Wanted to Knock Down Huawei. It's Only Getting Stronger," *The Wall Street Journal*, July 29, 2024, https://www.wsj.com/business/telecom/huawei-china-technology-us-sanctions-76462031?st=rag4lv7lobh913m.

4. Liza Lin and Raffaele Huang, "Huawei Bounces Back from U.S. Sanctions as Profit Doubles," *The Wall Street Journal*, March 29, 2024, https://www.wsj.com/business/earnings/huaweis-2023-net-profit-more-than-doubled-as-revenue-rose-b1b6d971; Michelle Toh, "Huawei Was Once Fighting for Its Survival. It Is Back to Nearly $100 Billion in Revenue in 2023," CNN, December 29, 2023, https://www.cnn.com/2023/12/29/tech/huawei-revenue-rebound-china-intl-hnk/index.html#:~:text=Huawei%20is%20back%20to%20nearly,US%2DChina%20tensions%20%7C%20CNN%20Business; Aaron Mc Nicholas, "ASML's Powder Keg," *The Wire China*, August 4, 2024, https://www.thewirechina.com/2024/08/04/asmls-powder-keg-asml-china/; Ian King and Debby Wu, "Huawei Building Secret Network for Chips, Trade Group Warns," Bloomberg, August 22, 2023, https://www.bloomberg.com/news/articles/2023-08-23/huawei-building-secret-chip-plants-in-china-to-bypass-us-sanctions-group-warns; Dylan Patel, Jeff Koch, and Sravan Kundojjala, "Fab Whack-A-Mole: Chinese Companies Are Evading U.S. Sanctions," *SemiAnalysis*, October 28, 2024, https://semianalysis.com/2024/10/28/fab-whack-a-mole-chinese-companies/; Peter Elstrom and Allen Wan, "China Secretly Transforms Huawei into Most Powerful Chip War Weapon," Bloomberg, December 1, 2023, https://www.bloomberg.com/graphics/2023-china-huawei-semiconductor/?sref=QmOxnLFz.

5. Raymond Zhong, "Huawei's Sales Jump Despite Trump's Blacklisting," *The New York Times*, July 30, 2019, https://www.nytimes.com/2019/07/30/technology/huawei-trump.html.

6. Liza Lin, "How Huawei Rose from the Ashes After Being Scorched by the U.S." *The Wall Street Journal—China Newsletter*, July 30, 2024, https://wsjchina.cmail20.com/t/d-e-ehllro-djtkukjlit-r/; Lin, Woo, and Huang, "The U.S. Wanted to Knock Down Huawei. It's Only Getting Stronger."

7. Yuan Gao and Debby Wu, "China's Chip Advances Stall as US Curbs Hit Huawei AI Product," Bloomberg, November 19, 2024, https://www.bloomberg.com/news/articles/2024-11-19/china-s-chip-advances-stall-as-us-curbs-hit-huawei-ai-product.

8. "Why America's Controls on Sales of AI Tech to China Are So Leaky," *The Economist*, January 21, 2024, https://www.economist.com/business/2024/01/21/why-americas-controls-on-sales-of-ai-tech-to-china-are-so-leaky.

9. Karen Freifeld, "Exclusive: TSMC Could Face $1 Billion or More Fine From US Probe, Sources Say," Reuters, April 8, 2025, https://www.reuters.com/technology/tsmc-could-face-1-billion-or-more-fine-us-probe-sources-say-2025-04-08/; Eliot Chen and Noah Berman, "TSMC's Huawei Headache," *The Wire China*, October 31, 2024, https://www.thewirechina.com/2024/10/31/tsmcs-huawei-headache-xiamen-sophgo-bitmain-sophon-tsmc/; Mackenzie Hawkins, "Huawei Technologies' Latest AI Chips Were Produced by TSMC," Bloomberg, October 22, 2024, https://www.bloomberg.com/news/articles/2024-10-22/huawei-technologies-latest-ai-processors-were-made-by-tsmc?sref=QmOxnLFz; Karen Freifeld and Fanny Potkin, "Exclusive: US Ordered TSMC to Halt Shipments to China of Chips Used in AI Applications," Reuters, November 10, 2024, https://www.reuters.com/technology/us-ordered-tsmc-halt-shipments-china-chips-used-ai-applications-source-says-2024-11-10/.

10. Lingling Wei, "Xi Prepares China for 'Extreme' Scenarios, Including Conflict with the West," *The Wall Street Journal*, June 12, 2023, https://www.wsj.com/world/chinas-xi-jinping-plays-up-possibility-of-worsening-tensions-with-the-west-aac2dff8.

11. Hung Tran, "Dual Circulation in China: A Progress Report," Atlantic Council, October 24, 2022, https://www.atlanticcouncil.org/blogs/econographics/dual-circulation-in-china-a-progress-report/.

12. Tai Ming Cheung, *Innovate to Dominate: The Rise of the Chinese Techno-Security State* (Ithaca, NY: Cornell University Press, 2022), 301–302; Will Hunt, "Sustaining U.S. Competitiveness in Semiconductor Manufacturing," Center for Security and Emerging Technology, January 2022, https://cset.georgetown.edu/publication/sustaining-u-s-competitiveness-in-semiconductor-manufacturing/; Patel, Koch, and Kundojjala, "Fab Whack-A-Mole: Chinese Companies Are Evading U.S. Sanctions."

13. Ming-Jer Chen, Y. P. Chan, and Mary Summers Whittle, "Wingtech Acquires Nexperia: Will the New Company Fly?," Darden Case No. UVA-S-0329, SSRN, September 30, 2020, https://ssrn.com/abstract=3614451; Karen Freifeld and David Shepardson, "Exclusive: Latest US Clampdown on China's Chips Hits Semiconductor Toolmakers," Reuters, December 2, 2024, https://www.reuters.com/technology/latest-us-strike-chinas-chips-hits-semiconductor-toolmakers-2024-12-02/.

14. Toby Sterling, "Dutch Government Took Control of Nexperia over Fears It Was Being Gutted—Sources," Reuters, October 27, 2025, https://www.reuters.com/sustainability/boards-policy-regulation/dutch-government-took-control-nexperia-over-fears-it-was-being-gutted-sources-2025-10-27/.

15. Meaghan Tobin and Xinyun Wu, "The Small Company in Europe Caught in the Big Trade War Between the U.S. and China," *The New York Times*, October 16, 2025, https://www.nytimes.com/2025/10/16/business/nexperia-netherlands-us-china.html.

16. Lily Ottinger, "Spies in NY, Ballpoint Industrial Policy, Chinese vs. US Horror Movies, Chinese Football," *ChinaTalk*, September 13, 2024, https://www.chinatalk.media/p/spies-in-ny-

ballpoint-industrial; Yu Jie, "China's New Scientists," Chatham House, July 24, 2023, https://www.chathamhouse.org/2023/07/chinas-new-scientists/breaking-chinas-technological-chokepoints.

17. Ben Murphy, "Chokepoints: China's Self-Identified Strategic Technology Import Dependencies," Center for Security and Emerging Technology, May 2022, https://cset.georgetown.edu/publication/chokepoints/.

18. Zhang Tong, "'Milestone' for China's Gas Turbine Technology as It Closes Gap with the West," *South China Morning Post*, October 9, 2024, https://archive.ph/48trA; "China Starts New Locally-Manufactured Gas Turbine for First Time," Bloomberg, December 2, 2025, https://www.bloomberg.com/news/articles/2025-12-02/china-starts-new-locally-manufactured-gas-turbine-for-first-time.

19. Iris Deng, "Huawei's HarmonyOS Unseats Apple's iOS to Become China's No 2 Mobile Operating Platform," *South China Morning Post*, June 15, 2024, https://www.scmp.com/tech/big-tech/article/3266747/huaweis-harmonyos-unseats-apples-ios-become-chinas-no-2-mobile-operating-platform.

20. Xinmei Shei, "China Touts Home-Grown Chip Lithography Machines amid Semiconductor Self-Sufficiency Drive," *South China Morning Post*, September 15, 2024, https://www.scmp.com/tech/tech-war/article/3278597/china-touts-home-grown-chip-lithography-machines-amid-semiconductor-self-sufficiency-drive; Fanny Potkin, "Exclusive: How China Built Its 'Manhattan Project' to Rival the West in AI Chips," Reuters, December 17, 2025, https://www.reuters.com/world/china/how-china-built-its-manhattan-project-rival-west-ai-chips-2025-12-17/.

21. Brian Spegele, "The Fortress That China Built for Its Battle With America," *The Wall Street Journal*, May 21, 2025, https://www.wsj.com/world/china/china-us-technology-economy-advancements-bb8d7439; Lingling Wei, "Beijing's 'Made in China' Plan Is Narrowing Tech Gap, Study Finds," *The Wall Street Journal*, May 5, 2025, https://www.wsj.com/world/china/beijings-made-in-china-plan-is-narrowing-tech-gap-study-finds-24ecbd34.

22. Dannie Peng, "Shadow Agents in China's Talent War Help Recruit Outstanding Overseas Scientists to Boost Self-Reliance," *South China Morning Post*, September 4, 2023, https://www.scmp.com/news/china/science/article/3233108/shadow-agents-chinas-talent-war-help-recruit-outstanding-overseas-scientists-boost-self-reliance; Stephen Chen, "China's Brain Drain to the US is Ending, Thanks to Higher Salaries and Donald Trump," *South China Morning Post*, September 6, 2018, https://www.scmp.com/news/china/science/article/2163001/chinas-brain-drain-us-ending-thanks-higher-salaries-and-donald; "China Has Become a Scientific Superpower," *The Economist*, June 12, 2024, https://www.economist.com/science-and-technology/2024/06/12/china-has-become-a-scientific-superpower.

23. Christina Lu and Anusha Rathi, "Chinese Scientists Are Leaving the United States," *Foreign Policy*, July 13, 2023, https://foreignpolicy.com/2023/07/13/chinese-scientists-united-states-research-tech-academia-china-initiative/.

24. Andrew Jones, "China's Enormous FAST Radio Telescope Is Getting Even Bigger (Video)," *Space*, September 30, 2024, https://www.space.com/china-fast-radio-telescope-24-new-dishes; Sean Gorman, "America Is Losing Its GPS Dominance to China's BeiDou Satnav," *SpaceNews*, April 8, 2024, https://spacenews.com/america-losing-gps-dominance-china-beidou-satnav/; Ciel Qi, "China's Quantum Ambitions: A Multi-Decade Focus on Quantum Communications," *Yale Journal of International Affairs*, February 2, 2024, https://www.yalejournal.org/publications/chinas-quantum-ambitions; "China Has Become a Scientific Superpower."

25. Jeffrey Ding, "Panic Buying, Speculative Booms, and Whack-A-Mole—What Lengths Will Chinese Companies Go to Get an NVIDIA A100 Chip?," *ChinAI* newsletter, February 26, 2024, https://chinai.substack.com/p/chinai-255-panic-buying-speculative.

26. Raffaele Huang, "The Underground Network Sneaking Nvidia Chips Into China," *The Wall Street Journal*, July 2, 2024, https://www.wsj.com/tech/the-underground-network-sneaking-nvidia-chips-into-china-f733aaa6; Erich Grunewald and Tim Fist, "Countering AI Chip Smug-

gling Has Become a National Security Priority," CNAS, June 11, 2025, https://www.cnas.org/publications/reports/countering-ai-chip-smuggling-has-become-a-national-security-priority; Dave Smith, "Nvidia Is Officially the World's First $5 Trillion Company. CEO Jensen Huang Says It's on Track for 'Half a Trillion Dollars' in Revenue," *Fortune*, October 29, 2025, https://fortune.com/2025/10/29/nvidia-first-5-trillion-company-ceo-jensen-huang-500-billion-revenue-blackwell-rubin-gpus-china/.

27. Ana Swanson and Claire Fu, "With Smugglers and Front Companies, China Is Skirting American A.I. Bans," *The New York Times*, August 4, 2024, https://www.nytimes.com/2024/08/04/technology/china-ai-microchips.html; "C4ADS Supports New York Times Investigation To Identify Banned Nvidia Chips in China," *C4ADS*, August 5, 2024, https://c4ads.org/news/c4ads-nyt-nvidia/; Stu Woo, "How Trump's U-Turn on Nvidia Chips Changes the Game for China's AI," *The Wall Street Journal*, December 9, 2025, https://www.wsj.com/tech/ai/china-ai-nvidia-trump-chip-export-ban-lift-c3d457c.

28. Eduardo Baptista, Fanny Potkin, and Karen Freifeld, "Chinese Entities Turn to Amazon Cloud and Its Rivals to Access High-End US Chips, AI," Reuters, August 23, 2024, https://www.reuters.com/technology/chinese-entities-turn-amazon-cloud-its-rivals-access-high-end-us-chips-ai-2024–08–23/; Victor Gevers, "China's Secret Backdoor to U.S. AI Tech," *Everything About AI*, August 26, 2024, https://about-ai.victorgevers.com/chinas-secret-backdoor-to-u-s-ai-tech-9bcaa9b47406.

29. Cheng Ting-Fang, "How China's Tech Ambitions Slip Through the U.S. Export Control Net," *Nikkei Asia*, October 20, 2023, https://asia.nikkei.com/Business/Business-Spotlight/How-China-s-tech-ambitions-slip-through-the-U.S.-export-control-net; Heekyong Yang, Fanny Potkin, and Karen Freifeld, "Chinese Firms Stockpile High-End Samsung Chips as They Await New US Curbs, Say Sources," Reuters, August 6, 2024 https://www.reuters.com/technology/chinese-firms-stockpile-high-end-samsung-chips-they-await-new-us-curbs-say-2024–08–06/.

30. Liza Lin, "China Intensifies Push to 'Delete America' from Its Technology," *The Wall Street Journal*, March 7, 2024, https://www.wsj.com/world/china/china-technology-software-delete-america-2b8ea89f?mod=djemchina.

31. Coco Feng, "'Are Your Chips Made Locally?' Xi Jinping Prods Construction Vehicle Maker About Semiconductors amid Self-Sufficiency Push," *South China Morning Post*, March 9, 2023, https://www.scmp.com/tech/policy/article/3212932/are-your-chips-made-locally-xi-prods-construction-vehicle-maker-about-semiconductors-amid-self; Shunsuke Tabeta, "Chinese Auto Chips Only: Inside Xi's Self-Sufficiency Campaign," *Nikkei Asia*, March 15, 2023, https://asia.nikkei.com/Business/Tech/Semiconductors/Chinese-auto-chips-only-Inside-Xi-s-self-sufficiency-campaign.

32. Alexandra Stevenson, "China Hits Dozens of U.S. Companies with Trade Controls," *The New York Times*, January 2, 2025, https://www.nytimes.com/2025/01/02/business/china-us-companies-entity-list.html.

33. Nick Rosenberger, "Big Change Is Coming to Boise Fast as 2 Tech Giants Invest $16 Billion. What to Expect," *Idaho Statesman*, March 22, 2024, https://www.idahostatesman.com/news/business/article286973665.html; "Micron Technology Stock Analysis: Losses Decrease, and Shares Gradually Rise," *RoboMarkets*, 2023, https://www.robomarkets.com.cy/blog/stock-market/analysis/micron-technology-stock-analysis-losses-decrease-and-shares-gradually-rise/.

34. Paul Mozur and John Liu, "With Ban on Micron, China Escalates Microchip Clash With U.S." *The New York Times*, May 22, 2023, https://www.nytimes.com/2023/05/22/business/micron-technology-china-ban.html.

35. Anirban Sen, "Intel Scraps $5.4 Bln Tower Deal After China Review Delay," Reuters, August 16, 2023, https://www.reuters.com/technology/intel-walk-away-54-bln-acquisition-tower

-semiconductor-sources-2023–08–16/; Raffaele Huang and Liza Lin, "Nvidia Probed in China over Possible Antimonopoly Violations," *The Wall Street Journal*, December 9, 2024, https://www.wsj.com/tech/china-nvidia-monopoly-probe-antitrust-da4f3d1f.

36. "China to Limit Antimony Exports in Latest Critical Mineral Curbs," Reuters, August 15, 2024, https://www.reuters.com/world/china/china-limit-antimony-exports-latest-critical-mineral-curbs-2024–08–15/; Jon Emont and Gavid Bade, "China's Lock on Rare Earths Dictated Path Toward Trade Truce," *The Wall Street Journal*, June 11, 2025, https://www.wsj.com/world/asia/trump-china-rare-earths-0cbf3f3c; Maggie Eastland, "US Formally Delays Affiliates Rule for One Year in US-China Pact," Bloomberg, November 10, 2025, https://www.bloomberg.com/news/articles/2025–11–10/us-formally-delays-affiliates-rule-for-one-year-in-us-china-pact.

37. Ana Swanson and Meaghan Tobin, "China's Rare Earth Restrictions Aim to Beat U.S. at Its Own Game," *The New York Times*, October 16, 2025, https://www.nytimes.com/2025/10/16/business/economy/china-rare-earths-supply-chain.html.

38. Simone McCarthy, "China Launches Satellites to Rival SpaceX's Starlink in Boost for Its Space Ambitions," CNN, August 9, 2024, https://www.cnn.com/2024/08/09/china/china-satellite-qianfan-g60-starlink-intl-hnk/index.html; Selam Gebrekidan, "This Was Supposed to Be the Year China Started Catching up with SpaceX," *The New York Times*, July 23, 2025, https://www.nytimes.com/interactive/2025/07/23/world/asia/starlink-spacex-musk-china-satellites.html.

39. Andrew Jones, "Chinese Rocket Stage Breaks Up into Cloud of More Than 700 Pieces of Space Debris," *SpaceNews*, August 9, 2024, https://spacenews.com/chinese-rocket-stage-breaks-up-into-cloud-of-more-than-700-pieces-of-space-debris/; Cissy Zhou, "China Races to Outflank Elon Musk's Starlink Satellite Internet Service," *Nikkei Asia*, October 11, 2024, https://asia.nikkei.com/spotlight/policy-asia/china-races-to-outflank-elon-musk-s-starlink-satellite-internet-service.

40. Walter Isaacson, "'How Am I in This War?': The Untold Story of Elon Musk's Support for Ukraine," *The Washington Post*, September 7, 2023, https://www.washingtonpost.com/opinions/2023/09/07/elon-musk-starlink-ukraine-russia-invasion/.

41. Christian Davenport and Joseph Menn, "Musk Refused to Allow Ukraine's Military to Use Starlink to Attack Russian Fleet," *The Washington Post*, September 11, 2023, https://www.washingtonpost.com/technology/2023/09/07/ukraine-starlink-musk-biography/.

42. Ibrahim Naber, "What Happened after Elon Musk Took the Russian Army Offline," *Politico*, February 25, 2026, https://www.politico.com/news/2026/02/25/elon-musk-russian-army-starlink-00793742.

43. Steven Feldstein, *The Rise of Digital Repression: How Technology Is Reshaping Power, Politics, and Resistance* (Oxford: Oxford University Press, 2021), 50; Starlink (@Starlink), "Starlink is connecting more than 5M people with high-speed internet across 125 countries, territories and many other markets. Thank you to all of our customers around the world!" X, February 28, 2025, 2:41 p.m., https://x.com/Starlink/status/1895559922991153367.

44. "China's Ground Segment: Building the Pillars of a Great Space Power," China Aerospace Studies Institute, March 1, 2021, https://www.airuniversity.af.edu/CASI/Display/Article/2517757/chinas-ground-segment-building-the-pillars-of-a-great-space-power/; William J. Dobson, Tarek Masoud, and Christopher Walker, eds., *Defending Democracy in an Age of Sharp Power* (Baltimore: Johns Hopkins University Press, 2023); Juliana Suess, "Guo Wang: China's Answer to Starlink?," Royal United Services Institute, May 3, 2023, https://www.rusi.org/explore-our-research/publications/commentary/guo-wang-chinas-answer-starlink.

17. Will China's Strategy Work?

1. Lingling Wei, "Wall Street Puts a 'Sell' on Its China Holdings," *The Wall Street Journal,* December 7, 2023, https://www.wsj.com/world/china/china-wall-street-investment-decline-3359f202; James T. Areddy and Liyan Qi, "China Signals Spying Fears amid Probe of Consulting Firms," *The Wall Street Journal,* May 8, 2023, https://www.wsj.com/articles/china-confirms-campaign-against-alleged-spying-by-foreign-consulting-companies-fef48e16; Lingling Wei and Stella Yifan Xie, "China's 40-Year Boom Is Over. What Comes Next?," *The Wall Street Journal,* August 20, 2023, https://www.wsj.com/world/china/china-economy-debt-slowdown-recession-622a3be4.

2. Lingling Wei and Jason Douglas, "Why China Is Starting a New Trade War," *The Wall Street Journal,* August 22, 2024, https://www.wsj.com/world/china/china-trade-war-xi-manufacturing-49f81f68; Karen Gilchrist, "EU's von der Leyen Echoes Yellen's Calls for Tough Stance on Chinese Overcapacity," CNBC, April 09, 2024, https://www.cnbc.com/2024/04/09/eus-von-der-leyen-echoes-yellens-calls-for-tough-stance-on-chinese-overcapacity.html.

3. Michael R. Gordon, "China's Newest Nuclear Submarine Sank, Setting Back Its Military Modernization," *The Wall Street Journal,* September 26, 2024, https://www.wsj.com/world/china/chinas-newest-nuclear-submarine-sank-setting-back-its-military-modernization-785b4d37?&-mod=djemchina.

4. Christopher J. Mihal, "Understanding the People's Liberation Army Rocket Force," *Military Review,* September 2021, https://www.armyupress.army.mil/Journals/Military-Review/English-Edition-Archives/China-Reader-Special-Edition-September-2021/Mihal-PLA-Rocket-Force/; Peter Martin and Jennifer Jacobs, "US Intelligence Shows Flawed China Missiles Led Xi to Purge Army," Bloomberg, January 06, 2024, https://www.bloomberg.com/news/articles/2024–01–06/us-intelligence-shows-flawed-china-missiles-led-xi-jinping-to-purge-military?sref=QmOxnLFz.

5. Sam Bresnick, "China's Military AI Roadblocks: PRC Perspectives on Technological Challenges to Intelligentized Warfare," Center for Security and Emerging Technology, June 2024, https://cset.georgetown.edu/publication/chinas-military-ai-roadblocks/; Seth G. Jones, "China Is Ready for War," *Foreign Affairs,* October 2, 2024, https://www.foreignaffairs.com/china/china-ready-war-america-is-not-seth-jones.

6. See, for example, Helen Toner, Jenny Xiao, and Jeffrey Ding, "The Illusion of China's AI Prowess," *Foreign Affairs,* June 2, 2023, https://www.foreignaffairs.com/china/illusion-chinas-ai-prowess-regulation-helen-toner.

7. Meaghan Tobin and Lyric Li, "Ernie, What Is Censorship? China's Chatbots Face Additional Challenges," *The Washington Post,* February 4, 2023, https://www.washingtonpost.com/world/2023/02/24/china-baidu-ernie-chatbot-chatgpt/; Steven Feldstein, "The Consequences of Generative AI for Democracy, Governance and War," *Survival* (October–November 2023): 123.

8. Liza Lin, "China Puts Power of State Behind AI—and Risks Strangling It," *The Wall Street Journal,* July 16, 2024, https://www.wsj.com/tech/china-puts-power-of-state-behind-aiand-risks-strangling-it-f045e11d.

9. "Why Chinese AI Has Stunned the World," *The Economist,* January 23, 2025, https://www.economist.com/briefing/2025/01/23/chinas-ai-industry-has-almost-caught-up-with-americas.

10. "Ban the H20: Competing in the Inference Age," *ChinaTalk,* March 7, 2025, https://www.chinatalk.media/p/ban-the-h20-competing-in-the-inference.

11. Marc Andreessen (@pmarca), "Deepseek R1 is AI's Sputnik moment," X, January 26, 2025, 5:16 p.m., https://archive.ph/lIWNv; "Why Chinese AI Has Stunned the World."

12. "Why Chinese AI Has Stunned the World"; David Sacks (@DavidSacks), "A BEST CASE SCENARIO FOR AI? The Doomer narratives were wrong . . ." X, August 9, 2025, 2:13 PM, https://x.com/davidsacks/status/1954244614304739360.

13. Jordan Schneider and Lily Ottinger, interview with Lennart Heim and Chris Miller, *ChinaTalk* podcast, January 6, 2025, https://www.chinatalk.media/p/ai-geopolitics-in-the-age-of-test; Sem Eifling, "China's Biggest AI Model Is Challenging American Dominance," *Rest of World*, September 23, 2024, https://restofworld.org/2024/alibaba-qwen-ai-model/; Ben Jiang, "Alibaba's AI Model Outperforms Chinese Rivals, Ranks Just Behind OpenAI, Anthropic," *South China Morning Post*, July 11, 2024, https://www.scmp.com/tech/big-tech/article/3270079/alibabas-ai-model-outperforms-chinese-rivals-ranks-just-behind-openai-anthropic; Jordan Schneider et al., "Deepseek: The Quiet Giant Leading China's AI Race," *ChinaTalk*, November 27, 2024, https://www.chinatalk.media/p/deepseek-ceo-interview-with-chinas.

14. Henry Farrell and Abraham Newman, *Underground Empire: How America Weaponized the World Economy* (New York: Random House, 2023), 37.

15. Site visit to Equinix data center facility, Ashburn, VA, January 10, 2025; Antonio Olivo and William Neff, "Our Digital Lives Need Massive Data Centers. What Goes On Inside Them?," *The Washington Post*, September 17, 2024, https://www.washingtonpost.com/dc-md-va/interactive/2024/data-centers-tour-northern-virginia/.

16. Dan Swinhoe, "Washington Commanders File for Permission to Build Data Centers on Ashburn Training Ground," *Data Center Dynamics*, June 18, 2024, https://www.datacenterdynamics.com/en/news/washington-commanders-file-for-permission-to-build-data-centers-on-ashburn-training-ground/.

17. Dylan Patel, Daniel Nishball, and Jeremie Eliahou Ontiveros, "Multi-Datacenter Training: OpenAI's Ambitious Plan to Beat Google's Infrastructure," *SemiAnalysis*, September 4, 2024, https://semianalysis.com/2024/09/04/multi-datacenter-training-openais/; Caroline O'Donovan, "Fighting Back Against Data Centers, One Small Town at a Time," *The Washington Post*, October 5, 2024, https://www.washingtonpost.com/technology/2024/10/05/data-center-protest-community-resistance/.

18. Patel, Nishball, and Ontiveros, "Multi-Datacenter Training."

19. Dylan Patel, Jeremie Eliahou Ontiveros, and Sravan Kundojjala, "Fab Whack-A-Mole: Chinese Companies Are Evading U.S. Sanctions," *SemiAnalysis*, October 28, 2024, https://semianalysis.com/2024/10/28/fab-whack-a-mole-chinese-companies/; Che Pan, "Tech War: China Sees Glut of AI Data Centres as GPU Mismatches Exacerbate Weak Demand," *South China Morning Post*, October 11, 2024, https://www.scmp.com/tech/article/3281894/tech-war-china-sees-glut-ai-data-centres-gpu-mismatches-exacerbate-weak-demand. Caiwei Chen, "China Built Hundreds of AI Data Centers to Catch the AI Boom. Now Many Stand Unused," *MIT Technology Review*, March 26, 2025, https://www.technologyreview.com/2025/03/26/1113802/china-ai-data-centers-unused/.

20. Pan, "Tech War;" "Bits, Bytes, and Bottlenecks: A Story of Two Data Center Corridors," Special Competitive Studies Project newsletter, November 20, 2024, https://scsp222.substack.com/p/bits-bytes-and-bottlenecks-a-story.

21. Patel, Ontiveros, and Kundojjala, "Fab Whack-A-Mole."

22. Tim Wu, "Could America Win the AI Race but Lose the War?," *Financial Times*, December 13, 2025, https://www.ft.com/content/12581344–6e37–45a0-a9d5-e3d6a9f8d9ba.

23. Xiaoying You, "China leads research in 90% of crucial technologies—a dramatic shift this century," *Nature*, December 12, 2025, https://www.nature.com/articles/d41586–025–04048–7; Jennifer Wong Leung and Stephan Robin, "Critical Technology Tracker," Australian Strategic Policy Institute, accessed December 16, 2025, https://www.aspi.org.au/programs/critical-technology-tracker/.

24. Ian Clay and Robert D. Atkinson, "Wake Up, America: China Is Overtaking the United States in Innovation Capacity," Information Technology and Innovation Foundation, January 23, 2023, https://itif.org/publications/2023/01/23/wake-up-america-china-is-overtaking-the-united-states-in-innovation-capacity/.

25. Karen Chiu, "The Story of Drone Pioneer DJI," *South China Morning Post,* October 2, 2018, https://www.scmp.com/abacus/who-what/what/article/3028241/story-drone-pioneer-dji.

26. Paul Mozur, Aaron Krolik, and Keith Bradsher, "As War in Ukraine Grinds On, China Helps Refill Russian Drone Supplies," *The New York Times,* March 21, 2023, https://www.nytimes.com/2023/03/21/business/russia-china-drones-ukraine-war.html; Nessa Anwar, "World's Largest Drone Maker Is Unfazed—Even If It's Blacklisted by the U.S.," CNBC, February 7, 2023, https://www.cnbc.com/2023/02/08/worlds-largest-drone-maker-dji-is-unfazed-by-challenges-like-us-blacklist.html; Kate Kelly, "A Chinese Firm Is America's Favorite Drone Maker. Except in Washington," *The New York Times,* April 25, 2024, https://www.nytimes.com/2024/04/25/us/politics/us-china-drones-dji.html.

27. Cate Cadell, "Drone Company DJI Obscured Ties to Chinese State Funding, Documents Show," *The Washington Post,* February 1, 2022, https://www.washingtonpost.com/national-security/2022/02/01/china-funding-drones-dji-us-regulators/.

28. Joyu Wang, "Taiwan Wants a Drone Army—but China Makes the Drones It Wants," *The Wall Street Journal,* June 23, 2024, https://www.wsj.com/world/taiwan-drone-stockpiling-efforts-bcf70ef5.

29. Dana Nickel, "FCC Bans Foreign-Made Drones over National Security, Spying Concerns," *Politico,* December 22, 2025, https://www.politico.com/news/2025/12/22/fcc-drone-ban-dji-00703742; Anthony Cuthbertson, "US Bans World's Biggest Drone Maker over National Security Concerns," *The Independent,* December 23, 2025, https://www.yahoo.com/news/articles/us-bans-world-biggest-drone-113753190.html.

30. Clay and Atkinson, "Wake Up, America."

31. William C. Hannas and Huey-Meei Chang, "Chinese Technology Transfer," in *China's Quest for Foreign Technology: Beyond Espionage* (New York: Routledge, 2021), 14–15.

32. Ben Murphy et al., "Xi Jinping: 'Strive to Become the World's Primary Center for Science and High Ground for Innovation,'" *DigiChina,* March 18, 2021, https://digichina.stanford.edu/work/xi-jinping-strive-to-become-the-worlds-primary-center-for-science-and-high-ground-for-innovation/.

33. Austin Ramzy and Keith Zhai, "Xi and Putin Rekindle 'Strategic Bromance' in Russia," *The Wall Street Journal,* March 21, 2023, https://www.wsj.com/articles/xi-and-putin-rekindle-strategic-bromance-in-russia-5f9f96dd; Charlie Campbell, "Vladimir Putin's Pomp-Filled China Trip Underscores the Limits of Western Pressure," *Time,* May 17, 2024, https://time.com/6979248/putin-russia-xi-china-western-pressure-analysis/.

34. Ramzy and Zhai, "Xi and Putin Rekindle 'Strategic Bromance' in Russia"; Evan Osnos, "China's Age of Malaise," *The New Yorker,* October 23, 2023, https://www.newyorker.com/magazine/2023/10/30/chinas-age-of-malaise; Alexander Gabuev, "Why China Is Sabotaging Ukraine," *Foreign Affairs,* June 14, 2024, https://www.foreignaffairs.com/china/why-china-sabotaging-ukraine-peace-agreement.

18. An Axis of Disruptors?

1. Andrea Kendall-Taylor and Richard Fountaine, "The Axis of Upheaval," *Foreign Affairs,* April 23, 2024, https://www.foreignaffairs.com/china/axis-upheaval-russia-iran-north-korea-taylor-fontaine; Special Competitive Studies Project, "The Existential Challenge of the Axis of Disruptors," in *Vision for Competitiveness: Mid-Decade Opportunities for Strategic Victory,* May 2024, https://www.scsp.ai/reports/vision/.

2. Interview with representatives from the French Ministry of Foreign Affairs, Zoom, May 6, 2024; Thomas Graham, "What Russia Really Wants," *Foreign Affairs,* October 9, 2023, https://www.foreignaffairs.com/asia/what-russia-really-wants; Bruce McClintock, Jeffrey W. Hor-

nung, and Katherine Costello, "Russia's Global Interests and Actions," RAND, June 15, 2021, https://www.rand.org/pubs/perspectives/PE327.html.

3. Hal Brands, "The New Autocratic Alliances," *Foreign Affairs*, March 29, 2024, https://www.foreignaffairs.com/united-states/new-autocratic-alliances; Oriana Skylar Mastro, *Upstart: How China Became a Great Power* (Oxford: Oxford University Press, 2024), 66; Eugene Rumer, "Taiwan and the Limits of the Russia-China Friendship," Carnegie Endowment for International Peace, September 3, 2024, https://carnegieendowment.org/research/2024/08/taiwan-and-the-limits-of-the-russia-china-friendship?lang=en.

4. Mastro, *Upstart*, 72; Rachel Cheung, "Oriana Skylar Mastro on How China Gained Power," *The Wire China*, September 29, 2024, https://www.thewirechina.com/2024/09/29/oriana-skylar-mastro-on-how-china-gained-power/; International Republic Institute, "The Authoritarian Nexus—How Russia and China Undermine Democracy Worldwide," April 18, 2024, https://www.iri.org/resources/the-authoritarian-nexus-how-russia-and-china-undermine-democracy-worldwide/.

5. Olena Bilousova et al., "Challenges of Export Controls Enforcement," Yermak-McFaul International Working Group on Russia Sanctions and KSE Institute, January 2024, https://kse.ua/wp-content/uploads/2024/01/Challenges-of-Export-Controls-Enforcement.pdf; Dan De Luce and Owen Hayes, "In Ukraine War, China Is Helping Tilt Momentum in Russia's Favor, Top U.S. Spy Says," NBC News, May 2, 2024, https://www.nbcnews.com/politics/national-security/china-helping-russia-momentum-ukraine-war-top-us-spy-rcna150437.

6. Rumer, "Taiwan and the Limits of the Russia-China Friendship"; Phelim Kine, "China's Russia Support Strategy," *Politico: China Watcher* newsletter, February 22, 2024, https://www.politico.eu/newsletter/china-watcher/chinas-russia-support-strategy/; Maxim Trudolyubov, "China's Balancing Act Between the U.S. and Russia," Wilson Center, May 24, 2024, https://www.wilsoncenter.org/blog-post/chinas-balancing-act-between-us-and-russia.

7. Austin Ramzy, "Putin Casts Russia and China as Defenders of Stability," *The Wall Street Journal*, May 16, 2024, https://www.wsj.com/world/putin-beijing-visit-xi-support-3b0531c3; Eugene Rumer, "The United States and the 'Axis' of Its Enemies: Myths vs. Reality," Carnegie Endowment for International Peace, November 25, 2024, https://carnegieendowment.org/research/2024/11/the-united-states-and-the-axis-of-its-enemies-myths-vs-reality; Laura Bicker, "China-Russia Relations: What Is Xi Jinping Prepared to Pay for Putin's War?," BBC, May 15, 2024, https://www.bbc.com/news/world-asia-china-69013520.

8. Kine, "China's Russia Support Strategy"; Nathaniel Sher, "Behind the Scenes: China's Increasing Role in Russia's Defense Industry," *Carnegie Politika*, May 6, 2024, https://carnegieendowment.org/russia-eurasia/politika/2024/05/behind-the-scenes-chinas-increasing-role-in-russias-defense-industry?lang=en; Trudolyubov, "China's Balancing Act Between the U.S. and Russia."

9. Rumer, "Taiwan and the Limits of the Russia-China Friendship."

10. Stuart Lau, "US Accuses China of Giving 'Very Substantial' Help to Russia's War Machine," *Politico*, September 10, 2024, https://www.politico.eu/article/united-states-accuse-china-help-russia-war-kurt-campbell/.

19. "We Are Strong People and Can Cope with Any Challenge"

1. "Vladimir Putin Fires a New Missile to Amplify His Nuclear Threats," *The Economist*, November 21, 2024, https://www.economist.com/europe/2024/11/21/vladimir-putin-fires-a-new-missile-to-amplify-his-nuclear-threats.

2. "Putin Touts Russia's New Missile and Delivers a Menacing Warning to NATO," Associated Press, November 21, 2024, https://apnews.com/article/russia-ukraine-intercontinental-missile-war-putin-d50183ccfc28b10c71e93f3e68159a61; Marc Santora et al., "With Use of New Missile, Rus-

sia Sends a Threatening Message to the West," *The New York Times*, November 21, 2024, https://www.nytimes.com/2024/11/21/world/europe/russia-ballistic-missile-ukraine-war.html; David L. Stern, Siobhán O'Grady, and Ellen Francis, "Russia Seeks to Intimidate Ukraine with New Missile, Officials Say," *The Washington Post*, November 22, 2024, https://www.washingtonpost.com/world/2024/11/22/ukraine-russia-missiles-war/.

3. "Putin Touts Russia's New Missile and Delivers a Menacing Warning to NATO."

4. "Putin Dismisses 'Stupid' Western Sanctions 'Blitzkrieg,'" Reuters, June 17, 2022, https://www.reuters.com/world/europe/putin-dismisses-stupid-western-sanctions-blitzkrieg-2022–06–17/.

5. "Responding to Two Years of Russia's Full-Scale War on Ukraine and Navalny's Death," US Department of State, February 23, 2024, archived February 28, 2024, archive.today, https://archive.ph/w9KUC.

6. Oriana Skylar Mastro, "China Isn't the Only Agent of Chaos; Russia Is, Too," *Foreign Affairs*, November/December 2024, https://www.foreignaffairs.com/united-states/china-agents-chaos-russia-mastro; interview with a senior Commerce Department official, Washington, DC, April 30, 2024.

7. "Heller Continues to Invest in Changzhou—Production Site to Be Significantly Expanded," Heller, June 26, 2023, https://www.heller.biz/en/news/heller-continues-to-invest-in-changzhou-production-site-to-be-significantly-expanded/.

8. Peter Apps, "The Western Industrial Components Rebuilding Russia's Military," Reuters, August 16, 2024, https://www.reuters.com/world/europe/western-industrial-components-rebuilding-russias-military-2024–08–16/.

9. Kevin Wolf, interview conducted by author, Zoom, March 25, 2024.

10. Parts of this section reflect ideas from a previously co-published article by the author: Steve Feldstein and Fiona Brauer, "Why Russia Has Been So Resilient to Western Export Controls," Carnegie Endowment for International Peace, March 11, 2024, https://carnegieendowment.org/research/2024/03/why-russia-has-been-so-resilient-to-western-export-controls; Frederick Wehrey, "Russia's Uneven Influence in the Maghreb," Carnegie Endowment for International Peace, October 2024, https://carnegieendowment.org/research/2024/10/russias-uneven-influence-maghreb?lang=en; Solomon Ekanem, "Top 10 Ports in Africa Based on Trade Volume," *Pulse Nigeria*, April 1, 2024, https://www.pulse.ng/business/domestic/top-ports-in-africa/ozl715p.

11. Paul Mozur, Aaron Krolik, and Adam Satariano, "Chinese Traders and Moroccan Ports: How Russia Flouts Global Tech Bans," *The New York Times*, December 19, 2023, https://www.nytimes.com/2023/12/19/technology/russia-flouts-global-tech-bans.html.

12. Mozur, Krolik, and Satariano, "Chinese Traders and Moroccan Ports."

13. Georgi Kantchev, Paul Hannon, and Laurence Norman, "How Sanctioned Western Goods Are Still Flowing into Russia," *The Wall Street Journal*, May 14, 2023, https://www.wsj.com/world/how-sanctioned-western-goods-are-still-flowing-into-russia-916db262.

14. "A wide range of services for the import and export of goods," IMEX Expert website, accessed January 3, 2026, https://archive.ph/wip/Ptljq; Kantchev, Hannon, and Norman, "How Sanctioned Western Goods Are Still Flowing into Russia."

15. Apps, "The Western Industrial Components Rebuilding Russia's Military."

16. Author interview with a senior Commerce Department official, April 30, 2024.

17. Eric Bazail-Eimil, Robbie Gramer, and Suzanne Lynch, "A Fresh Push to Hobble Russia's War Economy," *Politico: National Security Daily* newsletter, October 24, 2024, https://www.politico.com/newsletters/national-security-daily/2024/10/24/a-fresh-push-to-hobble-russias-war-economy-00183727; Georgi Kantchev and Lingling Wei, "China Is Studying Russia's Sanctions

Evasion to Prepare for Taiwan Conflict," *The Wall Street Journal*, December 1, 2024, https://www.wsj.com/world/china/china-is-studying-russias-sanctions-evasion-to-prepare-for-taiwan-conflict-5665f508.

18. Mathieu Boulègue et al., "Assessing Russian Plans for Military Regeneration," Chatham House, July 2024, https://www.chathamhouse.org/sites/default/files/2024-07/2024-07-09-as sessing-russian-plans-military-boulegue-et-al.pdf.

19. "How Russia's Fatalities Compare with Ukraine's," *The Economist*, February 23, 2026, https://www.economist.com/graphic-detail/2026/02/23/how-russias-fatalities-compare-with-ukraines; Daria Dmytriieva, "Russia's Losses in Ukraine as of March 17," RBC-Ukraine, March 17, 2026, https://newsukraine.rbc.ua/news/russia-s-losses-in-ukraine-as-of-march-17-1773669682.html; Matthew Luxmoore, "Russia Taps Soviet-Era Stockpiles as Armored-Vehicle Supply Dwindles," *The Wall Street Journal*, December 21, 2024, https://www.wsj.com/world/russia-taps-soviet-era-stockpiles-as-armored-vehicle-supply-dwindles-6f8f8df2;

20. Nan Tian et al., "Trends in World Military Expenditure, 2023," Stockholm International Peace Research Institute, April 2024, https://www.sipri.org/sites/default/files/2024-04/2404_fs_milex_2023.pdf; "Export Controls: A Key G-7 Tool to Halt Russia's War, Working Group Paper #20," International Working Group on Russian Sanctions, June 12, 2024, https://sanctions.kse.ua/wp-content/uploads/2024/06/working_paper_20_export_controls_june_12_2024_final_update.pdf; Dan Sabbagh, "Global Defence Spending Rises 9% to Record $2.2tn," *The Guardian*, https://www.theguardian.com/world/2024/feb/13/global-defence-spending-rises-9-per-cent-to-record-22tn-dollars.

21. Andrew Roth, "'A Lot Higher than We Expected': Russian Arms Production Worries Europe's War Planners," *The Guardian*, February 15, 2024, https://www.theguardian.com/world/2024/feb/15/rate-of-russian-military-production-worries-european-war-planners.

22. Roth, "A Lot Higher than We Expected."

23. "Kh-101/Kh-102," CSIS Missile Defense Project, Center for Strategic and International Studies, October 26, 2017, last modified April 23, 2024, https://missilethreat.csis.org/missile/kh-101-kh-102/; Jack Watling and Gary Somerville, "A Methodology for Degrading the Arms of the Russian Federation," Royal United Services Institute, June 26, 2024, https://www.rusi.org/explore-our-research/publications/occasional-papers/methodology-degrading-arms-russian-federation.

24. Steven Feldstein, "The Larger Geopolitical Shift Behind Iran's Drone Sales to Russia," Carnegie Endowment for International Peace, October 26, 2022, https://carnegieendowment.org/posts/2022/10/the-larger-geopolitical-shift-behind-irans-drone-sales-to-russia?lang=en.

25. Inst for Science (@TheGoodISIS), "A key detail in the new Zvezda video is that Alabuga now claims to be producing nine-times more drones than it planned 'at one time,' which can reasonably be interpreted as the first years of its" X, July 22, 2025, 12:34 p.m., https://x.com/TheGoodISIS/status/1947696765521605116; C. J. Chivers, "The Weapon That Terrorizes Ukrainians by Night," *The New York Times*, June 25, 2025, https://www.nytimes.com/2025/06/25/magazine/ukraine-russia-war-drone-shahed-iran.html; Mary Ilyushina, "Russia's deadly drone industry upgraded with Iran's help, report says," *The Washington Post*, May 29, 2025, https://www.washingtonpost.com/world/2025/05/29/russia-iran-drone-cooperation-industry/.

26. Ilyushina, "Russia's Deadly Drone Industry Upgraded with Iran's Help;" Omar Al-Ghusbi and Conrad Rousseau, "Airborne Axis: Inside the Deal That Brought Iranian Drone Production to Russia," *C4ADS*, 2025, https://c4ads.org/wp-content/uploads/2025/05/SaharaThunder-FinalLayout.pdf.

27. Dara Massicot, "Russian Military Reconstitution: 2030 Pathways and Prospects," Carnegie Endowment for International Peace, September 12, 2024, https://carnegieendowment.org/research/2024/09/russian-military-reconstitution-2030-pathways-and-prospects?lang=en.

28. Joby Warrick and Serhiy Morgunov, "Far from the Front Lines, a Spy War Rages over Russian Weapons," *The Washington Post*, December 5, 2024, https://www.washingtonpost.com/national-security/2024/12/05/russian-sanctions-weapons-ukraine-war/; Andrew E. Kramer, "Russia Unleashes Largest Drone Assault of War, Setting Government Building Ablaze," *The New York Times*, September 7, 2025, https://www.nytimes.com/2025/09/07/world/europe/russian-strike-kyiv-ukraine.html.

29. Mark Krutov, Sergei Dobrynin, and Yauhen Lehalau, "Inside Rubicon, The Elite Russian Drone Unit Wreaking Havoc On Ukraine's Troops," RFE/RL, September 17, 2025, https://www.rferl.org/a/russia-drone-rubicon-secret-ukraine-war/33532804.html.

30. Maria Snegovaya et al., "Back in Stock?," Center for Strategic and International Studies, April 2024, https://csis-website-prod.s3.amazonaws.com/s3fs-public/2024–04/240419_Snegovaya_Backin_Stock.pdf?VersionId=rwHuy82sf7y5TEoD8sRJLGF3lYmeGAnL; Mathieu Boulègue et al., "Assessing Russian Plans for Military Regeneration," Chatham House, July 9, 2024, DOI: 10.55317/9781784136178, p. 3.

31. Alexander H. Montgomery and Amy J. Nelson, "Ukraine and the Kinzhal: Don't Believe the Hypersonic Hype," Brookings, May 23, 2023, https://www.brookings.edu/articles/ukraine-and-the-kinzhal-dont-believe-the-hypersonic-hype/; Andrew W. Reddie, "Why the Alarm over Russia's Use of Hypersonic Missiles in Ukraine Is Misplaced," *Bulletin of the Atomic Scientists*, March 31, 2023, https://thebulletin.org/2023/03/why-the-alarm-over-russias-use-of-hypersonic-missiles-in-ukraine-is-misplaced/.

32. Ann M. Simmons, "Spy Mania Sows Fear Among Russia's Scientists," *The Wall Street Journal*, October 2, 2024, https://www.wsj.com/world/russia/putin-russia-scientist-arrests-ae6d9ed2.

33. Simmons, "Spy Mania Sows Fear Among Russia's Scientists;" Sébastian Seibt, "Why Russia Is Accusing Its Own Scientists of Treason," *France24*, May 9, 2024, https://www.france24.com/en/europe/20240905-why-russia-is-accusing-its-own-scientists-of-treason.

34. Peter Suciu, "Russia's Su-57 Fifth-Generation Fighter Mocked on Chinese Social Media," *Forbes*, November 4, 2024, https://www.forbes.com/sites/petersuciu/2024/11/04/russias-su-57-fifth-generation-fighter-mocked-on-chinese-social-media/; Clash Report (@clashreport), "Lots of screws: Closer look at Russia's 5th generation Su-57 fighter jet at China's Zhuhai Airshow," X, November 4, 2024, 6:45 a.m., https://archive.ph/BEZdP.

35. Christian Shepherd and Lyric Li, "With New Stealth Fighter, China Displays Ambitions to Challenge U.S.," *The Washington Post*, November 12, 2024, https://www.washingtonpost.com/world/2024/11/12/china-j-35a-stealth-fighter-jet-russia/.

36. Kenneth Chang, "Russian Launch Site Mishap Shows Perilous State of Storied Space Program," *The New York Times*, December 1, 2025, https://www.nytimes.com/2025/12/01/science/russia-base-damage-iss-launch-collapse.html; Alexandra Prokopenko, "Russia's New Budget Is a Blueprint for War, Despite the Cost," *Financial Times*, October 7, 2024, https://www.ft.com/content/0778af05-fde1–47e3–90e4-bf0b2411d155.

20. Military Innovations Spread Across the Globe

1. Katrina Manson, "Low-Cost Warfare: US Military Battles with 'Costco Drones,'" *Financial Times*, January 5, 2022, https://www.ft.com/content/aef5901e-4b9c-4561-a559-a6b7197bafe1.

2. Gordon Lubold, "How an Iranian-Backed Militia Ties Down U.S. Naval Forces in the Red Sea," *The Wall Street Journal*, June 12, 2024, https://www.wsj.com/world/middle-east/how-an-iranian-backed-militia-ties-down-u-s-naval-forces-in-the-red-sea-3821056c; Lynsey Chutal, "Here's What We Know About the Oil Tanker Stuck in the Red Sea," *The New York Times*, August 29, 2024, https://www.nytimes.com/2024/08/29/world/middleeast/oil-tanker-red-sea-yemen-houthi-attack.html; Joshua Keating, "How a Yemeni Rebel Group Is Creating Chaos in the Global Economy," *Vox*, January 12, 2024, https://www.vox.com/world-politics/24010092/houthis-red-sea-shipping-yemen-israel-gaza; Colin Demarest, "Exclusive: Houthi Arsenal Shocks the Pen-

tagon's Top Weapons Buyer," *Axios*, November 14, 2024, https://www.axios.com/2024/11/14/laplante-houthi-antiship-missiles-iran.

3. Benoit Faucon and Thomas Grove, "Russia Provided Targeting Data for Houthi Assault on Global Shipping," *The Wall Street Journal*, October 24, 2024, https://www.wsj.com/world/russia-provided-targeting-data-for-houthi-assault-on-global-shipping-eabc2c2b; Emily O. Goldman and Leslie C. Eliason, "Introduction: Theoretical and Comparative Perspectives on Innovation and Diffusion," in *The Diffusion of Military Technology and Ideas*, eds. Emily O. Goldman and Leslie C. Eliason (Stanford: Stanford University Press, 2003), 18–19.

4. Dov Lieber, Alexander Ward, and Laurence Norman, "Why the U.S. and Israel Struck When They Did: A Chance to Kill Iran's Leaders," *The Wall Street Journal*, February 28, 2026, https://www.wsj.com/world/middle-east/why-the-u-s-and-israel-struck-iran-when-they-did-a-chance-to-kill-its-leaders-bodbbc88.

5. Kim Barker, "Ukraine Helps U.S. Bases in the Mideast With Stopping Drones," *The New York Times*, March 9, 2026, https://www.nytimes.com/2026/03/09/world/middleeast/ukraine-shahed-drone-middle-east.html.

6. Rayhan Uddin, "Iranian Drones Cost a Fraction of Air Defences. How Long Can Gulf States Last?," *Middle East Eye*, March 2, 2026, https://www.middleeasteye.net/news/iranian-drones-fraction-air-defences-how-long-gulf-last; Jeff Wise, "We're in a New Kind of War," *New York*, March 11, 2026, https://archive.ph/1CZfI; David Sanger et al., "Entering War's Third Week, Trump Faces Stark Choices," *The New York Times*, March 15, 2026, https://www.nytimes.com/2026/03/15/us/politics/trump-stark-choices-iran-war.html

7. Michael R. Gordon, Daniel Michaels, and Alistair MacDonald, "U.S. Army Sends Ukraine-Tested Drones to Hit Iran's Drones," *The Wall Street Journal*, March 7, 2026, https://www.wsj.com/world/u-s-army-sends-ukraine-tested-drones-to-hit-irans-drones-1e4c983a.

8. Michael Schwirtz, "To Fight Iran's Drones, U.S. Taps Ukraine's Hard-Earned Knowledge," *The New York Times*, March 13, 2026, https://www.nytimes.com/2026/03/13/world/middleeast/ukraine-drone-knowledge.html.

9. Everett M. Rogers, *Diffusion of Innovations*, 5th ed. (New York: Free Press, 2003), 11.

10. Michael Horowitz, *The Diffusion of Military Power: Causes and Consequences for International Politics* (Cambridge, MA: Harvard University Press, 2007), 8, 19.

11. Goldman and Eliason, "Introduction: Theoretical and Comparative Perspectives on Innovation and Diffusion," 18; Andrea Gilli and Mauro Gilli, "Why China Has Not Caught Up Yet: Military-Technological Superiority and the Limits of Imitation, Reverse Engineering, and Cyber Espionage," *International Security* 43, no. 3 (2018): 141–42; Robert Gilpin, *War and Change in World Politics* (Cambridge: Cambridge University Press, 1981).

12. Audrey Kurth Cronin, *Power to the People: How Open Technological Innovation Is Arming Tomorrow's Terrorists* (New York: Oxford University Press, 2020), 19, 23.

13. Cronin, *Power to the People*, 34.

14. Cronin, *Power to the People*, 12.

15. Michael C. Horowitz, "Battles of Precise Mass," *Foreign Affairs*, October 22, 2024, https://www.foreignaffairs.com/world/battles-precise-mass-technology-war-horowitz.

16. Cronin, *Power to the People*, 13.

17. Thomas L. Friedman, "Inside America's Shadow War with Iran," *The New York Times*, March 5, 2024, https://www.nytimes.com/2024/03/05/opinion/biden-iran-middle-east.html.

18. Joshua Keating, "America No Longer Has a Monopoly on Deadly Drones," *Vox*, January 29, 2024, https://www.vox.com/2024/1/29/24055046/jordan-drone-strike-troop-deaths-proliferation.

19. Friedman, "Inside America's Shadow War with Iran."

20. Kyle Mizokami, "The CIA's Blade-Wielding 'Flying Ginsu' Missile Strikes Again," *Popular Mechanics*, December 10, 2019, https://www.popularmechanics.com/military/weapons/a30175425/cia-blade-missile/.

21. Goldman and Eliason, "Introduction: Theoretical and Comparative Perspectives on Innovation and Diffusion," 18–19.

21. Israel's Culture of Innovation

1. Steve Hendrix and Joby Warrick, "Who Was the Iranian Nuclear Scientist Killed in Tehran?," *The Washington Post*, November 27, 2020, https://www.washingtonpost.com/world/iran-fakhrizadeh-who-is-nuclear/2020/11/27/cffdb6a8-30ce-11eb-9dd6-2d0179981719_story.html; Ronen Bergman and Farnaz Fassihi, "The Scientist and the A.I.-Assisted, Remote-Control Killing Machine," *The New York Times*, September 18, 2021, https://www.nytimes.com/2021/09/18/world/middleeast/iran-nuclear-fakhrizadeh-assassination-israel.html.

2. Bergman and Fassihi, "The Scientist and the A.I.-Assisted, Remote Control Killing Machine." I incorporated a similar accounting of Israel's use of a robot machine gun in the following article: Steven Feldstein, "AI in War: Can Advanced Military Technologies Be Tamed Before It's Too Late?," *Bulletin of the Atomic Scientists*, January 11, 2024, https://thebulletin.org/2024/01/ai-in-war-can-advanced-military-technologies-be-tamed-before-its-too-late/.

3. Robert Gandt, "The Czech Fighter That Helped Israel Win Its War of Independence," *Smithsonian Magazine*, September 2019, https://www.smithsonianmag.com/air-space-magazine/czech-knife-180972958/; Shashank Joshi, "How Israel Fights," *Foreign Affairs*, March–April 2024, https://www.foreignaffairs.com/reviews/how-israel-fights; Edward N. Luttwak and Eitan Shamir, *The Art of Military Innovation: Lessons from the Israel Defense Forces* (Cambridge, MA: Harvard University Press, 2023), 24.

4. Timothy D. Hoyt, "Revolution and Counter-Revolution," in *The Diffusion of Military Technology and Ideas*, eds. Emily O. Goldman and Leslie C. Eliason (Stanford: Stanford University Press, 2003), 182–85; "Harpoon," Naval Air Systems Command, accessed March 14, 2025, https://www.navair.navy.mil/product/Harpoon.

5. Hoyt, "Revolution and Counter-Revolution," 186–87; David B. Ottaway, "War and Missiles," *The Washington Post*, June 13, 1982, https://www.washingtonpost.com/archive/politics/1982/06/14/war-and-missiles/12b4abf6–9aa5–44a2–990f-33d08a99dcf4/; John F. Kreis, "Unmanned Aircraft in Israeli Air Operations," *Air Power History* 37, no. 4 (1990): 48.

6. Joshi, "How Israel Fights."

7. Jon Swaine et al., "How Hamas Exploited Israel's Reliance on Tech to Breach Barrier on Oct. 7," *The Washington Post*, November 17, 2023, https://www.washingtonpost.com/investigations/2023/11/17/how-hamas-breached-israel-iron-wall/; Andrei Popuviciu and Lubna Masarwa, "Gaza: What the Iron Wall Built by Israel Means for Besieged Palestinians," *Middle East Eye*, December 11, 2021, https://www.middleeasteye.net/news/israel-iron-wall-gaza-palestinians-siege.

8. Swaine et al., "How Hamas Exploited Israel's Reliance on Tech to Breach Barrier on Oct. 7"; Ronen Bergman and Adam Goldman, "Israel Knew Hamas's Attack Plan More Than a Year Ago," *The New York Times*, November 30, 2023, https://www.nytimes.com/2023/11/30/world/middleeast/israel-hamas-attack-intelligence.html.

9. Swaine et al., "How Hamas Exploited Israel's Reliance on Tech to Breach Barrier on Oct. 7."

10. Ariel (Eli) Levite, "How Was Israel Caught Off-Guard?," *War on the Rocks*, February 22, 2024, https://warontherocks.com/2024/02/how-was-israel-caught-off-guard/.

11. Yuval Abraham, "'A Mass Assassination Factory': Inside Israel's Calculated Bombing of Gaza," *+972 Magazine*, November 30, 2023, https://www.972mag.com/mass-assassination-factory-israel-calculated-bombing-gaza/; Elizabeth Dwoskin, "Israel Built an 'AI Factory' for War. It Unleashed it in Gaza," *The Washington Post*, December 29, 2024, https://www.washingtonpost.com/technology/2024/12/29/ai-israel-war-gaza-idf/.

12. Feldstein, "AI in War: Can Advanced Military Technologies Be Tamed Before It's Too Late?"; Patrick Kingsley et al., "Israel Loosened Its Rules to Bomb Hamas Fighters, Killing Many More Civilians," *The New York Times*, December 26, 2004, https://www.nytimes.com/2024/12/26/world/middleeast/israel-hamas-gaza-bombing.html; Dwoskin, "Israel Built an 'AI Factory' for War"; Abraham, "'A Mass Assassination Factory.'"

13. Dwoskin, "Israel Built an 'AI Factory' for War"; Yuval Abraham, "'Lavender': The AI Machine Directing Israel's Bombing Spree in Gaza," *+972 Magazine*, April 3, 2004, https://www.972mag.com/lavender-ai-israeli-army-gaza/.

14. Abraham, "'Lavender'"; Dwoskin, "Israel Built an 'AI Factory' for War."

15. Abraham, "'Lavender.'"

16. Michael N. Schmitt, "Israel—Hamas 2024 Symposium—The Gospel, Lavender, and the Law of Armed Conflict," *Articles of War*, June 28, 2024, https://lieber.westpoint.edu/gospel-lavender-law-armed-conflict/.

17. Kingsley et al., "Israel Loosened Its Rules to Bomb Hamas Fighters."

18. Tara Copp, Elizabeth Dwoskin, and Ian Duncan, "Anthropic's AI Tool Claude Central to U.S. Campaign in Iran, Amid a Bitter Feud," *The Washington Post*, March 4, 2026, https://www.washingtonpost.com/technology/2026/03/04/anthropic-ai-iran-campaign/.

22. Drones and Geopolitics

1. Leo Sands, "Who Are the Key Players in the Syrian War?," *The Washington Post*, December 8, 2024, https://www.washingtonpost.com/world/2024/12/02/syria-war-rebels-hts-assad-key-figures/.

2. Samia Nakhoul et al., "Assad's Final Hours in Syria: Deception, Despair and Flight," Reuters, December 13, 2024, https://www.reuters.com/world/middle-east/assads-final-hours-syria-deception-despair-flight-2024-12-13/; Ben Hubbard et al., "Deception and Betrayal: Inside the Final Days of the Assad Regime," *The New York Times*, December 21, 2024, https://www.nytimes.com/2024/12/21/world/middleeast/assad-regime-syria-final-days.html.

3. Hubbard et al., "Deception and Betrayal"; Ragip Soylu, "Has Ukraine Helped the Syrian Rebel Offensive in Aleppo?," *Middle East Eye*, December 2, 2024, https://www.middleeasteye.net/news/has-ukraine-helped-syrian-rebel-offensive-aleppo.

4. Charles Lister, "Syria—Nov 26–Dec 3, 2024," *Syria Weekly* newsletter, December 3, 2024, https://www.syriaweekly.com/p/special-syria-nov-26-dec-3-2024?open=false#%C2%A7special-hts-and-opposition-offensive-changes-the-game-as-assads-regime-crumbles-in-northwestern-syria.

5. Don Rassler, "The Islamic State and Drones," Combating Terrorism Center at West Point, July 2018, https://ctc.westpoint.edu/wp-content/uploads/2018/07/Islamic-State-and-Drones-Release-Version.pdf; Rueben Dass, "Hayat Tahrir al-Sham's Drone Force," *Lawfare*, December 13, 2024, https://www.lawfaremedia.org/article/hayat-tahrir-al-sham-s-drone-force.

6. Soylu, "Has Ukraine Helped the Syrian Rebel Offensive in Aleppo?"; David Ignatius, "Syrian Rebels Had Help from Ukraine in Humiliating Russia," *The Washington Post*, December 10, 2024, https://www.washingtonpost.com/opinions/2024/12/10/ukraine-syria-russia-war/; Kateryna Zakharchenko, "Ukrainian Special Forces and Syrian Rebels Decimate Russian Mercenaries in Syria," *Kyiv Post*, June 3, 2024, https://www.kyivpost.com/post/33695.

7. Armed Conflict Location & Event Data Project (ACLED), accessed January 7, 2026, www.acleddata.com.

8. Armed Conflict Location & Event Data Project (ACLED), accessed January 7, 2026, www.acleddata.com.

9. Armed Conflict Location & Event Data Project (ACLED), accessed January 7, 2026, www.acleddata.com.

10. D. Max Ferguson and Russell Lemler, "Understanding the Counterdrone Fight: Insights from Combat in Iraq and Syria," Modern War Institute, May 14, 2024, https://mwi.westpoint.edu/understanding-the-counterdrone-fight-insights-from-combat-in-iraq-and-syria/; "Unmanned Aircraft System Airspace Integration Plan," Department of Defense, March 2011, archived January 21, 2016, Internet Archive, https://web.archive.org/web/20160121155841/http://www.acq.osd.mil/sts/docs/DoD_UAS_Airspace_Integ_Plan_v2_(signed).pdf; Audrey Kurth Cronin, *Power to the People: How Open Technological Innovation Is Arming Tomorrow's Terrorists* (New York: Oxford University Press, 2020), 207; Marc Santora et al., "A Thousand Snipers in the Sky: The New War in Ukraine," *The New York Times*, March 3, 2025, https://www.nytimes.com/interactive/2025/03/03/world/europe/ukraine-russia-war-drones-deaths.html.

11. Eleanor Watson, "3 U.S. MQ-9 Reaper Drones, Worth About $30 Million Each, Have Crashed In or Near Yemen Since November," CBS News, April 26, 2024, https://www.cbsnews.com/news/3-u-s-mq-9-reaper-drones-30-million-crash-yemen/.

12. Michael Marrow, "The Army's 'Most Challenging' Unmanned Threat? Group 3 Drones," *Breaking Defense*, October 17, 2024, https://breakingdefense.com/2024/10/the-armys-most-challenging-unmanned-threat-group-3-drones/.

13. Paul K. Kerr, "U.S.-Proposed Missile Technology Control Regime Changes," Congressional Research Service, May 10, 2022, https://www.congress.gov/crs-product/IF11069.

14. Mike Stone, "Exclusive: Trump to Reinterpret 1987 Missile Treaty to Sell Heavy Attack Drones Abroad," Reuters, September 5, 2025, https://www.reuters.com/business/aerospace-defense/trump-reinterpret-1987-missile-treaty-sell-heavy-attack-drones-abroad-2025–09–05/.

15. Osman Okyay, interview conducted by author, June 14, 2024.

16. Stephen Witt, "The Turkish Drone That Changed the Nature of Warfare," *The New Yorker*, May 9, 2022, https://www.newyorker.com/magazine/2022/05/16/the-turkish-drone-that-changed-the-nature-of-warfare.

17. Witt, "The Turkish Drone That Changed the Nature of Warfare."

18. Serhat Süha Çubukçuoğlu, interview conducted by the author, May 30, 2024; "The SIPRI Top 100 Arms-Producing and Military Services Companies in the World, 2024," Stockholm International Peace Research Institute, December 2025, https://www.sipri.org/visualizations/2025/sipri-top-100-arms-producing-and-military-services-companies-world-2024.

19. Asli Aydintaşbaş and Cinzia Bianco, "Useful Enemies: How the Turkey-UAE Rivalry Is Remaking the Middle East," European Council on Foreign Relations, March 15, 2021, https://ecfr.eu/publication/useful-enemies-how-the-turkey-uae-rivalry-is-remaking-the-middle-east/; Ali Bakir, "The UAE Just Received Twenty Drones from Turkey. What's the Backstory?," Atlantic Council, November 16, 2022, https://www.atlanticcouncil.org/blogs/menasource/the-uae-just-received-twenty-drones-from-turkey-whats-the-backstory/; Ralph Jennings, "Why Chinese Weapons Exports Are Declining Despite Loyal Clientele," Voice of America, October 5, 2021, https://www.voanews.com/a/chinese-weapons-exports-decline/6258163.html; Erik Lin-Greenberg, "The Dawn of Drone Diplomacy," *Foreign Affairs*, December 20, 2022, https://www.foreignaffairs.com/ukraine/dawn-drone-diplomacy.

20. Data compiled from the Stockholm International Peace Research Institute (SIPRI) Arms Transfer Database, "SIPRI Arms Transfer Database," SIPRI, accessed February 17, 2025, https://www.sipri.org/databases/armstransfers.

21. Arda Mevlütoğlu et al., "Adapting Security: The Intersection of Turkiye's Foreign Policy and Defence Industrialisation," International Institute for Strategic Studies, June 2024, https://www.iiss.org/research-paper/2024/06/adapting-security-the-intersection-of-turkiyes-foreign-policy-and-defence-industrialisation/.

22. The passage on the Sudan war reflects ideas from a previously published article—Steve Feldstein, "'Good enough' drones have become geopolitical chips," *Foreign Policy*, November 17, 2025, https://foreignpolicy.com/2025/11/17/drones-geopolitical-chips-sudan-warfare-russia-ukraine/; Nicholas Kristof, "Where there's no debate about genocide—and no response, either," *The New York Times*, August 30, 2025, https://www.nytimes.com/2025/08/30/opinion/sudan-genocide-famine.html; "Crisis in Sudan: What Is Happening and How to Help," International Rescue Committee, November 5, 2025, https://archive.ph/OjV3O; Ayesha Roscoe and Robbie Griffiths, "UN Humanitarian Chief: World Needs to 'Wake Up' and Help Stop Violence in Sudan," NPR, December 7, 2025, https://www.npr.org/2025/12/07/nx-s1–5634741/un-official-says-war-torn-darfur-sudan-is-the-epicenter-of-suffering.

23. Katharine Houreld and Elizabeth Dwoskin, "A Turkish Arms Firm Helped Fuel Sudan's Brutal Civil War, Records Show," *The Washington Post*, March 7, 2025, https://www.washingtonpost.com/world/2025/03/07/sudan-war-turkey-baykar-rsf-saf/.

24. Houreld and Dwoskin, "A Turkish Arms Firm Helped Fuel Sudan's Brutal Civil War, Records Show"; "Security Council Extends Sanctions Regime Against Sudan, Unanimously Adopting Resolution 2750 (2024)," United Nations Security Council, September 11, 2024, https://press.un.org/en/2024/sc15817.doc.htm.

25. Declan Walsh, "The Gold Rush at the Heart of a Civil War," *The New York Times*, February 7, 2025, https://www.nytimes.com/2024/12/11/world/africa/sudan-gold-rush-heart-civil-war.html; Jared Malsin, Benoit Faucon, and Robbie Gramer, "How UAE Arms Bolstered a Sudanese Militia Accused of Genocide," *The Wall Street Journal*, October 28, 2025, https://www.wsj.com/world/how-u-a-e-arms-bolstered-a-sudanese-militia-accused-of-genocide-781b9803; Patrick Wintour, "As Criticism Grows, Is UAE Ready to Walk Away from Sudan's RSF Militia?," *The Guardian*, November 4, 2025, https://www.theguardian.com/world/2025/nov/04/sudan-rsf-militia-uae-united-arab-emirates; Adya Madhavan, "The Sky's the Limit: Understanding China's UCAV Exports," Takshashila Institution, April 2025, https://static1.squarespace.com/static/618a55c4cb03246776b68559/t/67f767b05bfb4733d4765a2b/1744267199627/Takshashila+Discussion+Document+2025–08_AM.pdf.

26. Alexandra Sharp, "The Fall of El Fasher Marks Another Dark Turn in Sudan's Civil War," *Foreign Policy*, November 4, 2025, https://foreignpolicy.com/2025/11/04/sudan-el-fasher-rsf-famine-killings-sexual-violence-guterres/; Danai Nesta Kupemba and Chris Ewokor, "Drone Strike in Besieged Sudan City Kills Dozens," BBC, October 11, 2025, https://www.bbc.com/news/articles/c3rvrn8req40.

27. Alia Shoaib, "Ukraine's Drones Are Becoming Increasingly Ineffective as Russia Ramps Up Its Electronic Warfare and Air Defenses," *Business Insider*, July 3, 2022, https://www.businessinsider.com/drones-russia-ukraine-war-electronic-warfare-2022–7.

28. Agnes Helou, "With Turkish Drones in the Headlines, What Happened to Ukraine's Bayraktar TB2s?," *Breaking Defense*, October 6, 2023, https://breakingdefense.com/2023/10/with-turkish-drones-in-the-headlines-what-happened-to-ukraines-bayraktar-tb2s/.

29. Tim Zadorozhnyy, "Ukraine On Track to Receive Total of 3 Million FPV Drones in 2025, Defense Minister Says," The Kyiv Independent, December 24, 2025, https://kyivindependent.com/ukraine-on-track-to-receive-total-of-3-million-fpv-drones-in-2025/; "Sergiy Boyev at the OFDEF Conference: Air Defense, Drones, and Extended-Range Munitions Among Ukraine's Top Priorities for 2026," Ministry of Defense of Ukraine, January 26, 2026, https://mod.gov.ua/en/news/sergiy-boyev-at-ofdef-conference-air-defense-drones-and-extended-range-munitions-among-ukraine-s-top-priorities-for-2026.

30. Cronin, *Power to the People*, 379.

23. The Emiratis' Bid to Become an AI Power

1. Eric Pace, "Abu Dhabi Oil Money Developing the Desert," *The New York Times*, January 2, 1976, https://www.nytimes.com/1976/01/02/archives/abu-dhabi-oil-money-developing-the-desert-abu-dhabis-oil-revenues.html; Nicolas Parasie, "How Dubai Went from Desert Village to Global Player," *The Wall Street Journal*, April 24, 2016, https://www.wsj.com/articles/how-dubai-went-from-desert-village-to-global-player-1461550185; Eliot Brown and Berber Jin, "The 'Spy Sheikh' Taking the AI World by Storm," *The Wall Street Journal*, February 26, 2025, https://www.wsj.com/tech/ai/abu-dhabi-spy-sheikh-ai-c4a9d48c.

2. Brown and Jin, "The 'Spy Sheikh' Taking the AI World by Storm"; Ben Bartenstein, "How to Get a Meeting with the UAE's $1.5 Trillion Man," Bloomberg, April 24, 2024, https://www.bloomberg.com/news/features/2024–04–24/uae-how-to-get-a-meeting-with-sheikh-tahnoon-abu-dhabi-s-1–5-trillion-man; Bradley Hope, "A Spymaster Sheikh Controls a $1.5 Trillion Fortune. He Wants to Use It to Dominate AI," *Wired*, January 14, 2025, https://www.wired.com/story/uae-intelligence-chief-ai-money/.

3. Brown and Jin, "The 'Spy Sheikh' Taking the AI World by Storm"; Hope, "A Spymaster Sheikh Controls a $1.5 Trillion Fortune."

4. Malcolm Moore and Chloe Cornish, "How Abu Dhabi's National Oil Company Is Planning for the Energy Transition," *Financial Times*, October 16, 2024, https://www.ft.com/content/9983be98–8a6e-4edd-be28–3eb866a9f6c7; "UAE National Strategy for Artificial Intelligence 2031," National Program for Artificial Intelligence, 2018, https://ai.gov.ae/wp-content/uploads/2021/07/UAE-National-Strategy-for-Artificial-Intelligence-2031.pdf; Yousef Al Otaiba, interview conducted by the author, Washington, DC, September 12, 2024.

5. Hope, "A Spymaster Sheikh Controls a $1.5 Trillion Fortune."

6. Ben Bartenstein et al., "Gulf Royal's $1.5 Trillion Empire Draws Bankers and Billionaires," Bloomberg, September 5, 2023, https://www.bloomberg.com/news/articles/2023-09-05/abu-dhabi-royal-sheikh-tahnoon-is-among-the-world-s-most-influential-dealmakers; "First Abu Dhabi Bank PJSC," Reuters, accessed March 7, 2025, https://www.reuters.com/mar kets/companies/fab.ad/financials/balance-sheet-annual; "Ranking," Global SWF, accessed March 10, 2025, https://globalswf.com/ranking; "Royal Group LLC (Royal Group)," Sovereign Wealth Fund Institute, accessed March 7, 2025, https://www.swfinstitute.org/pro file/5d3fd22b46d64f65f9d59d79.

7. Simeon Kerr, "Sheikh Tahnoon Named Chair of $790bn Abu Dhabi Sovereign Wealth Fund," *Financial Times*, March 9, 2023, https://www.ft.com/content/9f965e4f-2447–4f28–8b18–75ddeee36c3e; Bartenstein et al., "Gulf Royal's $1.5 Trillion Empire Draws Bankers and Billionaires," Bloomberg, September 5, 2023, https://www.bloomberg.com/news/articles/2023-09-05/abu-dhabi-royal-sheikh-tahnoon-is-among-the-world-s-most-influential-dealmakers; Mark Mazzetti and Edward Wong, "Inside U.S. Efforts to Untangle an A.I. Giant's Ties to China," *The New York Times*, November 27, 2023, https://www.nytimes.com/2023/11/27/us/politics/ai-us-uae-china-security-g42.html.

8. Michael Peel and Simeon Kerr, "UAE's Top AI Group Vows to Phase Out Chinese Hardware to Appease US," *Financial Times*, December 7, 2023, https://www.ft.com/content/6710c259–0746–4e09–804f-8a48ecf50ba3; Paul Mozur and David E. Sanger, "Microsoft Makes High-Stakes Play in Tech Cold War with Emirati A.I. Deal," *The New York Times*, April 16, 2024, https://www.nytimes.com/2024/04/16/technology/microsoft-g42-uae-ai.html.

9. Edward Wong, Mark Mazzetti, and Paul Mozur, "Lawmakers Push U.S. to Consider Trade Limits with A.I. Giant Tied to China," *The New York Times*, January 9, 2024, https://www.nytimes.com/2024/01/09/us/politics/ai-china-uae-g42.html.

10. Wong, Mazzetti, and Mozur, "Lawmakers Push U.S. to Consider Trade Limits with A.I. Giant Tied to China"; Mark Bergen, Marissa Newman, and Ben Bartenstein, "Caught Between the US and China, a Powerful AI Upstart Chooses Sides," Bloomberg, April 25, 2024, https://www.bloomberg.com/news/articles/2024–04–25/g42-allies-with-us-over-china-in-global-ai

-conflict?sref=QmOxnLFz; Mazzetti and Wong, "Inside U.S. Efforts to Untangle an A.I. Giant's Ties to China."

11. Brown and Jin, "The 'Spy Sheikh' Taking the AI World by Storm"; Bergen, Newman, and Bartenstein, "Caught Between the US and China, a Powerful AI Upstart Chooses Sides"; Ben Bartenstein et al., "G42 Made Secret Pact with US to Divest from China Before Microsoft Deal," Bloomberg, April 16, 2024, https://www.bloomberg.com/news/articles/2024-04-16/g42-made-secret-pact-with-us-to-divest-from-china-before-microsoft-deal.

12. Bartenstein et al., "G42 Made Secret Pact with US to Divest from China Before Microsoft Deal"; Peel and Kerr, "UAE's Top AI Group Vows to Phase Out Chinese Hardware to Appease US"; Dong Cao, Ben Bartenstein, and Mackenzie Hawkins, "Abu Dhabi Investment Firm Lunate Takes On AI Firm G42's China Fund," Bloomberg, July 8, 2024, https://www.bloomberg.com/news/articles/2024-07-08/abu-dhabi-investment-firm-lunate-takes-on-ai-firm-g42-s-china-fund.

13. Yousef Al Otaiba, interview conducted by the author, Washington, DC, September 12, 2024.

14. Elizabeth Dwoskin, Ellen Nakashima, Nitasha Tiku, and Cat Zakrzewski, "How the Authoritarian Middle East Became the Capital of Silicon Valley," *The Washington Post*, May 14, 2024, https://www.washingtonpost.com/technology/2024/05/14/middle-east-ai-tech-companies-saudi-arabia-uae/; Chloe Cornish, Andrew England, and George Hammond, "US Seeks Alliance with Abu Dhabi on Artificial Intelligence," *Financial Times*, April 20, 2024, https://www.ft.com/content/843796a6-191c-4828-8bc9-d6648e4b460e.

15. Mozur and Sanger, "Microsoft Makes High-Stakes Play in Tech Cold War with Emirati A.I. Deal"; Mark Bergen et al., "Microsoft AI Deal With UAE's G42 at Risk over National Security Fears," Bloomberg, July 2, 2024, https://www.bloomberg.com/news/articles/2024-07-02/microsoft-ai-deal-with-uae-s-g42-at-risk-over-national-security-fears?sref=QmOxnLFz; Chloe Cornish, "UAE Seeks 'Marriage' with US over Artificial Intelligence Deals," *Financial Times*, June 4, 2024, https://www.ft.com/content/b72ba623-8e3c-4d29-837b-33f6b8407fb8.

16. Gregory C. Allen et al., "The United Arab Emirates' AI Ambitions," Center for Strategic and International Studies, January 24, 2025, https://www.csis.org/analysis/united-arab-emirates-ai-ambitions.

17. Yousef Al Otaiba, interview conducted by the author, Washington, DC, September 12, 2024.

18. Bergen et al., "Microsoft AI Deal With UAE's G42 at Risk Over National Security Fears."

19. Dina Bass and Farah Elbahrawy, "Microsoft, G42 Announce Geothermal-Powered Data Center in Kenya," Bloomberg, May 22, 2024, https://www.bloomberg.com/news/articles/2024-05-22/microsoft-g42-announce-1-billion-geothermal-data-center-in-kenya?sref=QmOxnLFz; Bergen et al., "Microsoft AI Deal with UAE's G42 at Risk over National Security Fears."

20. Alexis Akwagyiram and Kelsey Warner, "Africa's Forgotten Battleground in the US-China Tech War," Semafor, September 22, 2025, https://www.semafor.com/article/09/22/2025/kenya-us-china-tech-war-ai-data-center.

21. Eric Lipton, "The Trumps Get Richer—The Morning Newsletter," *The New York Times*, May 14, 2025, https://www.nytimes.com/2025/05/14/briefing/trump-family-business.html; https://www.wsj.com/world/middle-east/trump-family-business-gulf-06792821?mod=article_inline; Jonathan Karl and Katherine Faulders, "Trump Administration Poised to Accept 'Palace in the Sky' as a Gift for Trump from Qatar: Sources," ABC News, May 11, 2025, https://abcnews.go.com/Politics/trump-administration-poised-accept-palace-sky-gift-trump/story?id=121680511. Sam Kessler et al., "'Spy Sheikh' Bought Secret Stake in Trump Company," *The Wall Street Journal*, January 31, 2026, https://www.wsj.com/politics/policy/spy-sheikh-secret-stake-trump-crypto-tahnoon-ea4d97e8.

22. Eliot Brown and Amrith Ramkumar, "How Gulf Sheikhs Played Their Trump Cards Into a

Massive AI Chip Deal," *The Wall Street Journal*, May 16, 2025, https://www.wsj.com/tech/how-gulf-sheikhs-played-their-trump-cards-into-a-massive-ai-chip-deal-b048a3f9.

23. Jonathan Swan, Vivian Nereim, and Luke Broadwater, "Saudis Greet Trump to His Tastes, With Pomp, Opulence and Real Estate," *The New York Times*, May 13, 2025, https://www.nytimes.com/2025/05/13/world/middleeast/trump-saudi-arabia.html; Michael Birnbaum and Matt Viser, "In Mideast, Trump Cast Himself as U.S. Pitchman. What Did He Sell?," *The Washington Post*, May 16, 2025, https://www.washingtonpost.com/politics/2025/05/16/trump-middle-east-trip/.

24. Antonia Hitchens, "Donald Trump's Tariff Dealmaker-in-Chief," *The New Yorker*, July 21, 2025, https://www.newyorker.com/magazine/2025/07/28/donald-trumps-tariff-dealmaker-in-chief; Alexander Cornwell, "US-UAE Multi-Billion Dollar AI Data Campus Deal Far from Finalised, Sources Say," Reuters, June 6, 2025, https://www.reuters.com/business/finance/us-uae-multi-billion-dollar-ai-data-campus-deal-far-finalised-sources-say-2025–06–06/; Jordan Schneider, "'Just count the server racks,' . . ." *ChinaTalk* newsletter, May 21, 2025, https://www.chinatalk.media/p/about-that-ai-middle-east-deal; "Department of Commerce Announces Recission of Biden-Era Artificial Intelligence Diffusion Rule, Strengthens Chip-Related Export Controls," Bureau of Industry & Security, May 13, 2025, https://www.bis.gov/press-release/department-commerce-announces-recission-biden-era-artificial-intelligence-diffusion-rule-strengthens-chip.

25. David Sacks (@davidsacks47), "On top of being a historic diplomatic and economic success, President Trump's trip to the Middle East was also a game-changer in the global AI race," X, May 17, 2025, 8:26 a.m., https://x.com/davidsacks47/status/1923716659497283709; Eric Lipton et al., "Anatomy of Two Giant Deals: The U.A.E. Got Chips. The Trump Team Got Crypto Riches," *The New York Times*, September 15, 2025, https://www.nytimes.com/2025/09/15/us/politics/trump-uae-chips-witkoff-world-liberty.html; Eric Lipton and David Yaffe-Bellany, "Trump Aides' Involvement in Giant Chips and Crypto Deals Draws Ethics Scrutiny," *The New York Times*, September 24, 2025, https://www.nytimes.com/2025/09/24/us/politics/trump-witkoff-sacks-saudi-chips-crypto.html; Cecilia Kang et al., "Silicon Valley's Man in the White House is Benefiting Himself and His Friends," *The New York Times*, November 30, 2025, https://www.nytimes.com/2025/11/30/technology/david-sacks-white-house-profits.html.

26. Hope, "A Spymaster Sheikh Controls a $1.5 Trillion Fortune."

27. Hope, "A Spymaster Sheikh Controls a $1.5 Trillion Fortune;" "United Arab Emirates," *Freedom on the Net 2024*, 2024, https://freedomhouse.org/country/united-arab-emirates/freedom-net/2024.

28. Kai Ryssdal and Sean McHenry, *Marketplace* podcast, August 29, 2024, https://www.marketplace.org/story/2024/08/29/china-chip-black-market-semiconductor-nvidia-chipmaker-superchip; Eduardo Baptista, Fanny Potkin, and Karen Freifeld, "Chinese Entities Turn to Amazon Cloud and Its Rivals to Access High-End US Chips, AI," Reuters, August 23, 2024, https://www.reuters.com/technology/chinese-entities-turn-amazon-cloud-its-rivals-access-high-end-us-chips-ai-2024–08–23/; Aaron Mak, "Why Smuggle Chips When You Can Remote In?," *Politico: Digital Future Daily* newsletter, July 3, 2025, https://www.politico.com/newsletters/digital-future-daily/2025/07/03/who-needs-to-smuggle-chips-when-you-can-remote-in-00439547.

29. Allen et al., "The United Arab Emirates' AI Ambitions"; "Freedom on the Net 2024: The Struggle for Trust Online," Freedom House, 2024, https://freedomhouse.org/sites/default/files/2024–10/FREEDOM-ON-THE-NET-2024-DIGITAL-BOOKLET.pdf; Sheena Chestnut Greitens and Isaac Kardon, "Playing Both Sides of the U.S.-Chinese Rivalry," *Foreign Affairs*, March 15, 2024, https://www.foreignaffairs.com/united-states/playing-both-sides-us-chinese-rivalry; Author interview with a senior Biden administration official, May 14, 2024.

30. Yousef Al Otaiba, interview conducted by the author, Washington, DC, September 12, 2024.

24. Power Struggle

1. Antony J. Blinken, "Remarks to the Press," briefing, US Department of State, October 17, 2022, https://2021–2025.state.gov/secretary-antony-blinken-remarks-to-the-press-3/.
2. Nectar Gan and Eric Cheung, "China's Military Launches Live-Fire Exercise in Escalation of Blockade Drills Near Taiwan," CNN, April 2, 2025, https://edition.cnn.com/2025/04/01/china/china-taiwan-drills-live-fire-escalation-intl-hnk/index.html.
3. Joshua Posaner, "China's 'Mind-Boggling' Space Capabilities Worry US, Says Space Force Chief," *Politico*, October 31, 2024, https://www.politico.eu/article/china-mind-boggling-space-capabilities-worry-us-says-space-force-chief-b-chance-saltzman-russia-nasa/.
4. Jacco van Loon, "The US Is Now at Risk of Losing to China in the Race to Send People Back to the Moon's Surface," *The Conversation*, October 25, 2024, https://theconversation.com/the-us-is-now-at-risk-of-losing-to-china-in-the-race-to-send-people-back-to-the-moons-surface-241716; Joey Roulette, "NASA Announces Further Delays in Artemis Moon Missions," Reuters, December 6, 2024, https://www.reuters.com/technology/space/nasa-announces-further-delays-artemis-moon-missions-2024-12-05/.
5. Peng Zhang, "Does China's DeepSeek-V3 Make the Computing Power Advantages of US AI Companies Less Important?," *Geopolitechs*, December 29, 2024, https://www.geopolitechs.org/p/does-chinas-deepseek-v3-make-the.
6. Jordan Schneider and Lily Ottinger, "AGI and the Future of Warfare," *ChinaTalk*, March 28, 2025, https://www.chinatalk.media/p/agi-and-the-future-of-warfare.
7. Robert D. Atkinson, "China Is Rapidly Becoming a Leading Innovator in Advanced Industries," Information Technology and Innovation Foundation, September 16, 2024, https://itif.org/publications/2024/09/16/china-is-rapidly-becoming-a-leading-innovator-in-advanced-industries/.
8. Martijn Rasser et al., "The American AI Century: A Blueprint for Action," Center for a New American Security, December 17, 2019, https://www.cnas.org/publications/reports/the-american-ai-century-a-blueprint-for-action.
9. "China and America Are Racing to Develop the Best AI. But Who Is Ahead in Using It?," *The Economist*, April 3, 2025, https://www.economist.com/business/2025/04/03/china-and-america-are-racing-to-develop-the-best-ai-but-who-is-ahead-in-using-it.
10. "Global Innovation Index," WIPO, 2025, https://www.wipo.int/en/web/global-innovation-index (accessed November 28, 2025); Soumitra Dutta, Bruno Lanvin, and Sacha Wunsch-Vincent, eds., "Global Innovation Index 2020: Who Will Finance Innovation?," Cornell University, INSEAD, and the World Intellectual Property Organization, 2020, https://www.wipo.int/edocs/pubdocs/en/wipo_pub_gii_2020.pdf; Soumitra Dutta, Bruno Lanvin, Lorena Rivera León, and Sacha Wunsch-Vincent, eds., "Global Innovation Index 2024: Unlocking the Promise of Social Entrepreneurship," World Intellectual Property Organization, 2024, https://www.wipo.int/web-publications/global-innovation-index-2024/assets/67729/2000%20Global%20Innovation%20Index%202024_WEB3lite.pdf; "China and America Are Racing to Develop the Best AI."
11. Keith Bradsher, "'The Tsunami Is Coming': China's Global Exports Are Just Getting Started," *The New York Times*, April 7, 2025, https://www.nytimes.com/2025/04/07/business/china-manufacturing-exports-trump-tariffs.html; Sanya Jain, "Viral Video Shows BYD Mega Factory in China That's Bigger than San Francisco, 10 Times the Size of Tesla Gigafactory," *Hindustan Times*, March 20, 2025, https://www.hindustantimes.com/trending/viral-video-shows-byd-mega-factory-in-china-that-s-bigger-than-san-francisco-10-times-the-size-of-tesla-gigafactory-101742450040329.html; Iris Deng, "Huawei Completes US$1.4 Billion Campus in Shanghai 'with 100 Cafes' to Lure Foreign Talent," *South China Morning Post*, July 15, 2024, https://www.scmp.com/tech/big-tech/article/3270536/huawei-completes-us14-billion-campus-shanghai-100-cafes-lure-foreign-talent.

12. Joseph Ataman and Richard Quest, "Europe 'Not in the AI race Today,' French President Macron Says," CNN, February 9, 2025, https://www.cnn.com/2025/02/09/europe/france-macron-europe-ai-race-intl/index.html; Pablo Chavez, "France Pursues an AI 'Third Way,'" Center for European Policy Analysis, February 13, 2025, https://cepa.org/article/france-pursues-an-ai-third-way/; Sam Schechner and Asa Fitch, "France Taps Nuclear Power in Race for AI Supremacy," *The Wall Street Journal*, February 10, 2025, https://www.wsj.com/tech/ai/france-taps-nuclear-power-for-new-ai-training-cluster-a7804107.

13. Marissa Newman et al., "Saudis Plan $100 Billion AI Powerhouse to Rival UAE Tech Hub," Bloomberg, November 6, 2024, https://www.bloomberg.com/news/articles/2024-11-06/saudis-plan-100-billion-ai-powerhouse-to-rival-uae-s-tech-hub.

14. Marc Santora et al., "A Thousand Snipers in the Sky: The New War in Ukraine," *The New York Times*, March 3, 2025, https://www.nytimes.com/interactive/2025/03/03/world/europe/ukraine-russia-war-drones-deaths.html; Chivers, "How Suicide Drones Transformed the Front Lines in Ukraine."

15. Shashank Joshi, "AI on the Battlefield: A Fait Accompli?," *Survival* 67, no. 2 (2025), 120, https://www.tandfonline.com/doi/abs/10.1080/00396338.2025.2481775.

16. Carl von Clausewitz, *On War*, trans. and eds. Michael Howard and Peter Paret (Princeton, NJ: Princeton University Press, 1976), 101; James O'Donnell, "Generative AI Is Learning to Spy for the US Military," *MIT Technology Review*, April 11, 2025, https://www.technologyreview.com/2025/04/11/1114914/generative-ai-is-learning-to-spy-for-the-us-military/.

17. O'Donnell, "Generative AI Is Learning to Spy for the US Military."

18. Joshi, "AI on the Battlefield: A Fait Accompli?"

19. Arthur Holland Michel, "Known Unknowns: Data Issues and Military Autonomous Systems," United Nations Institute for Disarmament Research, 2021, https://unidir.org/wp-content/uploads/2023/05/Holland_KnownUnknowns_20210517_0.pdf; Emelia S. Probasco et al., "AI for Military Decision-Making," Center for Security and Emerging Technology, April 2025, https://cset.georgetown.edu/wp-content/uploads/CSET-AI-for-Military-Decision-Making.pdf. See also Steven Feldstein and Brian (Chun Hey) Kot, "Two Deeply Troubling Trends from Ukraine's Year of War," Carnegie Endowment for International Peace, February 22, 2023, https://carnegieendowment.org/posts/2023/02/two-deeply-troubling-trends-from-ukraines-year-of-war?lang=en.

20. Yvonne McDermott, Alexa Koenig, and Daragh Murray, "Open Source Information's Blind Spot: Human and Machine Bias in International Criminal Investigations," *Journal of International Criminal Justice* 19, no. 1 (March 2021): 85–105, https://academic.oup.com/jicj/article/19/1/85/6214619; Probasco et al., "AI for Military Decision-Making."

21. Yuval Abraham, "'Lavender': The AI Machine Directing Israel's Bombing Spree in Gaza," *+972 Magazine*, April 3, 2024, https://www.972mag.com/lavender-ai-israeli-army-gaza/.

22. David Ignatius, "A Cry for the Refugees of Emptied Nagorno-Karabakh: 'We Are Nobody,'" *The Washington Post*, March 11, 2024, https://www.washingtonpost.com/opinions/2024/03/11/david-ignatius-nagorno-karabakh-refugees-azerbaijan-enclave/; "New Report: Azerbaijani Regime Ethnically Cleansed Nagorno-Karabakh According to International Fact-Finding Mission," Freedom House, November 11, 2024, https://freedomhouse.org/article/new-report-azerbaijani-regime-ethnically-cleansed-nagorno-karabakh-according-international.

23. "Political Declaration on Responsible Military Use of Artificial Intelligence and Autonomy," US Department of State, November 27, 2024, https://www.state.gov/bureau-of-arms-control-deterrence-and-stability/political-declaration-on-responsible-military-use-of-artificial-intelligence-and-autonomy; Chivers, "How Suicide Drones Transformed the Front Lines in Ukraine."

24. "About the Leahy Law," US Department of State, January 20, 2025, accessed April 28, 2025,

https://web.archive.org/web/20250204112133/https://www.state.gov/bureau-of-democracy-human-rights-and-labor/releases/2025/01/leahy-law-fact-sheet.

25. Eric Lipton, "As A.I.-Controlled Killer Drones Become Reality, Nations Debate Limits," *The New York Times*, November 21, 2023, https://www.nytimes.com/2023/11/21/us/politics/ai-drones-war-law.html.

26. Casey Newton, "America Crafts an AI Action Plan," *Platformer* newsletter, March 18, 2025, https://www.platformer.news/ai-action-plan-submissions-meta-google-openai-anthropic/?ref=platformer-newsletter.

27. Lolita C. Baldor, "Hegseth Says He Fired the Top Military Lawyers Because They Weren't Well Suited for the Jobs," Associated Press, February 24, 2025, https://apnews.com/article/pentagon-hegseth-firing-chairman-lawyers-6bead3346b1210e45e77648e6cbc3599; John Ismay and Azmat Khan, "Hegseth Cuts Pentagon Work on Preventing Civilian Harm," *The New York Times*, March 4, 2025, https://www.nytimes.com/2025/03/04/us/politics/hegseth-pentagon-civilian-harm.html; Courtney Albon, "Hegseth Cuts Pentagon's Testing Office in Half, Appoints New Leader," DefenseNews, May 28, 2025, https://www.defensenews.com/pentagon/2025/05/28/hegseth-directs-reorg-of-pentagon-testing-office-appoints-new-leader/.

25. Tech War in the Age of Trump

1. Joseph R. Biden, "State of the Union Address," The White House, March 1, 2022, https://bidenwhitehouse.archives.gov/state-of-the-union-2022/; Thomas Carothers and Frances Brown, "Democracy Policy Under Biden: Confronting a Changed World," Carnegie Endowment for International Peace, February 6, 2024, https://carnegieendowment.org/research/2024/02/democracy-policy-under-biden-confronting-a-changed-world?lang=en.

2. Max Boot, "Trump's Emerging Foreign Policy Is a Disturbing 19th-Century Throwback," *The Washington Post*, April 7, 2025, https://www.washingtonpost.com/opinions/2025/04/07/donald-trump-foreign-policy/.

3. Lauren Weber, "Can Trump's Tariff Offensive Deliver New American Jobs?," *The Wall Street Journal*, April 11, 2025, https://www.wsj.com/economy/trump-tariffs-manufacturing-jobs-impact-8b25b257.

4. Mitch Ambrose, "NSF Stands Up Directorate for Technology, Innovation, and Partnerships," American Institute of Physics, March 17, 2022, https://ww2.aip.org/fyi/2022/nsf-stands-directorate-technology-innovation-and-partnerships; "NSF Announces 7 New National Artificial Intelligence Research Institutes," National Science Foundation, May 4, 2023, https://www.nsf.gov/news/nsf-announces-7-new-national-artificial; "Democratizing the Future of AI R&D: NSF to Launch National AI Research Resource Pilot," National Science Foundation, January 24, 2024, https://www.nsf.gov/news/democratizing-future-ai-rd-nsf-launch-national-ai; "Fact Sheet: President Biden's 2025 Budget Invests in Science and Technology to Power American Innovation, Expand Frontiers of What's Possible," The White House, March 13, 2024, https://bidenwhitehouse.archives.gov/ostp/news-updates/2024/03/13/fact-sheet-president-bidens-2025-budget-invests-in-science-and-technology-to-power-american-innovation-expand-frontiers-of-whats-possible/.

5. Jeffrey Mervis, "Exclusive: NSF Director to Resign amid Grant Terminations, Job Cuts, and Controversy," *Science*, April 24, 2025, https://www.science.org/content/article/nsf-director-resign-amid-grant-terminations-job-cuts-and-controversy; Corbin Hiar, "National Science Foundation Mass Firings Go Beyond Trump's Orders, Sparking Outrage," *Scientific American*, February 20, 2025, https://www.scientificamerican.com/article/national-science-foundation-mass-firings-go-beyond-trumps-orders-sparking/; Meryl Kornfield, Hannah Natanson, and Laura Meckler, "The Federal Government Is Paying More Than 154,000 People

Not to Work," *The Washington Post*, July 31, 2025, https://www.washingtonpost.com/politics/interactive/2025/white-house-budget-federal-government/; Jeremy B. Merrill, Kati Perry and Jacob Bogage, "The White House's plan to downsize the federal government, in charts," *The Washington Post*, July 16, 2025, https://www.washingtonpost.com/politics/interactive/2025/white-house-budget-federal-government/; "Fiscal Year 2026 Discretionary Budget Request," Executive Office of the President, Office of Management and Budget, May 2, 2025, https://www.whitehouse.gov/wp-content/uploads/2025/05/Fiscal-Year-2026-Discretionary-Budget-Request.pdf; "NSF Grant Terminations in 2025," Grant Witness, accessed December 8, 2025, https://grant-witness.us/nsf-data.html.

6. Matteo Wong, "A Disaster for American Innovation," *The Atlantic*, April 11, 2025, https://www.theatlantic.com/technology/archive/2025/04/trump-jeopardizing-ai-boom/682404/; "Fact Sheet: President Donald J. Trump Takes Action to Enhance America's AI Leadership," The White House, January 23, 2025, https://www.whitehouse.gov/fact-sheets/2025/01/fact-sheet-president-donald-j-trump-takes-action-to-enhance-americas-ai-leadership/.

7. Patricia Cohen, "Trump's Trade War Risks Forfeiting America's Economic Primacy," *The New York Times*, April 3, 2025, https://www.nytimes.com/2025/04/03/business/trump-tariffs-dollar.html; G. John Ikenberry, *Liberal Leviathan: The Origins, Crisis, and Transformation of the American World Order* (Princeton, NJ: Princeton University Press, 2011), 2–3.

8. Nicholas Mulder, "The Paradox of Trump's Economic Weapon," *Foreign Affairs*, March 6, 2025, https://www.foreignaffairs.com/united-states/paradox-trumps-economic-weapon.

9. Mulder, "The Paradox of Trump's Economic Weapon."

10. Andrew H. Kydd, *Trust and Mistrust in International Relations* (Princeton, NJ: Princeton University Press, 2005), 7; Aaron Hoffman, "A Conceptualization of Trust in International Relations," *European Journal of International Relations* 8, no. 3 (September 2002): 377; Damien Cave, "How Trump Supercharged Distrust, Driving U.S. Allies Away," *The New York Times*, March 31, 2025, https://www.nytimes.com/2025/03/31/world/trump-foreign-policy-trust.html; Paul Slovic, "Perceived Risk, Trust, and Democracy," *Risk Analysis* 13, no. 6 (December 1993): 675–82.

11. Adam S. Posen, "The new economic geography," *Foreign Affairs*, September/October 2025, https://www.foreignaffairs.com/united-states/new-economic-geography-posen; Mark Carney, "Davos 2026: Special Address by Mark Carney, Prime Minister of Canada," World Economic Forum, January 20, 2026, transcript of speech delivered at the World Economic Forum Annual Meeting, Davos, https://www.weforum.org/stories/2026/01/davos-2026-special-address-by-mark-carney-prime-minister-of-canada/.

12. Suzanne Lynch, "Biden–von der Leyen 2024?," *Politico*, April 30, 2023, https://www.politico.eu/article/joe-biden-ursula-von-der-leyen-2024-elections-european-commission/.

13. Lynch, "Biden–von der Leyen 2024?"

14. Anthony Faiola, Steve Hendrix, and Ellen Francis, "After Trump's Broad Tariffs, Europe Reels from the Loss of an Old Ally," *The Washington Post*, April 3, 2025, https://www.washingtonpost.com/world/2025/04/03/trump-tariffs-eu-europe/; Ursula von der Leyen, "2025 State of the Union Address by President von der Leyen," transcript of speech delivered at the European Parliament, Strasbourg, France, September 9, 2025, https://ec.europa.eu/commission/presscorner/detail/ov/SPEECH_25_2053.

15. The White House, "Winning the Race: America's AI Action Plan," July 2025, https://www.whitehouse.gov/wp-content/uploads/2025/07/Americas-AI-Action-Plan.pdf; Daniella Cheslow and Giselle Ruhiyyih Ewing, "How a New Russia-China-US Network Could Work," *Politico: National Security Daily* newsletter, December 10, 2025, https://www.politico.com/newsletters/national-security-daily/2025/12/10/how-a-new-russia-china-us-network-could-work-00685342; Stu Woo, "How Trump's U-Turn on Nvidia Chips Changes the Game for

China's AI," *The Wall Street Journal*, December 9, 2025, https://www.wsj.com/tech/ai/china-ai-nvidia-trump-chip-export-ban-lift-c3d457c1; David Sanger, "Trump's Nvidia Chip Deal Reverses Decades of Technology Restrictions," *The New York Times*, December 9, 2025, https://www.nytimes.com/2025/12/09/us/politics/trump-nvidia-ai-chips-china.html.

16. Ana Swanson, "With China Truce, U.S. National Security Controls Now Appear Up for Negotiation," *The New York Times*, October 30, 2025, https://www.nytimes.com/2025/10/30/us/politics/china-trump-trade-security-chips-minerals.html; Jeffrey Goldberg and Shane Harris, "Here Are the Attack Plans That Trump's Advisers Shared on Signal," *The Atlantic*, March 26, 2025, https://www.theatlantic.com/politics/archive/2025/03/signal-group-chat-attack-plans-hegseth-goldberg/682176/.

17. Anton Trojanovski, "Trump's Security Strategy Focuses on Profit, Not Spreading Democracy," *The New York Times*, December 5, 2025, https://www.nytimes.com/2025/12/05/us/politics/trump-national-security-strategy.html; "National Security Strategy of the United States of America," The White House, November 2025, https://www.whitehouse.gov/wp-content/uploads/2025/12/2025-National-Security-Strategy.pdf; National Security Strategy of the United States of America, The White House, December 2017, https://trumpwhitehouse.archives.gov/wp-content/uploads/2017/12/NSS-Final-12–18–2017–0905.pdf; Henry Farrell, "America Has Identified Its Greatest Enemy: Western Europe," *Programmable Mutter* newsletter, December 6, 2025, https://www.programmablemutter.com/p/america-has-identified-its-final.

18. Ishan Tharoor, "A Line in the Sand Moment for Europe and the U.S.," *The Washington Post*, December 7, 2025, https://www.washingtonpost.com/world/2025/12/07/europe-united-states-national-security/; Martin Wolf, "China Senses an Opportunity in Trump's Cultural Revolution," *Financial Times*, April 1, 2025, https://www.ft.com/content/80ab4ac8–0Decdeb-4deb-b54a-7d0d06d12f1c.

19. Kurt M. Campbell and Rush Doshi, "Underestimating China," *Foreign Affairs*, May–June 2025, https://www.foreignaffairs.com/china/underestimating-china.

20. Jeremy W. Peters and Andrea Fuller, "How Universities Became So Dependent on the Federal Government," *The New York Times*, April 18, 2025, https://www.nytimes.com/2025/04/18/us/trump-universities.html.

21. Ali Swenson, "Trump, a Populist President, Is Flanked by Tech Billionaires at His Inauguration," Associated Press, January 20, 2025, https://apnews.com/article/trump-inauguration-tech-billionaires-zuckerberg-musk-wealth-0896bfc3f50d941d62cebc3074267ecd.

22. Arjun Kharpal, "Tesla Sales Plunge 40% in Europe as Chinese EV Rival BYD's Triple," CNBC, August 28, 2025, https://www.cnbc.com/2025/08/28/tesla-europe-sales-plunge-40percent-chinese-ev-rival-byd-up-225percent.html; William Echikson, "Europe's Tech Dilemma: An Olive Branch or a Bazooka," Center for European Policy Analysis, April 11, 2025, https://cepa.org/article/europes-tech-dilemma-an-olive-branch-or-a-bazooka/.

23. Pieter Haeck, "EU-US Rift Triggers Call for Made-in-Europe Tech," *Politico*, March 10, 2025, https://www.politico.eu/article/push-for-eurostack-as-eu-us-tech-tensions-grow/.

24. Peters and Fuller, "How Universities Became So Dependent on the Federal Government."

25. Angela Shen and Jordan Schneider, "Trump's Crackdown on Foreign Students Is a gift to China," *The Washington Post*, April 21, 2025, https://www.washingtonpost.com/opinions/2025/04/21/trump-student-visas-china-crackdown/; Chan Ho-Him, "China Rolls Out Its Version of the H-1B Visa to Attract Foreign Tech Workers," Associated Press, November 10, 2025, https://apnews.com/article/china-us-talent-tech-visa-h1b-4c5792cf1b9fec9a846ca002f9e383ae.

26. Paul Krugman, "Trump's Cultural Revolution," Paul Krugman newsletter, April 21, 2025, https://paulkrugman.substack.com/p/trumps-cultural-revolution.

27. Amy Zegart, "The Crumbling Foundations of American Strength," *Foreign Affairs*, August 20, 2024, https://www.foreignaffairs.com/united-states/crumbling-foundations-american-strength-amy-zegart.

28. Zegart, "The Crumbling Foundations of American Strength"; Andrew Dessler, "Make China Great Again? How Trump's Attacks on Universities Will Backfire," *Bulletin of the Atomic Scientists*, April 18, 2025, https://thebulletin.org/2025/04/make-china-great-again-how-trumps-attacks-on-universities-will-backfire/.

29. David Pierson and Damien Cave, "China Wants Countries to Unite Against Trump, but Is Met with Wariness," *The New York Times*, April 17, 2025, https://www.nytimes.com/2025/04/17/world/asia/china-trump-tariffs-trade.html.

30. Darren J. Lim, "Chinese Economic Coercion during the THAAD Dispute," The ASAN Forum, December 28, 2019, https://theasanforum.org/chinese-economic-coercion-during-the-thaad-dispute/; Rory Smith and Tariq Panja, "The Erasure of Mesut Özil," *The New York Times*, October 26, 2020, https://www.nytimes.com/2020/10/26/sports/soccer/mesut-ozil-arsenal-china.html.

31. Matthew Reynolds and Matthew P. Goodman, "China's Economic Coercion: Lessons from Lithuania," Center for Strategic and International Studies, May 6, 2022, https://www.csis.org/analysis/chinas-economic-coercion-lessons-lithuania.

32. Vivian Wang, "China Warns Countries Not to Team Up with U.S. Against It on Trade," *The New York Times*, April 21, 2025, https://www.nytimes.com/2025/04/21/world/asia/china-tariffs-trade-warning.html.

33. Phelim Kine, "Pentagon Follies, Beijing Jollies," *Politico: China Watcher* newsletter, April 24, 2025, https://www.politico.eu/newsletter/china-watcher/pentagon-follies-beijing-jollies/.

26. Managing the Future

1. Peter Guest, "Mykhailo Fedorov Is Running Ukraine's War Like a Startup," *Wired*, July 25, 2023, https://www.wired.com/story/ukraine-runs-war-startup/; "Air Raid Alert Map of Ukraine," *Ukrzen*, https://alerts.in.ua/en; John Leicester, "Mark Hamill Lends 'Star Wars' Voice to Ukrainian Air-Raid App," Associated Press, March 28, 2023, https://apnews.com/article/russia-ukraine-star-wars-luke-hamill-app-08ec03bf1a2c9c0378857090079f00f9.

2. Robert Daly and Robert Litwak, "How to Manage and De-Risk an Emerging Cold War II with China," *Bulletin of the Atomic Scientists*, June 26, 2024, https://thebulletin.org/2024/06/how-to-manage-and-de-risk-an-emerging-cold-war-ii-with-china/; Dean A. Wilkening and Ken Watman, "Nuclear Deterrence in a Regional Context," RAND, 1995, https://www.rand.org/pubs/monograph_reports/MR500.html?utm_source=substack&utm_medium=email; Nicholas Welch and Kyle Chan, "Weapons of Cold War 2.0 + 'People's War' Invasion Fleet," *ChinaTalk*, April 25, 2025, https://www.chinatalk.media/p/cold-war-20-tools-peoples-war-invasion.

3. Daly and Litwak, "How to Manage and De-Risk an Emerging Cold War II with China."

4. Welch and Chan, "Weapons of Cold War 2.0 + 'People's War' Invasion Fleet."

5. Michiaki Matsushima, "Yuval Noah Harari: 'How Do We Share the Planet with This New Superintelligence?'" *Wired*, April 1, 2025, https://www.wired.com/story/questions-answered-by-yuval-noah-harari-for-wired-ai-artificial-intelligence-singularity/; Sam Meacham, "A Race to Extinction: How Great Power Competition Is Making Artificial Intelligence Existentially Dangerous," *Harvard International Review*, September 8, 2023, https://hir.harvard.edu/a-race-to-extinction-how-great-power-competition-is-making-artificial-intelligence-existentially-dangerous/; Dan Hendrycks, Eric Schmidt, and Alexandr Wang, "Superintelligence Strategy: Expert Version," *arXiv*, April 14, 2025, https://arxiv.org/abs/2503.05628.

6. Joshua Rovner, *Strategy and Grand Strategy* (New York: Routledge, 2025), 141–42.

7. David Rising, "US Defense Secretary Says War with China Neither Imminent nor Unavoidable, Stressing Need for Talks," Associated Press, June 1, 2024, https://apnews.com/article/shangrila-dialogue-austin-china-taiwan-philippines-defense-46868bec3799f5043276d9fdeca62a41.

8. Trevor Hunnicutt and Jeff Mason, "Biden Calls Xi a Dictator After Carefully Planned Summit," Reuters, November 16, 2023, https://www.reuters.com/world/biden-calls-xi-dictator-after-carefully-planned-summit-2023-11-16/; Lauren Egan and Phelim Kine, "Biden's Final Meeting with Xi Jinping Reaps Agreement on AI and Nukes," *Politico*, November 16, 2024, https://www.politico.com/news/2024/11/16/biden-xi-jinping-ai-00190025.

9. Robert Daly and Robert Litwak, "How to Manage and De-Risk an Emerging Cold War II with China," *Bulletin of the Atomic Scientists*, June 26, 2024, https://thebulletin.org/2024/06/how-to-manage-and-de-risk-an-emerging-cold-war-ii-with-china/.

10. Lauren Kahn, "Ground Rules for the Age of AI Warfare," *Foreign Affairs*, June 6, 2023, https://www.foreignaffairs.com/world/ground-rules-age-ai-warfare.

11. Steven Feldstein, "The Consequences of Generative AI for Democracy, Governance and War," *Survival* 65, no. 5 (September 2023): 117–42, https://www.tandfonline.com/doi/full/10.1080/00396338.2023.2261260; Paul Scharre, "AI's Gatekeepers Aren't Prepared for What's Coming," *Foreign Policy*, June 19, 2023, https://foreignpolicy.com/2023/06/19/ai-regulation-development-us-china-competition-technology/.

12. Steven Feldstein, "AI in War: Can Advanced Military Technologies Be Tamed Before It's Too Late?," *Bulletin of the Atomic Scientists*, January 11, 2024, https://thebulletin.org/2024/01/ai-in-war-can-advanced-military-technologies-be-tamed-before-its-too-late/.

13. Sheera Frenkel and Natan Odenheimer, "Israel's A.I. Experiments in Gaza War Raise Ethical Concerns," *The New York Times*, April 25, 2025, https://www.nytimes.com/2025/04/25/technology/israel-gaza-ai.html.

14. Volodymyr Zelenskyy, "Speech by the President During the General Debate of the UN General Assembly in New York," transcript of speech delivered at the United Nations, New York, NY, September 24, 2025, https://www.president.gov.ua/en/news/vistup-prezidenta-pid-chas-zagalnih-debativ-generalnoyi-asam-100349; "Political Declaration on Responsible Military Use of Artificial Intelligence and Autonomy," US Department of State, November 9, 2023, https://www.state.gov/political-declaration-on-responsible-military-use-of-artificial-intelligence-and-autonomy-2/.

15. Keith Bradsher, "China Has an Army of Robots on Its Side in the Tariff War," *The New York Times*, April 23, 2025, https://www.nytimes.com/2025/04/23/business/china-tariffs-robots-automation.html.

16. Bradsher, "China Has an Army of Robots on Its Side in the Tariff War."

17. Gregory Zuckerman and Raffaele Huang, "DeepSeek Chief's Journey from Math Geek to Global Disruptor," *The Wall Street Journal*, January 28, 2025, https://www.wsj.com/tech/ai/liang-wenfeng-deepseek-ai-8bba3bb3?mod=livecoverage_web; "Why Chinese AI Has Stunned the World," *The Economist*, January 23, 2025, https://www.economist.com/briefing/2025/01/23/chinas-ai-industry-has-almost-caught-up-with-americas.

18. Kirsten Grind and Keach Hagey, "Why Did Facebook Fire a Top Executive? Hint: It Had Something to Do with Trump," *The Wall Street Journal*, November 11, 2018, https://www.wsj.com/articles/why-did-facebook-fire-a-top-executive-hint-it-had-something-to-do-with-trump-1541965245; Erin Griffith and Cade Metz, "A.I. Military Start-Up Anduril Close to Deal That Would Value It at $28 Billion," *The New York Times*, February 7, 2025, https://www.nytimes.com/2025/02/07/technology/anduril-funding-valuation-28-billion.html.

19. Guest, "Mykhailo Fedorov Is Running Ukraine's War Like a Startup."

20. Amanda Gerut, "Larry Ellison and Elon Musk 'Begged' Nvidia's Jensen Huang for More GPUs over a Fancy Sushi Dinner," *Fortune*, September 16, 2024, https://fortune.com/2024/09/16/larry-ellison-elon-musk-begged-nvidias-jensen-huang-more-gpus-fancy-sushi-dinner/.

Appendix II

1. Ben Murphy, "Chokepoints: China's Self-Identified Strategic Technology Import Dependencies," Center for Security and Emerging Technology, May 2022, https://cset.georgetown.edu/publication/chokepoints/.
2. Zijing Wu and Eleanor Olcott, "China Trials Its First Advanced Tools for AI Chipmaking," *Financial Times*, September 17, 2025, https://www.ft.com/content/8fd79522-e34f-4633-bc87-efoaae2d9159; Fanny Potkin, "Exclusive: How China Built Its 'Manhattan Project' to Rival the West in AI Chips," Reuters, December 17, 2025, https://www.reuters.com/world/china/how-china-built-its-manhattan-project-rival-west-ai-chips-2025-12-17/.
3. Adnan Memija, "ZTT Delivers First Batch of Submarine Cables for China's 1 GW Off-Shore Wind Project," offshoreWIND.biz, September 10, 2025, https://www.offshorewind.biz/2025/09/10/ztt-delivers-first-batch-of-submarine-cables-for-chinas-1-gw-offshore-wind-project/.
4. Yvette Wu, "Zhaosheng Microelectronics Pushes Forward with 3.475 Billion RMB Private Placement to Expand RF Chips Manufacturing," WalliSemi, October 16, 2025, https://www.wallisemi.com/blog/industry-insights/zhaosheng-microelectronics-pushes-forward-with-3–475-billion-rmb-private-placement-to-expand-rf-chips-manufacturing/; Antonio Hmaidi, "Huawei Is Quietly Dominating China's Semiconductor Supply Chain," MERICS, April 9, 2024, https://merics.org/en/report/huawei-quietly-dominating-chinas-semiconductor-supply-chain.
5. Murphy, "Chokepoints: China's Self-Identified Strategic Technology Import Dependencies."
6. "China Starts New Locally-Manufactured Gas Turbine for First Time," Bloomberg, December 2, 2025, https://www.bloomberg.com/news/articles/2025–12–02/china-starts-new-locally-manufactured-gas-turbine-for-first-time.
7. Stephen Chen, "Chinese Scientists Break Design 'Curse' That Killed US Navy's Drone Programme," *South China Morning Post*, July 23, 2025, https://archive.ph/dKHmH.
8. Jia Liu, "Big Bearing Makes Engineering Breakthrough," Chinese Academy of Sciences, December 16, 2022, https://english.cas.cn/newsroom/cas_media/202212/t20221216_325646.shtml.
9. Zhang Tong, "China's Fatigue-Free Alloy Breakthrough Paves Way for Aerospace Revolution," *South China Morning Post*, April 10, 2025, https://archive.ph/cp1ou.
10. Zhang Tong, "China's Heat-Defying Aluminium Alloys Are 6 Times Stronger than Previous Ones," *South China Morning Post*, May 21, 2024, https://archive.ph/PPxJz.
11. Amber Wang, "China Enters Hi-Tech Imaging Race with New 'Core' Electron Microscope," *South China Morning Post*, January 21, 2024, https://archive.ph/TNJX6.
12. Iris Deng, "Huawei's HarmonyOS Tops Apple's iOS as China's No 2 Mobile Platform," *South China Morning Post*, June 15, 2024, https://archive.ph/PFjvJ; Xinmei Shei, "Huawei's HarmonyOS Stays Ahead of Apple iOS in China," *South China Morning Post*, September 23, 2025, https://www.scmp.com/tech/big-tech/article/3326505/huaweis-harmonyos-maintains-lead-over-apple-ios-china-sixth-consecutive-month.
13. Murphy, "Chokepoints: China's Self-Identified Strategic Technology Import Dependencies."

Appendix III

1. Armed Conflict Location & Event Data Project (ACLED), accessed January 7, 2026, www.acleddata.com.
2. Armed Conflict Location & Event Data Project (ACLED), accessed January 7, 2026, www.acleddata.com.

Index

A100, 116
Abu Amama, 247
Abu Dhabi, 8, 248, 251
Abu Dhabi Investment Authority (ADIA), 253
Abu Nidal Organization, 232
ACLED. *See* Armed Conflict Location and Event Data
Adeyemo, Wally, 138
ADIA. *See* Abu Dhabi Investment Authority
Advanced Micro Devices (AMD), 142, 173
  x86, 166–67
Aegis radar system, 169
affiliate rule, 184
Afghanistan, 19
  United States at war with, 100
Africa, 187
AGI. *See* artificial general intelligence
AI. *See* artificial intelligence
Air Alert, 293
Airbnb, 54
Alabuga special economic zone, 213
Aleppo, 238
Alexander, Keith, 161
Ali Cloud, 256
Alibaba, 58, 95, 112, 267
Aliyev, Ilham, 37, 274
All Measures Short of War (Wright), 7
Allen, Gregory, 125–26
Alphabet (company), 56
AlphaZero, 252
Altman, Sam, 57, 63, 84–85, 125
Amazon, 42, 53, 134
Amazon Web Services, 50, 182, 261
AMD. *See* Advanced Micro Devices
America First movement, 265–66
American Dynamism fund, 81
Analog Devices, 142
Andreesen, Marc, 113
Andreesen Horowitz, 69, 77, 81, 86
Android, 177
Anduril Industries, *82*, 86, 87, 183, 301–2
  founding of, 79–80
  OpenAI and, 84–85
  Pentagon and, 78–79
Ansaldo, 176–77
Ant Group, 59
Anthropic, 85, 91, 110, 114, 191, 237
  Hegseth and, 62
APEC. *See* Asia-Pacific Economic Cooperation
Apple, 31, 53, 109, 113, 134, 177, 225
Applebaum, Anne, 56
Applied Intuition, *82*
Armed Conflict Location and Event Data (ACLED), 240, *241*
Armenia, 38, 210, 274–75
  Azerbaijan conflict with, 36–37
arms races, AI, 32–35
Arpacik, Cihat, 239
ARPANET, 30, 46, 225
Artemis III, 266–67
artificial general intelligence (AGI), 11, 31
  China and, 109
  infrastructure for, 55
  national security benefits of, 33
  potential impacts of, 265
  race to develop, 267–68
artificial intelligence (AI), 7
  arms race, 32–35, 296
  Biden on, 123
  China and, 109–10, 181–82, 190–93, 267–68
  drone piloting with, 74
  dual uses of, 33
  Google on, 85–86
  impact of, 34–35
  inference scaling of, 193
  in Iran, 41

artificial intelligence (AI) (*continued*)
  Karp on, 90–91
  military power and, 33, 90, 272
  private sector development of, 61*n*
  R&D, 281
  sovereign, 114–15
  Trump, D., and, 115–18, 258–59, 281
  in UAE, 252–54
  United States development of, 267–68
Artsakh, 37
Asia-Pacific Economic Cooperation (APEC), 159
  Obama and, 161
ASML, 111, 177, 184–85
Aspin, Les, 65
al-Assad, Bashar, 238
al-Assad, Maher, 238
Astranis, *82*
Athens, 98
Atkinson, Robert, 268
Atlantic Council, 134
Augustine, Norm, 65
Austin, Lloyd, 297
Australia Group, 141
Australian Strategic Policy Institute, 27–28
  Critical Technology Tracker, 196
Auterion, 74
authoritarianism, technology and, 34
automation bias, 273
Azerbaijan, 42, 274
  Armenia conflict with, 36
Azure, 84

Baidu, 95, 191
balance of power, 22–26
bandwagoning, 43
Bank of America, 68
Barinov, Dmytro, 76
Barkin, Noah, 130, 133
Bartiromo, Maria, 62
Battle of Agincourt, 27
Baums, Ansgar, 112
Baykar, 41–42, 244–45, 250
Bayraktar, Selçuk, 42, 245, 300–301
Beglaryan, Artak, 275
BeiDou, 179
Beijing, 7–8
Bekaa Valley, 232
Belarus, 49
Benaich, Nathan, 110
Bendett, Samuel, 249
Berlinska, Maria, 214
Best Buy, 165
Bhatia, Karan, 51
Bhattacharjee, Yudhijit, 168
Biari, Ibrahim, 298–99
bias
  automation, 273
  organizational, 274
Biddle, Stephen, 31, 89
Biden, Joe, 10, 21, 106, 109, 115, 175, 203
  on AI, 123
  on China, 95–96
  Leyen and, 283
  on Russia, 139, 210–11
  Trump, D., and, 155–56, 278–80
  on Ukraine, 206
  Xi Jinping and, 159–60, 297
Bier, David, 179
Big Data valley, 195
Big Fund, 156
bin Laden, Osama, 68
bin Salman, Mohammed, 258, 270–71
bin Zayed Al Nahyan, Khalifa, 252
bin Zayed Al Nahyan, Mohamed, 251–52
bin Zayed Al Nahyan, Tahnoon, 251, 301
Black Sea, 2, 76, 91
Black Sea fleet, 75, 186
BlackSea Technologies, 92
Blackwell, 117
Blank, Steve, 121
Blinken, Antony, 265
Boeing, 65, 168–69, 183
Boise State University, 183
Bolsonaro, Jair, 44
Bolton, John, 158
Boot, Max, 5, 29, 279
Boullenois, Camille, 178
Bowden, Mark, 66
Boyle, Katherine, 77
Bradford, Anu, 133
Brands, Hal, 20
  on China, 23–24
Brazil, 25
  China and, 44
  technological development in, 44
  United States and, 44
Bresnick, Sam, 190
Bretton Woods, 54, 265–66

BRICS, growth of, 25
Brin, Sergey, 66
British East India Company, 52
Brooks, Rosa, 8
Brown, Michael, 70, 86, 91, 164–65, 170
Brussels, 134, 137
Budnikov, Sergei, 214
Bulletin of the Atomic Scientists, 6
Burj Khalifa, 258
Burns, William, 17
Bush, George H. W., 16, 152
Bush, George W., 15, 285
Bush, Vannevar, 119
BYD, 113, 167, 269
ByteDance, 105

C4ADS, 181
CAC. *See* Cyberspace Administration of China
Caffarra, Cristina, 134
Californian Ideology, 52
Campbell, Kurt, 152, 286
Canada, 55
  GDP of, 156
capitalism, 123
Carlstrom, Greg, 223
Carnegie Endowment for International Peace, 90, 203
Carnegie Mellon, 30
Carney, Mark, 283
Carter, Ash, 66
Cato Institute, 179
Caucusus, 36
CCP. *See* Chinese Communist Party
Center for Naval Analyses, 249
Center for Strategic and International Studies, 125
Centers for Disease Control and Prevention, 69
Central Science and Technology Commission, 176
CEOs, 56
Chamber of Commerce, US, 178
Chang, Morris, 124–25, 135
Chang'e-4, 1
Changzhou, 208
ChatYuan, 190
Chelyabinsk, 3
Chelyabinsk Tractor Plant, 142
Chemical Weapons Convention, 43
Chen Jian, 151
Chen QiNan, 168
Chernihiv, 145
Chevron, 68
Chhabra, Tarun, 256
Chi Mak, 169
Chiang Kai-shek, 168
China, 1, 5, 111
  AGI and, 109
  AI and, 109–10, 181–82, 190–93, 267–68
  Biden on, 95–96
  Brands on, 23–24
  Brazil and, 44
  chip manufacturing, 97–98
  circumvention strategies of, 174
  corruption in, 189
  data centers in, 195–96
  drones produced by, 244
  as emerging superpower, 266
  EU and, 128–34
  EVs in, 196–97
  funding model of, 156–57
  GDP of, 24, 151–52
  growth of, 23–24, 128
  industrial production of, 107
  JVs in, 167
  Lithuania and, 291
  LLMs in, 190
  manufacturing in, 21
  MCF in, 28, 154–56
  military power of, 24–25, 69–70
  nuclear weapons of, 28
  overcapacity technique of, 188
  population of, 24
  Posen on, 24
  potential of war with, 80–81
  R&D of, 27–28, 178
  reshoring, 116–17
  Russia and, 129–30, *144*, 200, 201–5, 216–17
  satellite internet in, 185–87
  semiconductor industry in, 157–58, 159
  Soviet Union and, 162
  space efforts of, 1–2
  subsidies in, 174, 188–89
  Taiwan and, 135–36, 189, 266
  tariffs against, 95
  tech companies in, 174–79
  tech transfer strategy of, *169*

China (*continued*)
technology in, 8–9, 27–28, *153*
techno-security ambitions, 156–57
Tesla in, 167
Trump, D., on, 95, 158, 170
UAE and, 261
United States' rivalry with, 7–8, 18, 21, 22–23, 138, 170–71, 174, 176–78, 179–87, 266–69, 294–95, 297–99
in World Trade Organization, 188
China Dream, 171
Xi Jinping on, 151, 152–53, 187
China State Shipbuilding, 189
China United Heavy-Duty Gas Turbine Technology Co., 177
Chinese Communist Party (CCP), 18, 151, 152, 159, 163, 291
Ma and, 58–59
CHIPS Act, 106, 126, 136, 278
subsidies from, 123, 125
choke points, 127
CHPL. *See* Common High Priority List
Chung, Dongfan "Greg," 168
Churchill, Winston, 53
Cisco, 182
Clark, Jack, 114, 191
Clarke, Richard, 260
Claude (AI), 62, 85, 91, 181–82, 237
Clausewitz, Carl von, 272
Clinton, Hillary, 18, 79
closed innovation, 225
in Cold War, 30
Cloudflare, 53
CNC. *See* computer numerical control
The Code (O'Mara), 120
Cold War, 5, 6, 15, 19, 45, 110, 120, 152–53
closed innovation in, 30
end of, 65, 122
technology and, 225
Collaborative Combat Aircraft program, 79
color revolutions, 17
Common High Priority List (CHPL), 141–42
compute (resource), 57–58
computer numerical control (CNC), 207–8, 210
Conflict Insights Group, 247
corruption, in China, 189
COVID, 60, 128
Crimea, 49
Russian annexation of, 17–18
Cronin, Audrey Kurth, 30, 46, 122–23, 225, 250
CubicPV, 188
Çubukçuoğlu, Serhat Süha, 245
Cultural Revolution, 289
Cyber Command, 161
cyberattacks, 33
in Russia-Ukraine conflict, 50–51, 53
Cyberspace Administration of China (CAC), 183, 191

Da Jiang Innovations (DJI), 51, 70, 197–98, 237, 243
da Vinci, Leonardo, 29
Daly, Robert, 295
Dar al-Arqam, 248
Daraa, 238
Dark Matter, 254, 260
data centers
in China, 195–96
demand for, 194
size of, 194–95
in United States, 193–94
Davos, 58
DeepMind, 109
DeepSeek, 7, 110, 112–13, 191, 267, 301
Defense Innovation Unit, 66, 164, 273
Defense Production Act, 58
Defense Tech Sector, *81*
Dell, 182
Delta system, 91
Demarais, Agathe, 100
Democratic Republic of Congo, 100
Deng Xiaoping, 18–19, 164
reforms under, 151
deregulation, under Reagan, 107
Dessler, Andrew, 290
The Diffusion of Military Power (Horowitz), 31
Digital Silk Road (DSR), 186
Ding, Jeffrey, 24, 32
Ding Wenwu, 157
distillation, 114
DIU, 86
DJI. *See* Da Jiang Innovations
Dnipro, 206
DOGE, 265
Musk at, 279–80

Donbas, 49
Doshi, Rush, 106
Draghi, Mario, 134
Draper, Gaither, and Anderson, 121
Drezner, Daniel, 139, 147
drones, 3–4, 72, 197
AI piloting of, 74
categories of, *242*, 243
Chinese production of, 244
in conflicts, *241*
dual use of, 102
fatalities from, *242*, *311–12*
Geran-2 and Geran-3, 213–14
Ghost, 78
global orders for, *246*
Harop, 36–37
HTS, 249–50
HX-2, 83
innovation in, 270
Iranian production of, 212–13
Lancet, 146
MAGURA, 75–76
mothership, 73
MQ-9 Reaper drone, 42, 226
Orlan-10, 144
Russian production of, 212–14
Shaheen, 240
TB2 drones, 37, 41–42, 88, 300–301
Türkiye development of, 41–42, 245–47, 249–50
Ukraine's use of, 30, 102
Vampire, 73
DSR. *See* Digital Silk Road
dual circulation strategy, 174
Dubai, 112
Duggan, Daniel, 162
Dutch East India Company, 52

Echikson, William, 288
EcoFlow, 3
Economy, Elizabeth, 152
Egypt, 231–32
Eisenhower, Dwight D., 227
Elbit Systems, 41
electric scooters, 88, *88*
electric vehicles (EVs), 167
in China, 196–97
Eliason, Leslie, 222, 228
Ellison, Larry, 106, 302
Entity List, 175
Epirus, *82*
Equinix, 193–94
Erdoğan, Recept Tayyip, 245, 246
Erdoğan, Sümeyye, 245
escalation dominance, 295
Esper, Mark, 82
Ethiopia, 19, 261
European Union (EU), 95
China and, 128–34
subsidies in, 126
Trump, D., on, 131–33
United States and, 132–33
EuroStack, 133–34
Eutelsat OneWeb, 185
Everbright Group, 165
Evers, Miles M., 54
EVs. *See* electric vehicles
exchange rates, 54

F-35 fighter planes, 72, 80
facial recognition software, 34
Fakhrizadeh, Mohsen, 229, 230, *230*
Fannie Mae, 69
Farrell, Henry, 53, 110, 184–85
El Fasher, 248
FCC. *See* Federal Communications Commission
Federal Aviation Administration, 69
Federal Communications Commission (FCC), 198–99
Federal Reserve Bank of New York, 116
Federov, Mykhailo, 49, 302
Musk and, 50–51
feeder vessels, 209
Fei-Fei Li, 67
Feinberg, Steve, 92
financial crisis of 2008, 18
firearms, 29
Firefly Aerospace, *82*
Five Eyes, 162
force employment, 31–32
Ford, Henry, 52
Ford Motors, 50
Foreign Direct Product Rule, 171
Foreign Investment Risk Review Modernization Act, 170
Forever Wars, 100
fossil fuel subsidies, 106–7
Founders Fund, 79, 81, 86

Fourth Industrial Revolution, 153*n*, 187
  Xi Jinping on, 153
FPVs, 70, 72, 240
France, 270
Frederiksen, Mette, 132
Freedom on the Net, 261
Freilich, Charles, 234
Fujian, 28
Fujian Forecam Optics, 173
Fukushima, 126
Fukuyama, Francis, 15, 19
Fuller, Andrea, 287

G7. *See* Group of Seven
G20, 55
G42, 254–55
  Microsoft and, 255–57
Gabuev, Alexander, 200
Gates, Bill, 58
Gaza, 7, 42, 83, 233, 235, 241, 298–99
GDP. *See* gross domestic product
GE Vernova, 177
Gelsinger, Pat, 183–84
General Catalyst, 81
General Dynamics, 65
General Electric, 176–77
General Instrument, 135
Geneva Conventions, 297
George Washington University, 24
Georgetown University, 155, 176, 190
Georgia (country), 17, 210
Geran-2 drones, 213–14
Geran-3 drones, 213–14
Germany, 131
Gertz, Geoffrey, 54
Ghaffary, Shirin, 53
Ghost drones, 78
Gilpin, Robert, 224
Global Positioning System (GPS), 30, 73
  development of, 46
global power index, 22
Global Resilience fund, 81
Global South, 25
globalization, 45, 107, 210, 261
GLONASS system, 212
Goldman, Emily O., 222, 228
Google, 31, 56, 66
  on AI, 85–86
  on foreign policy, 52–53
Google Cloud, 67
Google Gemini, 110
Google Maps, 273
Gospel AI, 274
Goujon, Reva, 176
government guidance funds, 156
GPS. *See* Global Positioning System
GPT-NL, 114
Graff, Gerard de, 129
Grandin, Greg, 281
graphics processing units (GPUs), 194
Great Depression, 18
Greene, Diane, 67
Greer, Jamieson, 117
Grieco, Kelly, 223
gross domestic product (GDP)
  of Canada, 156
  of China, 24, 151–52
Group of Seven (G7), 40
  economies of, 55
  large company revenues and, 55
Guangdong, 73
The Guardian, 84
Guizhou, 195
Gulf War, 19
Gumstix, 144
gunpowder, 29

H20 chips, 61
H100, 116
Haines, Avril, 203
Haiyin Capital, 165
Hamas, 7, 233–34, 238, 274
  IDF conflicts with, 234–37
  Israeli conflict with, 42
Hamill, Mark, 293–94
Hammes, T. X., 90
Hangzhou, 112
Hannas, William, 199
Harari, Yuval Noah, 296
Harbin Electric Corp., 177
HarmonyOS, 177
Harop drones, 36–37
Hasbora, 35
Hayat Tahrir al-Sham (HTS), 238–39
  drone program of, 249–50
Hegseth, Pete, 85, 132
  Anthropic and, 62–63
  at Pentagon, 277
Heim, Lennart, 193
Helberg, Jacob, 81

Hellfire missile, 228
Helsing, 83
Henry V (King), 27
Hermeus, *82*
Hezbollah, 238
Hirschhorn, Eric, 127, 284
Hirschman, Albert, 98
Hoffman, Reid, 79
Homs, 238
Hope, Bradley, 253
Horowitz, Michael, 31, 72, 226
Houthis, attacks by, 221–22
Hrytsenko, Valentine, 293
Hsieh, Chang-Tai, 125
HTS. *See* Hayat Tahrir al-Sham
Huang, Alexander, 137
Huang, Jensen, 61, 116, 117, 302
Huaqiangbei, 179, *180*
Huawei, 43, 44, 95, 113, 130, 254, 269, 278
  blacklisting of, 170, 171
  revenue of, 172
  United States and, 171–72
Huawei Cloud, 256
Huey-Meei Chang, 199
Human Rights Watch, 275–76
The Human-Machine Team (Sariel), 236
Huttenlocher, Daniel, 33
HX-2 drones, 83

IBM, 182
ICE. *See* Immigration and Customs Enforcement
ICT. *See* information and communications technology
IDF. *See* Israeli Defense Forces
Idlib, 238
Ignatius, David, 240
Ikenberry, G. John, 15
Imex-Expert, 210
Immigration and Customs Enforcement (ICE), 280
India, 25
  population of, 40–41
Infineon, 142
Inflation Reduction Act, 278
information and communications technology (ICT), 166
Information Technology and Innovation Foundation, 268
innovation
  closed, 30, 225
  in drones, 270
  economic power and, 101
  fast followers, 41
  first movers in, 29–30
  geopolitical advantages and, 28–29
  in Israel, 230–32
  military power and, 29–30, 41–46
  open, 30
  through private sector, 30–31
  strategic, 103–4
  in UAE, 261
  in United States, 63–64
innovation through, public sector, 30–31
In-Q-Tel, 67
Integrated Silicon Solution Inc., 145
Intel, 61–62, 123, 142
International Monetary Fund, 6
Iran, 19
  AI in, 41
  drone production in, 212–13
  nuclear weapons of, 229–30
  United States war with, 222–23
Iranian Revolutionary Guard, 228
Iraq War, 15–16, 100
Iron Curtain, 7
Iron Wall, 33, 42
Isaacson, Walter, 186
ISIS, 100
Iskander-K cruise missile, 145
Islamic Revolutionary Guard, 222
Israel, 7, 19, 29, 36
  Hamas conflict with, 42
  innovation in, 230–32
  military power of, 232–33
  military spending of, 39
  technology in, 233–37
Israeli Defense Forces (IDF), 43, 231–32, 233, 298–99
  data collection by, 83–84
  Hamas conflicts with, 234–37
  Microsoft and, 83–84
IVPM, 198
Izmail, 75

J-35A, 216–17
James, Deborah Lee, 165
Japan, 5, 110, 126
JEDI, 67, 69

Jetson TX2 AI, 146–47
Jobs, Steve, 66
Johnson, Boris, 139
joint ventures (JVs), 167
Joint Warfighting Cloud Capability, 69
al-Jolani, Abu Mohammed, 239
Joshi, Shashank, 233, 272
JVs. *See* joint ventures

Kagame, Paul, 58
Kagan, Robert, 56
Kahn, Lauren, 298
Kalashnikov Concern, 210
Kalibr cruise missiles, 212–13
Kalout, Hussein, 44
Kania, Elsa, 155
Kantrowitz, Alex, 53
Karabakh, 275
Karp, Alex, 67–68
  on AI, 90–91
Karpathy, Andrej, 113
Kataib Hezbollah, 227–28
Kazakhstan, 210
Kazan, 25
Kendall, Frank, 60
Kennedy, Paul, 29
Kenya, 256
Khamenei, Ali, 222
Kharkiv, 129
Khrushchev, Nikita, 163
Kimi, 110
Kinzhal dagger missiles, 215
Kissinger, Henry, 33
Kmentt, Alexander, 276
Konza Technopolis, 256–57
Korean War, 120, 227
Korolev, Sergei, 216
Kratsios, Michael, 81, 114, 115
Krauthammer, Charles, 19
Kristol, William, 56
Krolikowski, Alanna, 2
Krugman, Paul, 289
Kundojjala, Sravan, 195
Kuomintang Party, 137
Kushner, Jared, 258
Kyiv, 3, 5, 53, 129, 143, 223–24, 293
Kyrgyzstan, 17, 210

L3 Harris, 183
Lamprecht, Karl, 131
Lancet drone, 146
LaPlante, William, 87, 221–22
large language models (LLMs)
  in China, 190
  in Netherlands, 114
  Singapore, 114
  in Taiwan, 114
Laskai, Lorand, 155
Lattice, 78
Lavender, 236, 264
Lavrov, Sergey, 139
Leahy Law, 276
leapfrogging moments, 187
Lebanon, 232
LEO satellites, 60–61
Leonard, Mark, 130, 131
Levite, Ariel, 234
Lew, jack, 100
Leyen, Ursula von der, 130, 189
  Biden and, 283
Li, Carrie, 300
Li, K. T., 135
Li, Lyric, 217
Liang Wenfeng, 112–13, 193, 301
liberalism, 19
Libya, 42, 100
Lin, Liza, 173
Lingling Wei, 113
Lin-Greenberg, Erik, 246
Lithuania, China and, 291
Litwak, Robert, 295
LLMs. *See* large language models
Lockheed Martin, 65, 72, 161, 183, 216–17
Luckey, Palmer, 79, 80–81, 301–2
Lukashevych, Ivan, 75–76
Lula da Silva, Luiz Inácio, 44
Lutnick, Howard, 117
Lynch, Justin, 247
Lysenko, Eduard, 87

M1 Abrams tank, 72
Ma, Jack
  CCP and, 58–59
  exile of, 59*n*
Macau University of Science and Technology, 1
Machiavelli, Niccolò, 29
Macron, Emmanuel, 270, 283
MAGA, 91
Maginot Line, 224

MAGURA drone, 75–76
Malhuret, Claude, 132
Maliuk, Vasyl, 2, 4
Mansoor, Ahmed, 260
Mao Zedong, 151, 159, 171, 289
Mar-a-Lago, 117, 132
Marczak, Bill, 253
Marietta, Martin, 65
Maslov, Anatoly, 215
Mastro, Oriana Skylar, 202
Mate 60 Pro, 172
Maven AI, 68–69
Maven Smart System, 91
Mavic, 70, 237
Maxwell, Liam, 50
MCF. *See* military-civil fusion
McNamara, Robert, 89
Mehmed II, 224
Mellanox Technologies, 184
Merops, 223
Meta, 53, 56, 79
MGX, 125, 258
Michel, Arthur Holland, 273
Michel, Charles, 83
Michelangelo, 29
Micron, 183–84
Microsoft, 11, 31, 57, 134, 177
  digital Geneva convention proposed by, 53
  G42 and, 255–57
  IDF and, 83–84
MicroStrategy, 254
military power
  AI and, 27, 33, 90
  of China, 24–25, 69–70
  force employment, 31–32
  history of, 28–29
  information in, 78
  innovation and, 29–30, 41–46
  of Israel, 39, 232–33
  of Russia, 203–4, 211
  technology and, 271–73
  of United States, 24–25
military-civil fusion (MCF), in China, 28, 154–56
Missile Technology Control Regime (MTCR), 244
Mitre, Jim, 33–34
Mitsubishi, 176
Mogelson, Luke, 30
Monteiro, Nelson, 252
Moonshot AI, 191
Morgan, J. P., 52
Morocco, Russia and, 208–9
Mossad, 229
mothership drones, 73
MQ-9 Reaper drone, 42, 226
MTCR. *See* Missile Technology Control Regime
Mulder, Nicholas, 282
Murdoch, Rupert, 106
Musk, Elon, 31, 63, 104, 185, 186, 245, 258, 288
  at DOGE, 279–80
  Federov and, 50–51
  on Starlink, 59–60

Nadella, Satya, 83
Nagorno-Karabakh, 36, 37, 240, 274
NAIRR. *See* National Artificial Intelligence Research Resource
Napoleon, 98
NASA, 1, 60
Nasirzadeh, Amir, 222
National Artificial Intelligence Research Resource (NAIRR), 61
National Defense University, 90, 235
National Science Foundation (NSF), 120
national security
  AGI and, 33
  tech companies and, 55–56, 58
  Trump, D., on, 285–86
National Security Agency, 66
National Security Council, 158
nationalism, 21
NATO, 44, 147, 202
  Putin and, 206
  Trump, D., on, 10
Navarro, Peter, 155, 158
Nazi Germany, 98
neoliberalism, end of, 21
Netanyahu, Benjamin, 222
Netherlands, 110
  LLMs in, 114
Neurala, 165
Neutrality Act, 231
Newman, Abraham, 53, 110
Newton, Casey, 85
Nexperia, 174–75
Nineteenth Party Congress, 164

Ningbo, 300
Nixon, Richard, 54
North Korea, 111, 147
Northrop Grumman, 65
NSF. *See* National Science Foundation
nuclear weapons, 102
  of China, 28
  of Iran, 229–30
Nvidia, 61, 62, 112, 114–15, 146, 179, 187, 196, 255, 259, 302
  Trump, D., and, 117–18

Obama, Barack, 62, 100, 110, 152, 278–79
  APEC and, 161
October War of 1973, 232
Oculus, 79
Odesa, 75
Office of Scientific Research and Development (OSRD), 119, 120
offsetting strategies, 43
Okyay, Osman, 244
Al Olama, Omar Sultan, 256
Olkaria, 257
O'Mara, Margaret, 66, 120
OneWeb, 60–61
Ontiveros, Jeremie Eliahou, 195
Ope, 62
open innovation, 30
OpenAI, 57, 110, 112, 113
  Anduril and, 84–85
Operation Gatekeeper, 181
Operational Test and Evaluation, 277
Oracle, 105, 125, 302
Oreshnik missile, 206–7
organizational bias, 274
Orlan-10 drone, 144
Osa quadcopters, 4–5
OSRD. *See* Office of Scientific Research and Development
Otaiba, Yousef Al, 252, 261

Padilla, Christopher, 285
Page, Larry, 66
Pakistan, 19
Palantir Technologies, 67–68, 86, 87
Palestine Liberation Organization (PLO), 232
Pardee Center, 22
Paris AI Action Summit, 276
Patel, Dylan, 195
Patriot interceptor, 223
PayPal, 79
Peace Research Institute, 19
Pearl, Matt, 69
Pearl Harbor, 5
Pegasus, 254
Peng Xiao, 54, 267
Pentagon, 5, 28, 60, 70, 77, 164, 243
  Anduril Industries and, 78–79
  critical technologies according to, 103
  Defense Innovation Unit, 66
  expenditures of, 72
  Hegseth at, 277
People's Liberation Army (PLA), 85–86, 154, 190
Peskov, Dmitry, 38
Peters, Jeremy W., 287
Phillips, 135
Pichai, Sundar, 56, 287
Pinky Wu, 300
pixel lock technology, 74
PLA. *See* People's Liberation Army
PLO. *See* Palestine Liberation Organization
Poghosyan, Benyamin, 38
Politburo, 176
political repression, technology and, 34
Politico, 114
population
  of China, 24
  of India, 40–41
Posen, Barry, 24
Pottinger, Matthew, 158
Predd, Joel B., 33–34
private sector
  AI development by, 61*n*
  innovation through, 30–31
PrivateBank, 50
Procter & Gamble, 68
Project Maven, 66
Project Raven, 260
Project Transcendence, 271
Prokopenko, Alexandra, 217
property rights, 54
ProSoft, 209
public sector, innovation through, 30–31

Putin, Vladimir, 6, 8, 17, 49, 71, 129, 137, 143
NATO and, 206
on Ukraine, 138–39
Xi Jinping and, 16, 199–200, 203

Qatar, 202, 257–58
Qianfan, 185
quadcopters, 240
Quadrilateral Security Dialogue, 10
Qwen, 110, 191

R1, 113
railways, 33
Raimondo, Gina, 96, 110, 123, 142, 155, 254, 255
RAND, 33–34, 121, 161
Rapid Support Forces (RSF), 247
Raspberry Pi Zero, 74
Ratner, Ely, 152
Raytheon, 65, 82, 183
R&D. *See* research and development
Reagan, Ronald, deregulation under, 107
Re:Build Manufacturing, *82*
Red Cell, 82
Red Sea, 221
Ren Zhengei, 172
research and development (R&D)
AI, 281
of China, 27–28, 178
funding trends in, *122*
under Trump, D., 288–90
United States federal, *280*
reshoring, 21
China and, 116–17
Reuters, 142
Rhode Island, 96
Rhodium Group, 176
Rice, Condoleezza, 17
rising tech powers
challenges from, 39–41
countries included in, *40*
defining, 39–40
fast followers, 41
Riyadh, 258
Rockwell Corporation, 168
Roeh-Yoreh, 234
Rogin, Josh, 86
Rollet, Charles, 198
Romanenko, Ihor, 214
Roosevelt, Franklin Delano, 119
Ross, Wilbur, 96
Rovner, Joshua, 296–97
Royal United Services Institute, 207
RSF. *See* Rapid Support Forces
Rubio, Marco, 117
Russia, 5, 9, 25
Biden on, 139, 210–11
China and, 129–30, *144*, 200, 201–5, 216–17
Crimea annexed by, 17–18
critical technology in, *144*
cyberattacks by, 50–51, 53
drone production in, 212–14
economy of, 211–12
manufacturing capacity of, 207–8
military power of, 203–4, 211
Morocco and, 208–9
sanctions against, 111–12, 140–48, 208–11
tech companies in, 56–57
Ukraine-Russia war, 2–4, 29–30, 49–51, 71–74, 87–88, 141, 186, 201, 214–15
United States and, 16–18
webs of intermediaries, 111–12
Russian Central Bank, 140
Russkiy Mir, 139
Ruto, William, 257
Rutte, Mark, 132
Rwanda, 19

Saakashvili, Mikheil, 17
Sacks, David, 79, 80, 115, 192, 259
Sadjadpour, Karim, 229
al-Saedi, Abu Baqir, 227–28
SAF. *See* Sudanese Armed Forces
Salesforce, 53
Saltzman, B. Chance, 266
Samsung, 123–24, 183
San Clemente, 78
sanctions, 99
impact of, 146–48
against Russia, 111–12, 140–48, 208–11
Sandkyan, 181
Sariel, Yossi, 83, 235–36
Saronic, 92
satellite internet, 185–87
Saudi Arabia, 25, 257–58
Savill, Matthew, 207

Scale AI, 81, *82*
Scharre, Paul, 298
Scherf, Gundbert, 83
Schimpf, Brian, 78
Schmidt, Eric, 33, 67, 96, 223
Schmitt, Michael, 236
Scholz, Olaf, 130, 131
Schwab, Klaus, 153*n*
Science and Technology (newspaper), 153*n*
Scott, Mark, 134
Sea Limited, 41
Security Council, UN, 15
semiconductor industry, 142–43
  in China, 157–58, 159
  in Soviet Union, 162–63
  in Taiwan, 135
  United States and, 123–26
Sentry, 78
September 11 attacks, 15
Sergey Kotov, 76
Shah, Raj, 66, 81
Shahed-136, 212–13, 222–23
Al-Shaheen Brigade, 239
Shaheen drones, 240
Shakhnazarov, Karen, 211
Shanahan, Jack, 91
Shanghai Jiao Tong University, 154
shareholder value, 56
Shenyang FC-31 stealth fighter, 161
Shenzhen, 73
Shepherd, Christian, 217
Sherman tanks, 50
Shield AI, *82*
Shield Capital, 81
shipbuilding, 107–8
Shiplyuk, Alexander, 215
Shotwell, Gwynne, 80–81, *81*
Siberia, 2
Siemens, 176
silicon shield theory, 135
Silicon Valley, 52, 63, 84, 122–23, 287
  history of, 121
  pro-military arm of, 67–70
Singapore, 8, 41
  LLMs in, 114
Singles Day, 58
SIPRI. *See* Stockholm International Peace Research Institute
Siri, 225
SK Hynix, 183
Skydio, 70, *82*, 85–86
Skynode S, 74
Slovic, Paul, 282–83
Small, Andrew, 128
smartphones, 30
SMIC, 173
Smith, Brad, 53, 256–57
Snowball units, 50
Snowden, Edward, 66
SoftBank, 125
solar panels, 3
Sophgo Technologies, 173
South Africa, 40
South China Sea, 18
sovereign AI, 114–15
Soviet Union, 30, 120, 122
  China and, 162
  fall of, 15, 274–75
  semiconductor industry in, 162–63
Soyuz rocket, 217
SpaceX, 11, 51, 60, 68, 86, 87, 104, 185
Spalding, Robert, 166
Sparta, 98
Spider's Web, 2–4
spyware, 34
  in UAE, 261
St. Petersburg, 140
Stalin, Joseph, 36, 216
Stanford, 120, 265
Stargate, 125, 259
Starlink, 51, 57, 75, 185, 186, 288
  Musk on, 59–60
Starshield, 60
steam engine, 29
Stephens, Trae, 68, 79, 87, 301
Stimson Center, 112, 223
Stockholm International Peace Research Institute (SIPRI), 246
Strait Thunder-2025A, 266
Su, Lisa, 166
submarines, 189
subsidies
  in China, 174, 188–89
  CHIPS Act, 123, 125
  in EU, 126
  fossil fuel, 127
Sudan, 247, 248
Sudanese Armed Forces (SAF), 247
Suess, Juliana, 187
Sugon Information Industry, 166

Sukho SU-57 stealth fighter, 216–17
Suleimani, Qassim, 228
Suleyman, Mustafa, 109
Sullivan, Jake, 98, 157–58, 285
Sunnyvale, 115–16
Surovikin Line, 129
SWIFT payments, 139
Syria, 100, 227
systemic rivalry, 128

Tai Ming Cheung, 155, 157
Taide, 114
Taipei, 61
Taiwan, 19, 60, 123–26, 159–60, 286
  China and, 135–36, 189, 266
  LLMs in, 114
  semiconductor industry in, 135
  United States and, 136–37
Taiyuan Satellite Launch Center, 185
Tan, Lip-Bu, 62
Tanger Med, 209
tariffs
  against China, 95
  as punishment, 98–101
  reciprocal, 95
  Trump, D., on, 95, 266
Tatarstan, 43, 111–12, 213
TB2 drones, 37, 41–42, 88, 300–301
tech companies
  in China, 174–79
  defense, *82*, 87–92
  national security and, 55–56, 58
  rising powers, *305–8*
  in Russia, 56–57
  Ukraine supported by, 50–53, 57
  in United States, 286–90
technology. *See also* innovation
  authoritarianism and, 34
  in Brazil, 44
  in China, 8–9, 27–28, *153*
  choke points, *309–10*
  Cold War and, 225
  control of, 10
  dual use, 31, 33, 226–27
  in Israel, 233–37
  military power and, 271–73
  Pentagon's critical, 103
  political repression via, 34
  in Russia, *144*
Tehran, 43
Tel Aviv, 8
Telegram, 8
Tencent, 191
Terman, Frederick, 120, 268
Tesla, 56, 104, 113, 288
  in China, 167
Teter, Tim, 116
Texas Instruments, 142
THAAD interceptors, 223, 291
Thiel, Peter, 67–68, 79, 301–2
Thousand Talents Plan, 178
Tiananmen Square, 190, 203
TikTok, 198
  algorithm, 105
  ownership of, 105
Tong Zhao, 203
Tooze, Adam, 188
ToTok, 254
Tower 22 attack, 227
Tower Semiconductor, 183–84
trace italienne, 29
Treasury Department, 20
Triebert, Christiaan, 71
Trudeau, Justin, 58
Trump, Donald, 5, 9, 25–26, 44, 82, 105, 127
  AI and, 115–18, 258–59, 281
  Biden and, 155–56, 278–80
  on China, 95, 158, 170
  coercive tactics of, 45, 61–62, 281–84
  on EU, 131–33
  fossil fuel subsidies and, 106–7
  market interventionism of, 63
  national security strategies of, 285–86
  on NATO, 10
  Nvidia and, 117–18
  R&D under, 288–90
  second term of, 278–79
  on tariffs, 95, 266
  UAE and, 257–62
  war on science, 280–81
  Xi Jingping and, 117–18, 158–59, 176, 184, 290–92
Trump, Donald, Jr., 257
Trump, Eric, 257
Tsai Ing-wen, 135
Tsinghua Unigroup, 157
TSMC, 41, 123–25, 146, 173
TU-22 bombers, 4
TU-95 bombers, 4

Tupolev, Andrei, 216
Türkiye, 36, 112
  drones developed by, 41–42, 245–47, 249–50
Tymofeyev, Artem, 3
Tyofeyeva, Kateryna, 3

UAE. *See* United Arab Emirates
UAVs. *See* unmanned aerial vehicles
Ukraine, 5, 5, 7–8, 17, 45, 271
  Biden on, 206
  casualty rates in, 89
  cyberattacks against, 50–51, 53
  drone attacks by, 30, 102
  Putin on, 138–39
  Russia war with, 2–4, 29–30, 49–51, 71–74, 87–88, 141, 186, 201, 214–15
  Western tech company support of, 50–53, 57
Ukrainian Aerial Reconnaissance Support Center, 214
Underground Empire (Farrell and Newman), 53
unipolarity, 16
Unit 8200, 235–36
United Arab Emirates (UAE), 25, 245–48, 250, 251
  AI in, 252–54
  China and, 261
  innovation in, 261–62
  spyware use in, 261
  Trump, D., and, 257–62
United States
  Afghanistan war, 100
  AI development in, 267–68
  allies of, 284–86
  Brazil and, 44
  challenges to, 185–87
  China's rivalry with, 7–8, 18, 21, 22–23, 138, 170–71, 174, 176–78, 179–87, 266–69, 294–95, 297–99
  data centers in, 193–94
  declining power of, 6
  EU and, 132–33
  federal R&D, *280*
  Huawei and, 171–72
  innovation in, 63–64
  Iran at war with, 222–23
  manufacturing in, 21
  military power of, 24–25
  Russia and, 16–18
  semiconductors and, 123–26
  shipbuilding in, 107–8
  Taiwan and, 136–37
  tech companies in, 286–90
  tech controls of, 179–82
  waning influence of, 25–26
  weaponized interdependence of, 110
  Xi Jinping and, 182
unmanned aerial vehicles (UAVs), 4, 248
Uralvagonzavod Corporation, 142
US Trade Representative (USTR), 170
U.S.-China Economic and Security Review Commission, 170
USTR. *See* US Trade Representative
Uyghurs, 254, 291
Uzbekistan, 210, 261

Vampire drones, 73
Vance, J. D., 276–77, 285
Vannevar Labs, 272–73
Versace, Max, 165
Virkkunen, Henna, 288
Volkswagen, 269
Von Kármán crater, 1
Vox, 53
Vysikan, Oleksandr, 145

Wall Street, 18
Wallace, Ben, 38
Walmart, 20
Wang, Frank, 197
War on Terror, 17
warfare, modern, 31–32
Warner, Mark, 56
Warsaw Pact, 49
Wassenaar Arrangement, 141
WeChat, 85
Wei, C. C., 136
Weiss, Jessica Chen, 169*n*
Wennink, Peter, 111
West Bank, 83
West Point, 236–37
Wilde, Gavin, 90
Wille, Belkis, 275–76
Wingtech, 175
WIPO. *See* World Intellectual Property Organization
Wolf, Kevin, 110, 147–48, 208
Wolfsburg, 269

World Bank, 6
World Economic Forum, 153*n*
World Intellectual Property Organization (WIPO), 39, 40, 268
World Liberty Financial, 257
World Trade Organization, China in, 188
World War II, 19, 29, 37, 107
Wozniak, Steve, 109
Wright, Thomas, 7
Wu, Tim, 196
Wuhan, 60
WZ-8, 69
WZ-9, 69

X (website), 56
X10D, 70
xAI, 62
Xi Jinping, 6, 8, 63, 130, 170–71, 175, 289
  Biden and, 159–60, 297
  on China Dream, 151, 152–53, 187
  on Fourth Industrial Revolution, 153–54
  government guidance funds used by, 156
  on MCF, 28, 154–56
  Putin and, 199–200, 203, 216
  repression under, 18
  Trump, D., and, 117–18, 158–59, 176, 184, 290–92
  on United States, 182
Xi'an Aircraft Industrial Corporation, 168
Xiaoying You, 196
Xinjiang, 18
Xuezheng Zhang, 175
Xuzhou Construction Machinery Group, 182

Y Combinator, 69
Yahoo, 66
Yang Jiechi, 18
Yanukovych, Viktor, 3
Yeltsin, Boris, 16
Yemen, 19
  civil war of 2015, 221
Yom Kippur War, 232
YouTube, 51
Yuan Yu, 190

Zeekr, 300
Zegart, Amy, 289
Zeiss, 131
Zelenograd, 163
Zelensky, Volodymyr, 2, 49, 56, 75, 299
Zhan Ketuan, 173
Zhang Xiaoping, 254
Zhang Youxia, 190
Zhejiang University, 112
Zhu Menghua, 1
Zhuhai, 161, 216
Zoellick, Robert, 18
ZTE, 43, 158, 159, 278
  blacklisting of, 170, 171
Zuckerberg, Mark, 66, 287

ABOUT THE AUTHOR

Carnegie Endowment for International Peace

Steve Feldstein is a senior fellow at the Carnegie Endowment for International Peace. An international expert on technology, democracy, and national security, he previously served as a senior official at the US State Department and for the Senate Foreign Relations Committee. His first book, *The Rise of Digital Repression,* won the 2023 Grawemeyer Award for Ideas Improving World Order. He has appeared on CNBC, ABC, CSPAN, NPR, and the BBC, and his writing has been featured in *The Atlantic, Foreign Affairs,* and *Lawfare,* among others.